PROBABILITY AND INFERENCE IN THE LAW OF EVIDENCE

The Uses and Limits of Bayesianism

BOSTON STUDIES IN THE PHILOSOPHY OF SCIENCE

Editor

ROBERT S. COHEN, *Boston University*

Editorial Advisory Board

ADOLF GRÜNBAUM, *University of Pittsburgh*
SYLVAN S. SCHWEBER, *Brandeis University*
JOHN J. STACHEL, *Boston University*
MARX W. WARTOFSKY, *Baruch College of the City University of New York*

VOLUME 109

K 2261 PRO

PROBABILITY AND INFERENCE IN THE LAW OF EVIDENCE

The Uses and Limits of Bayesianism

Edited by

PETER TILLERS
Benjamin N. Cardozo School of Law,
Yeshiva University, New York, U.S.A.

and

ERIC D. GREEN
Boston University School of Law, Boston, U.S.A.

KLUWER ACADEMIC PUBLISHERS
DORDRECHT / BOSTON / LONDON

Library of Congress Cataloging in Publication Data

Probability and inference in the law of evidence: the uses and limits of Bayesianism /
edited by Peter Tillers and Eric D. Green.
    p.   cm. —(Boston studies in the philosophy of science; v. 109)
    Grew out of the April 1986 Symposium on Probability and Inference in the
Law of Evidence, sponsored by Boston University School of Law and the Boston
Colloquium for the Philosophy of Science.
    Includes index..
    ISBN 90-277-2689-2
    1.   Evidence (Law)—Congresses.   2.   Burden of proof—Congresses.
3.   Inference (Logic)—Congresses.   4.   Bayesian statistical decision
theory—Congresses.   I.   Tillers, Peter.   II.   Green, Eric D.   III.   Boston
University. School of Law.   IV.   Symposium on Probability and Inference in the
Law of Evidence (1986: Boston, Mass.).   V.   Boston Colloquium for the
Philosophy of Science (1986).   VI.   Series.
Q174.B67 vol. 109
[K2261] 001'.01 s—dc 19 [347'.06] [001'.01 s] [342.76]   87-37664
                                                                                                                                                       CIP

ISBN 90-277-2689-2

Published by Kluwer Academic Publishers,
P.O. Box 17, 3300 AA Dordrecht, The Netherlands.

Kluwer Academic Publishers incorporates
the publishing programmes of
D. Reidel, Martinus Nijhoff, Dr W. Junk and MTP Press.

Sold and distributed in the U.S.A. and Canada
by Kluwer Academic Publishers,
101 Philip Drive, Norwell, MA 02061, U.S.A.

In all other countries, sold and distributed
by Kluwer Academic Publishers Group,
P.O. Box 322, 3300 AH Dordrecht, The Netherlands.

*The papers in this volume are reproduced here
with the kind permission of the editor-in-chief of the
Boston University Law Review*

All Rights Reserved
© 1988 by Kluwer Academic Publishers
No part of the material protected by this copyright notice may be reproduced or
utilized in any form or by any means, electronic or mechanical
including photocopying, recording or by any information storage and
retrieval system, without written permission from the copyright owner.

Printed in The Netherlands

To my mother.

P.T.

To my teachers.

E.D.G.

TABLE OF CONTENTS

PETER TILLERS & ERIC D. GREEN / Preface	ix
DAVID H. KAYE / Introduction. What is Bayesianism?	1
RONALD J. ALLEN / A Reconceptualization of Civil Trials . . .	21
RICHARD LEMPERT / The New Evidence Scholarship: Analyzing the Process of Proof	61
RONALD J. ALLEN / Analyzing the Process of Proof: A Brief Rejoinder	103
L. JONATHAN COHEN / The Role of Evidential Weight in Criminal Proof	113
DAVID H. KAYE / Do We Need a Calculus of Weight to Understand Proof Beyond a Reasonable Doubt?	129
LEA BRILMAYER / Second-Order Evidence and Bayesian Logic	147
ANNE W. MARTIN / A Comment in Defense of Reverend Bayes	169
DAVID H. KAYE / A First Look at "Second-Order Evidence"	177
GLENN SHAFER / The Construction of Probability Arguments .	185
DAVID A. SCHUM / Beating and Boulting an Argument	205
DAVID A. SCHUM / Probability and the Processes of Discovery, Proof, and Choice	213
WARD EDWARDS / Insensitivity, Commitment, Belief, and Other Bayesian Virtues, or, Who Put the Snake in the Warlord's Bed?	271
PETER TILLERS / Mapping Inferential Domains	277
WARD EDWARDS / Summing Up: The Society of Bayesian Trial Lawyers	337
Name Index	343

PREFACE

This book explores the nature of factual inference in adjudication. The book should be useful to students of law in Continental Europe as well as to students of Anglo-American law. While a good many countries do not use the sorts of rules of evidence found in the Anglo-American legal tradition, their procedural systems nevertheless frequently use a variety of rules and principles to regulate and structure the acquisition, presentation, and evaluation of evidence. In this sense, almost all legal systems have a law of proof.

This book should also be useful to scholars in fields other than law. While the papers focus on inference in adjudication, they deal with a wide variety of issues that are important in disciplines such as the philosophy of science, statistics, and psychology. For example, there is extensive discussion of the role of generalizations and hypotheses in inference and of the significance of the fact that the actors who evaluate data also in some sense constitute the data that they evaluate. Furthermore, explanations of the manner in which some legal systems structure fact-finding processes may highlight features of inferential processes that have yet to be adequately tackled by scholars in fields other than law.

The topic of uncertainty must loom large in any significant discussion of the nature of factual inference and it looms large in this book. The authors differ in their views of the properties of factual inference but with only one possible exception they assume that a degree of uncertainty infects all or practically all factfinding. There may have been times when people generally believed that certainty about facts is attainable. Today, however, it is hard to take seriously any theorizing that assumes that human beings can generally reach conclusions about facts that are entirely free of doubt.

This book focuses on the merits and demerits of using a particular structure – Bayesianism – for grappling with the problem of uncertainty in factfinding in adjudication. While Bayesianism takes many different forms, Bayesians do agree that uncertainty is not synonymous with disorder or irrationality and that there are coherent and rational ways of thinking about uncertainty. The debate in this book is not primarily if at all about the validity of Bayes' Theorem; even the most rigorous critics of Bayesianism do not argue that Bayes' Theorem is invalid. The issues take a more discrete form. The papers examine, for example, the question of whether there are other logically coherent portraits of the structure of rational inference, the question of whether there are certain features of proof processes that cannot be effectively portrayed by a Bayesian structure, and the question of the effect that variations in the dissection and partitioning of evidence have on the validity of Bayesian interpretations of problems of evidence and inference.

Lurking in the details of these interesting discussions is the more general question of the value of trying to use any kind of formal analysis to portray the nature of factual inference in adjudication. Some of the authors very much doubt that formal logical analysis has much value whereas a good many of the authors very much believe it does. The outcome of this debate may have some important practical consequences. We have the sense that skepticism about the reliability and predictability of adjudication is spreading rapidly in the United States and that an increasing number of informed observers are embracing the hard-bitten saying that trials are nothing more than "crap shoots." If we are right about this, it is a fair guess that a large part of the explanation for the emergence of what Professor Lempert calls the "new evidence scholarship" is this growing skepticism about the reliability of factfinding in adjudication. By the same token, if this new evidence scholarship proves to be a chimera, it is possible that factual adjudication will be viewed with an increasing degree of distrust and that proposals for the reform of proof processes will be met with an equal degree of distrust. If so, the broader matter at stake in this debate about Bayesianism is nothing less than the idea of rational adjudication. Adjudication, of course, serves a variety of values of a cultural and political nature in addition to the concern for reliable factfinding. But, whatever other purposes adjudication may serve, we cannot bring ourselves to believe that a system of adjudication is worth preserving if it is entirely indifferent to the realities that its own rules and principles proclaim are essential for the exercise of legal coercion.

This book grows out of the April 1986 Symposium on Probability and Inference in the Law of Evidence sponsored by Boston University School of Law and the Boston Colloquium for the Philosophy of Science. We would like to thank Dean William Schwartz of the Boston University School of Law and Professor Robert Cohen, Chair of the Colloquium, for their generous support. We would also like to thank Professors Aviam Soifer, Michael Graham, Adrian Zuckerman, David Schum, and William Twining who saw the need for the Symposium and contributed in many ways to its organization. We would like to thank the *Boston University Law Review*, which published a special issue of the *Review* containing all of the papers and comments from the Symposium [66 B.U. L. Rev. 377–952 (1986)]. Space and subject matter limitations unfortunately prevented us from including all of the excellent Symposium papers in this volume. This volume focuses on Bayesianism; readers interested in the broader subject of probability and inference in evidence would do well to consult the *Law Review* issue. Finally, we would like to thank the authors of all the Symposium papers. It is ultimately their intellectual effort that justifies the publication of this book.

Peter Tillers, *Rockport, Massachusetts*
Eric D. Green, *Boston, Massachusetts*
U.S.A.
July, 1987

D. H. KAYE*

INTRODUCTION
WHAT IS BAYESIANISM?

Thomas Jefferson once remarked that "[m]athematical reasonings and deductions are ... a fine preparation for investigating the abstruse speculations of the law."[1] Perhaps they are, but few lawyers have tried to use mathematics or logic not merely as propaedutics for the study of law, but as tools for explicating or criticizing legal doctrine. The papers in this volume, then, are exceptional. They discuss mathematical formulations of such matters as the probative value of courtroom evidence and the burden of persuasion, and they ask which such formulations (if any) best further our understanding of the rules of evidence and how jurors or jurists should apply these rules. All the papers are concerned with inference in the law of evidence, and all invoke the name "Bayes."

Why Bayes? Not the name, of course, for that is accidental. If the theorem published in 1763 that bears the name of the Thomas Bayes had been attributed to another author, it would have made no difference to the dissenting Reverend Bayes, who was already dead and buried. Neither would it have made a difference to us. Indeed, the question of who really discovered Bayes' theorem is not free from doubt.[2] But why have scholars of the law begun to speak of "Bayesianism" – whatever that may be?

In Anglo-American legal systems, trials are contests in which each side tells a story through the testimony of witnesses and other evidence. Counsel tend to structure these stories so that under the governing substantive law, they produce the legal consequences that their clients prefer. To this extent, trials are exercises in the confirmation and refutation of historical theories, much like scientific experiments are exercises in the confirmation or refutation of scientific theories. Inasmuch as some philosophers have found Bayes' theorem helpful in explicating the concepts of evidence and proof in science,[3] it is hardly surprising that some legal scholars would turn to the same source for inspiration. Trials also bear a superficial resemblance to the testing of statistical hypotheses, an analogy that some statisticians have exploited in discussing hypothesis testing.[4] Here too, Bayesian methods have been prominent.[5]

In short, the scholars of evidence law are hardly writing on a blank slate. To appreciate the appeal and limits of Bayesian ideas in the new evidence scholarship, one needs to understand what Bayes' theorem is, how it has matured into a school of statistical inference, and how it has provided a conceptual framework for understanding aspects of scientific inference. Although it may be foolhardy to attempt a brief description of the many facets of "Bayesianism," two possible prizes make this task worth trying. First, such a discussion may sharpen a few easily blurred distinctions for the segment of the legal community that is beginning to burrow more deeply into "Bayesianism" and its alternatives. In this regard, the major distinctions or controversies that one would do well to keep in mind include Bayesian versus other interpretations of probability, Bayesian versus other theories of statistical inference, and Bayesian versus other theories of inference generally. Second, a short explanation of the "Bayesian" approaches may serve as a vehicle for identifying the legal issues for philosophers, statisticians, psychologists, logicians and others who are well acquainted with Bayesian ideas generally and are curious about the influence that these ideas have had on theories of forensic proof.

I. THEORIES OF PROBABILITY

> The concept of probability is an unusually slippery and puzzling one. – J.L. Mackie[6]
> [N]early every adult speaker of English can use the word "probable" correctly. – Ian Hacking[7]

Day in and day out, attorneys, judges, jurors and witnesses speak of probabilities. No, the jury decides, it is not very likely that the fashionable physician murdered his elderly patients to expedite his receipt of their legacies. Yes, the prosecutor insists, the senator must have known that the cash contribution he accepted was the gift of a foreign national. No, the expert witness testifies, chances are that the plane crash that left the plaintiff a penniless widow was the result not of pilot error but of an unexpected summer squall.

Is there any connection between these informal probabilities and the probabilities that mathematicians and statisticians manipulate or apply? The authors whose work is represented here have divergent views on this fundamental question. To evaluate these views, a clear conception of mathematical probability and its interpretations is imperative. The topic could fill volumes. Indeed, it already has. Still, setting a few more remarks adrift in this sea of verbiage can do no noticeable harm. In any case, I shall delineate seven types or theories of probability and indicate which ones arguably apply to proof in the courtroom.[8]

A. Seven Types of Probability

Mathematical Probability. From the standpoint of pure mathematics, "there is no problem about probability: it is simply a non-negative, additive set function, whose maximum value is unity."[9] Mathematical probability obeys the many rules latent in this slightly facetious definition, and these axioms and theorems constitute a rich formal system. Yet, "probability" in this sense is sterile. Before we can know that this probability has anything to do with legal evidence, we must find some connection between the abstract probability that satisfies the mathematical stipulations and the "probability" that is spoken of in judicial opinions or lawyers' arguments. Mathematical probability is but an uninterpreted term in a powerful formalism.

Informal Probability. One way to transform "probability" into a term with a clear application to forensic proof is to argue that mathematical probability is a quantitative measure of the strength of a person's belief about the truth of propositions. John Venn, better known as an advocate of the frequentist conception of probability, appears to have resorted to this psychological interpretation when he speculated that it might be possible to compare probabilities derived from repeated outcomes with the "amount of belief" produced by "many conflicting arguments, and many analogies more or less remote."[10]

Of course, most people do not wander through life quantifying the intensity of all their convictions, and there is ample reason to doubt that even if these psychological "probabilities" were to be elicited, they would obey the rules of mathematical probability.[11] If these partial beliefs lack the requisite mathematical properties, then the term "probability" is a misnomer.[12]

Classical Probability. The "classical theory" of probability rests on the principle of insufficient reason, which holds that if there are mutually exclusive alternatives of the same form, backed by symmetrical reasons, then the probability of each of these alternatives is equal.[13] For example, since a coin has two sides, and since we have no reason to favor the side designated "heads" over the one designated "tails," the *a priori* probability that the outcome of a coin toss will be heads is one-half. Difficulties in framing such alternatives uniquely and in establishing the symmetrical quality of the reasons make this interpretation all but untenable.[14]

Frequentist Probability. The frequency theory identifies probability with some suitably defined relative frequency or a limiting value of such a frequency. One might say that the probability that a coin will turn up heads is the proportion of heads that would be observed in the limit, or in the long run. This interpretation is appealing in some domains, but it does not recognize probability as applied to unique events (except in a contrived and unfalsifiable way).[15] Since the events in dispute in most litigation are not subject to repetition, the theory has limited applicability in law.[16]

Logical Probability. The logical theory treats probabilities as expressing the degree of certainty that a particular body of evidence gives to the hypothesis in question.[17] For a deductively valid argument the probability of a conclusion conditioned on the premises is one. The premises E of an argument in inductive logic support the conclusion H to a lesser extent, and the inductive probability Pr(H|E) is a numerical statement of the degree of support.

The logical relation theory of probability, with its insistence that the probability of a hypothesis H must be judged in relation to some body of evidence E, is congenial to jurisprudential applications. It appears to be employed at least implicitly in the discussions of legal theorists.

Personal Probability. Informal probability, I suggested, was not a valid interpretation of "probability" because the degrees of belief that you and I hold do not conform to the mathematical theory of probability. However, if we view these informal probabilities as crude or typically error-prone approximations to a system of partial beliefs that *does* adhere to the probability calculus, the notion of personal, or subjective probability emerges.[18]

There are quite a few variations on personal probability theory.[19] At a minimum, this school of thought insists that partial beliefs be assigned to proportions "coherently." This much is needed to ensure not just any logically consistent set of partial beliefs, but a pattern that can serve as a *bona fide* interpretation of mathematical probability. "Coherence" is a technical term describing the completeness and structure of preferences. Given some seemingly innocuous postulates (like transitivity and irreflexivity) concerning preferences, it can be shown that partial beliefs satisfy all the rules of mathematical probability.[20]

Although by no means essential to establishing the requirement of coherence, an analysis of betting quotients for gambles commonly is used to define or motivate the constraints. Unless a bookie establishes his betting quotients coherently, a clever bettor can engage the bookie in a set of bets, such that even though the bookie judges each bet fair or favorable, he is bound to suffer a loss overall.[21] Whether this so-called Dutch book argument for coherence is compelling in the jurisprudential context has been the subject of much drum beating and teeth gnashing.[22]

Propensity Theory. The last theory of probability that I shall mention treats probability as a characteristic of an object or a set of observational conditions.[23] For instance, the probability that an atom of radium will emit an alpha particle in the next ten minutes is said to be an "objective chance" that applies to the single atom and does not vary with the observer's personal knowledge of nuclear physics.[24] The nature of chance is a perplexing metaphysical topic,[25] and the necessity of relying on the propensity interpretation in certain forensic applications has been debated.[26]

B. *Cross-links and "Bayesianism"*

This seven-fold taxonomy is more convenient than conceptually elegant. Mathematical probability, we saw, is a name for a measure that has particular properties. With varying degrees of success, the other six theories seek to interpret the mathematical formalism, but they are not as neatly compartmentalized as my cursory outline may have suggested. Thus, a propensity theorist may maintain that personal probability is fine as far it goes, but that there are further empirical constraints of rationality of partial beliefs.[27] A devotee of the logical school may contend that in addition to the inductive probability that applies to inductive arguments and that measures the degree of rational support that the premises give to the conclusion, there is an epistemic probability Pr(H), which measures the degree of rational belief in a proposition H, which may vary with the available stock of knowledge that a person happens to have, and which may change over time.[28] Even a personalist may adopt the view that in appropriate situations, limiting relative frequency determines the personal probability.[29] The fact that the *minimum* constraint on personal probability is coherence does not imply that it is the *only* requirement for a jurisprudentially useful probability.[30]

Given the cross-connectivity among the various theories of probability, it is not easy to say what "Bayesian probability" is.[31] Usually, the adjective is used as a synonym for the personal, or subjective interpretation. However, since personal probability can be conceived of either as an overarching theory that includes other theories or as an element of certain other theories, proponents and critics of "Bayesian probability" should specify which features of this type of probability they find attractive or odious.[32] While this attention to detail will not settle any arguments, it at least should clarify the crux of the disagreement.

II. Theories of Statistical Inference

> Ideally, we would like a method of inference which would allow us to compare, on some scale, the merits of different possible parameter values, or of simple rival hypotheses. – A.W.F. Edwards[33]

In addition to toying with phrases like "Bayesian probability" and "Bayesian distributions," the new evidence scholars speak of "Bayesian inference." This expression does have a fairly settled meaning in statistics, and an exposition of this type of inference may illuminate the legal discussions. By way of proem, I begin with a thumbnail sketch of the classical approach to such inference. I contrast this to Bayesian inference and mention what theory or theories of probability Bayesian inference necessitates.[34]

A. *Classical Inference*

A problem of statistical inference begins with data. There are two or more probability distributions that may have generated the data, and the problem is to make inferences about the unknown distribution. Consider the following illustration.[35] The Woozy Manufacturing Company (WOOZYMAN) uses a machine to produce some kind of electronic component, called a woozy. This machine makes woozies in big batches, and it has a fixed probability of producing a defective woozy each time it makes one. Suppose that we draw at random and test $n = 200$ woozies from each batch. The resulting data may be summarized by the numbers of defective woozies for the samples. Suppose, for instance, that there were $x_i = 60$ defectives for the ith sample, and that we wish to draw some inference about the proportion of defectives Θ_i in that batch.

The number of defectives x_i is an observation of a random variable whose probability distribution is described by a function $f(x_i|n, \Theta_i)$ that gives the probability of all possible values of x for a particular n (here 200) and Θ_i (an unknown parameter).[36] The statistical problem is to make some judgment about Θ_i.

The classical method relies exclusively on the sample data. The proportion of defectives for the sample is $x_i/n = 0.3$. If we conclude, on the strength of this sample statistic, that the proportion for the complete batch is around 0.3, then we are using the sample proportion as a point estimator of Θ_i. In classical inference, the sample proportion is considered a good estimator because it has certain properties, such as unbiasedness, consistency, or minimum variance. These properties make sense in the context of a view that sees the sample proportion of .3 as a single instance in a postulated infinite sequence of similar samples. The sample proportion x_i/n, in other words, is a value of a random variable whose probability distribution has a frequentist interpretation.

Of course, there are fancier (and more revealing) techniques of classical inference. Continuing to restrict ourselves to the sample data for the ith batch, we may compute the interval estimate (.236, .364) which has a "confidence coefficient" of 95 per cent. It is easy to misconstrue this estimate as having a .95 probability of covering the true value Θ_i. Yet, from the frequentist perspective, there is only a hypothetical, infinite collection of various intervals, of which 95 per cent cover Θ_i. The classical analysis does not permit us to state the chance that the present interval is one of the correct intervals.[37]

Another tool of classical statistical inference is hypothesis testing.[38] Suppose that WOOZYMAN has signed a contract that obligates him to pay a penalty when the proportion of defective woozies in a batch exceeds 24 per cent. Then he may wish to institute a quality control program involving random sampling. We may designate the claim that $\Theta_i \leq .24$ as the null hypothesis H_0 and the claim H_1 and $\Theta_i > .24$ as the alternative hypothesis. Since sample proportions x_i/n greatly in excess of .24 seem inconsistent with the claim that $\Theta_i \leq .24$, we say that we will reject H_0 if and only if the sample proportion exceeds some critical value. Classical statistics offers procedures for choosing this critical value that attend to the probability of rejecting H_0 when H_0 is true (a false alarm) and the probability of not rejecting H_0 when H_1 is true (a miss). For example, to keep the maximum risk of a false alarm to no more than .05, we should find the batch unacceptable whenever the sample proportion (of size $n = 200$) exceeds 0.29. Applying this rule, we would conclude that the batch we examined, for which $x_i/n = .30$, is unacceptable.

As with estimation, these test procedures look strictly to sample data, and the conditional probabilities for errors do not give the chances for a mistake with respect to any single batch. The properties of classical hypothesis tests are framed in terms of a sequence of similar situations. Thus, the .05 level for false alarms is an upper bound to the long-term proportion of batches *satisfying the claim of the null hypothesis* that will be rejected. Classical statistics cannot give the probability that the rejected batch actually has an unacceptably high proportion of defectives.

B. *Bayesian Inference*

Where classical inference restricts itself to sample data and does not lead to direct statements of the probability for the possible values of a parameter, Bayesian inference combines sample data with prior information to produce direct probability statements concerning unknown parameters. To solve WOOZYMAN's problem, we treat the unknown proportion Θ_i as the latest in a series of batches with proportions $\Theta_1, \Theta_2, \ldots, \Theta_i$. If these proportions were known (because we had exhaustively tested every previous batch[39]), we might describe them with a probability distribution $\pi(\Theta)$. Suppose, for instance, that previous values of Θ suggest a particular distribution $\pi(\Theta) = \beta(20, 80)$.[40] This prior distribution peaks at $\Theta = .194$. Consequently, before turning to the sample data on the ith proportion, we might take .194 as a prior point estimate of Θ_i.[41]

Now for the sample data. To simplify the notation, let me denote the sample proportion x_i/n as t_i. The observed value $t_i = 0.3$ seems atypical compared to the prior distribution $\pi(\Theta)$, which is centered at 0.194. One might think that the combined information from the previous batches and the latest sample data should yield an estimate Θ_i somewhere between its most likely prior value of 0.194 and the sample estimate t_i of 0.300.

This is what happens when one applies Bayes' rule. Bayes' rule shows us how to combine two ingredients – the prior distribution $\pi(\Theta)$ and something called a likelihood function – to arrive at the posterior probability density of Θ given the sample data. The likelihood function $l(\Theta|t_i)$ is computed as the probability of the given sample proportion t_i as a function of Θ,[42] but it measures the support that the statistic t_i lends to the possible values for Θ.[43] Bayes' rule says that the posterior distribution is proportional to the likelihood function times the prior distribution:

$$\pi(\Theta|t_i) \propto l(\Theta|t_i)\pi(\Theta).\text{[44]} \qquad (1)$$

In our example, multiplying the prior distribution $\pi(\Theta)$ by the likelihood function yields the posterior distribution $\beta(80, 220)$. With $\pi(\Theta|t_i)$ so specified, Θ is readily estimated and hypotheses about the value of Θ are easily tested.[45] For instance, since the peak of this distribution occurs at .265, the sample proportion .3 has shifted our point estimate of the proportion of defective woozies from the prior estimate of 0.194 to the posterior estimate of 0.265.[46]

The distinctive characteristic of Bayesian statistical inference is the use of Bayes' theorem to process the sample data in the light of previous experience.[47] In contrast to classical inference, this gives rise to posterior distributions that permit direct statements about the probability of particular hypotheses or estimates.

When Bayes' theorem is used with random variables for which a frequentist probability distribution exists, this procedure should be unobjectionable.[48] To make inferences about parameters that cannot be given a frequentist interpretation, however, we must assign probabilities in the sense of partial beliefs to the possible values of the parameter. For this reason, a broadly applicable theory of Bayesian inference presupposes a personal or logical interpretation of probability.[49]

C. *Bayesian Inference Writ Large*

The previous section described statistical inference as a process of choosing among the probability distributions that might have generated the data. The aim of statistical inference can also be described as an effort to make meaningful statements about assertions of the form "Θ is in S," where Θ stands for a true but unknown state of nature, and S is a subset of the possible states of nature.[50] Put this abstractly, theories of statistical inference are also theories of inductive reasoning. Bayes' rule[51] has played a large role in some theories of inductive logic because its "conditionalization" prescription for updating prior knowledge constitutes a rational way to assimilate new information into one's structure of beliefs.

In this regard, Bayes' rule is usually expressed as applying to a discrete set of mutually exclusive and collectively exhaustive hypotheses $H_1, \ldots, H_k$. If the prior probability of any such hypothesis H_i is $Pr(H_i)$ and the evidence for or against this hypothesis is E, then the posterior probability is proportional[52] to the prior probability times the likelihood of E when H_i is true:

$$Pr(H_i|E) \propto Pr(E|H_i) Pr(H_i). \qquad (2)$$

According to (2), we update our belief in H_i by conditioning on the evidence E. For any two particular hypotheses H_i and H_j, it follows that

$$\frac{Pr(H_i|E)}{Pr(H_j|E)} = \frac{Pr(E|H_i) Pr(H_i)}{Pr(E|H_j) Pr(H_j)}. \qquad (3)$$

In other words, the posterior odds are the product of the prior odds and the likelihood ratio $LR = PR(E|H_i)/Pr(E|H_j)$. This version of Bayes' theorem has had considerable exposure in the writings on legal evidence. The theorem has been advanced as a practical device for displaying the probative value of certain kinds of evidence,[53] as a heuristic device for understanding various evidentiary doctrines,[54] and as a standard against which to judge the inferential success of real or mock jurors.[55] As in other settings, the reliance on Bayes' rule has been challenged, and several of the papers in the present volume raise objections to conditionalization.

III. Decision Theory

> [A]nother type of measurement is possible in a decision problem: a measurement of utility. The two measurements, utility and probability, will then be combined to provide a coherent solution to problem of decision-making under uncertainty. – D.V. Lindley[56]

To this point, I have tried to clarify the nature of "Bayesianism" as it pertains to interpretations of probability and to inference and induction. There is yet a third sense in which "Bayesianism" pertains to the law of evidence. Bayesian statistical inference, we saw, takes a controversial step beyond classical inference by incorporating prior information or beliefs. Decision theory takes another step forward by incorporating into the analysis the consequences of actions.

WOOZYMAN, for instance, wanted to avoid shipping a batch with an unacceptably high proportion of defective woozies. As a result, we focused on the hypothesis H_0: $\Theta \leq .24$. The version of classical hypothesis testing known as Neyman–Pearson testing permits us to keep the risk of a false alarm below a fixed level and to maximize the detection probability at the preselected level. In practice, this level may be adjusted, somewhat subjectively, due to a recognition of the relative costs of false alarms and misses. For example, if the penalty WOOZYMAN pays is huge compared to the cost of not shipping, then the maximum false alarm probability will be set at a smaller level than if the costs are roughly equal.

Decision theory makes these costs explicit and uses them along with the prior information and the sample data to identify the optimal decision. To illustrate the basic concepts, we may consider the simplest class of decision problems – binary decisions with a single observation. The unknown state of nature Θ has only two possible values, Θ_0 and Θ_1. If Θ_0 is true but we accept Θ_1 (call this decision d_1), we have a false alarm. The resulting loss is some number $L(\Theta_0, d_1) = A$. If Θ_1 is true but we accept Θ_0 (a miss), the loss is $L(\Theta_1, d_0) = B$. To keep things especially simple, let us suppose that a correct decision brings no loss and no gain. The loss table below summarizes this:

$$\begin{array}{c} \quad \Theta_0 \quad \Theta_1 \\ \begin{array}{c} d_0 \\ d_1 \end{array} \begin{bmatrix} 0 & B \\ A & 0 \end{bmatrix} \end{array}$$

Table 1. A particular loss function $L(\Theta_i, d_j)$ for a binary decision

A modification of the WOOZYMAN problem exemplifies this situation. We imagine that WOOZYMAN operates a machine that has only two states. In state Θ_0 the machine churns out batches in which 20 per cent of the woozies are defective. In state Θ_1 the machine grinds out batches in which 30 per cent of the woozies are defective. If WOOZYMAN ships (this is decision d_0) a 30 per cent defective batch (Θ_1), he loses B dollars due to a penalty clause in his contract. If he retains (d_1) a 20 per cent defective batch (Θ_0), he loses A dollars in sales. To obtain some information, WOOZYMAN tests one woozy from a batch. Either the randomly sampled woozy is satisfactory (x = 0) or it is defective (x = 1). Under these conditions, the probability of x depends on Θ in a simple way: $\Pr(x = 0|\Theta_0) = .8$, $\Pr(x = 1|\Theta_0) = .2$; $\Pr(x = 0|\Theta_1) = .7$, $\Pr(x = 1|\Theta_1) = .3$.

We want the best decision rule δ, where δ depends on nothing more than the single observation of x. One conceivable decision rule would be always to ship. Another is to ship only if the test is satisfactory (x = 0). A third is to ship only the test indicates a defect (x = 1). A fourth is never to ship. We may label these particular rules δ_1 through δ_4, respectively, and, in general, we may speak of a decision rule $\delta(x)$ that tells us how to act when confronted with any sample value x.

To assess the relative merit of such rules, decision theory relies on the expected loss, that is, the mean loss with respect to the different data that might arise. This expected loss is known as the risk function $R[\delta(x), \Theta_i]$. As this notation suggests, each decision rule $\delta(x)$ has a corresponding risk function for each possible state of nature. To determine the risk function, for every Θ_i, we weigh the loss $L[\delta(x), \Theta_i]$ that could come about by the probability of the observation x that would lead to this loss, and add these products together.[57] For instance, if a batch contains 30 per cent defectives (Θ_1), then δ_2 will produce a loss of B dollars with probability .7 when x = 0, and no loss when x = 1; hence, $R[\delta_2(x), \Theta_0] = .7B$.[58]

Since the risk function is different for each unknown state of nature, it is far from clear which decision rule is best. If Θ_0 applies, then the "always ship" rule δ_1 minimizes the risk. But if Θ_1 holds, then the "never ship" rule $\delta 4$ is optimal. To reach a decision at this point, we would have to rely on some arbitrary criterion like the minimax principle, which instructs us to adopt the decision rule that minimizes the maximum possible risk.[59] Notice that so far, nothing in our analysis is "Bayesian." The probability distribution over which we average the losses L to obtain the risk R need not be "Bayesian," and we have made no use of Bayesian inference. But neither have we gotten very far.

The Bayesian resolution of the quandary resorts to a prior distribution $\pi(\Theta_i)$ for the states of nature. As with Bayesian inference, sometimes a frequentist interpretation of this prior distribution will be available, but in many circumstances the argument must be that the prior distribution represents partial beliefs about the possible states of nature. Regardless of the characterization of π, we can order the decision rules according to the expectation of their risks over the prior distribution of Θ. In short, we average the losses twice – first over the evidence x, then over the prior information. This results in a single number – the posterior expected loss $r(\delta, \pi)$ – for each decision rule. The Bayes decision rule is the one with the smallest expected posterior loss, and this one number is called the Bayes risk.[60] The mathematical expression for the Bayes risk can be rewritten to make the reliance on Bayes' theorem explicit.[61]

Having indicated the framework and normal recipe for Bayesian decision theory (BDT), it is time to ask why the criterion of minimizing expected posterior loss has some appeal. WOOZYMAN might like it because it produces the decision rule that, in the long run, will minimize actual losses. In law, however, this frequentist argument is unavailing. We do not relitigate the same case over and over, drawing the evidence according to a nicely structured prior distribution. To defend broadly the expected loss minimization criterion, we must enter the domain of utility theory. The normative treatment of preferences mentioned in connection with personal probability leads to a derivation of (1) personal probabilities that characterize partial beliefs, (2) cardinal utilities that order preferences for uncertain consequences, and (3) the prescription to maximize the expected utility of decisions, using the personal probabilities to form the expectation. This is the subjective expected utility theory of decision-making.[62] With the losses measured as disutilities, it yields BDT's directive to minimize expected loss.

In sum, decision theory, like Bayesian inference, is "Bayesian" to the extent that it relies on Bayes' theorem in finding the decision rule that minimizes the expected loss. It is also "Bayesian" in that the criterion of minimizing expected loss can be defended from the standpoint of personal probability.

This Bayesian decision theory, or some less elaborate variant on it, has proved fruitful in the study of the burden of persuasion. As with Bayesian inference, its usefulness in describing the decision-making of jurors and mock jurors has been the subject of many empirical studies.[63] Likewise, as the citations in several papers in this volume attest, many a law review article has drawn on the power of decision theory to analyze the burden of persuasion.

As the present collection also indicates, however, the value of decision theory in explicating the meaning of the burden of persuasion has been sharply questioned. A panoply of objections stem from what has come to be known, picturesquely, as the problem of naked statistical evidence.[64] This problem, which usually comes with a colorful wardrobe of hypothetical cases, receives more than passing attention in the following chapters. And, of course, there are other concerns with BDT as a tool for the study of rules of proof.

In sum, the new evidence scholars are fishing in deep waters. Something surely is flopping about in their nets, but the catch has yet to be cleaned, prepared and served as dish that delights every discerning palate – a true bouilliBayes.

NOTES

* Professor of Law, Arizona State University College of Law.

Stephen Fienberg and Brian Skyrms commented on a draft of this Introduction.

[1] Henry S. Randall, *The Life of Thomas Jefferson* 53 (1858) (letter to Bernard Moore, ca. 1765).

[2] *See* Stephen M. Stigler, *Who Discovered Bayes Theorem?*, 37 Am. Statistician 290 (1983) (estimating 3:1 odds that Nicholas Saunderson rather than Thomas Bayes penned the posthumously published essay). In any case, "Bayesian" concepts were ripe for discovery in the 18th Century. *See* Stephen M. Stigler, *Laplace's 1774 Memoir on Inverse Probability*, 1 Statistical Sci. 359 (1986).

[3] *See, e.g.*, Paul Horwich, *Probability and Evidence* (1982); Wesley C. Salmon, *The Foundations of Scientific Inference* (1967); Brian Skyrms, *Causal Necessity: A Pragmatic Investigation of the Necessity of Laws* (1980).

[4] J. Neyman & E.S. Pearson, *On the Problem of the Most Efficient Tests of Statistical Hypotheses*, 231 Phil. Trans. Roy. Soc. 289 (1933); I.J. Good, *Good Thinking: The Foundations of Probability and Its Applications* 11–12 (1983); William E. Feinberg, *Teaching the Type I and Type II Errors: The Judicial Process*, 25 Am. Statistician, June 1971, at 30.

[5] *See, e.g.*, George E.P. Box & George C. Tiao, *Bayesian Inference in Statistical Analysis* (1973); Harry V. Roberts, Bayesian Inference, in 1 Int'l Enc. Statistics 9, 15 (W. Kruskal & J. Tanur eds. 1978).

[6] J.L. Mackie, *Truth, Probability and Paradox: Studies in Philosophical Logic* 154 (1973).

[7] Logic of Statistical Inference 227 (1965).

[8] This typology is not exhaustive, and it contains several overlapping categories whose boundaries are not crystal clear. Furthermore, as always, there is more than one way to slice a salami. For other categorizations, *see, e.g.*, Vic Barnett, *Comparative Statistical Inference* 64–95 (2d ed. 1982); Terrence E. Fine, *Theories of Probability: An Examination of Foundations* (1973); I.J. Good, *supra* note 4, at 70–71; J.L. Mackie, *supra* note 6, at 154–88; Brian Skyrms, Choice and Chance: An Introduction to Inductive Logic 129, 205–15 (3d ed. 1986). I should acknowledge also that my assessments of the theories that I discuss are at least mildly opinionated and are not universally accepted.

[9] Henry E. Kyburg, Jr. & Howard E. Smokler, *Introduction*, in Studies in Subjective Probability 3, 4 (H.E. Kyburg & H.E. Smokler 2d ed. 1980).

[10] Venn wrote that "these proofs themselves may have mostly faded from my mind, but they will leave their effect behind them in a weak or strong conviction. At the time, therefore, I may still be able to say, with some degree of accuracy, though a very slight degree, what amount of belief I entertain upon the subject." John Venn, *The Logic of Chance*, ch. 6 (1888), reprinted as *The Subjective Side of Probability*, in Studies in Subjective Probability 43 (H. Kyburg & H. Smokler eds. 1964). Venn illustrated his suggestion with a forensic example. He wrote that one might compare the probability that a letter has been lost in the Post Office (which experience shows to be one in a million) to the probability that his servant stole the letter (which introspection might show to be greater than one in a million).

[11] *See, e.g.*, Daniel Kahneman, Paul Slovic & Amos Tversky, *Judgment Under Uncertainty: Heuristics and Biases* (1982); R.E. Nisbett, D.H. Krantz, D. Jepson & Z. Kunda, *The Use of Statistical Heuristics in Everyday Inductive Reasoning*, 90 Psych. Rev. 339 (1983).

[12] Such an ill-mannered set of partial beliefs "hardly deserves to be called a probability." I.J. Good, *supra* note 4, at 70. But *cf.* L.J. Cohen, *The Probable and the Provable* (1977) (distinguishing between "Pascalian" probability, which is what I have called mathematical probability, and "Baconian" probability, which has different mathematical properties).

[13] For a concise summary of the origins of the classical theory, *see* Max Black, *Probability*, in 6 Enc. Phil. 464, 474 (P. Edwards ed. 1968).

[14] But cf. Rudolph Carnap, *Logical Foundations of Probability* (2d ed. 1962); Harold Jeffreys, *Theory of Probability* (3d ed. 1967).

[15] But cf. W.C. Salmon, *supra* note 3, at 93 (relating a frequentist probability that does not apply to individual events to a "weight" that does).

[16] In some kinds of cases, a long run frequency may be the subject of proof. *E.g.*, Mapes Casino, Inc. v. Maryland Casualty Co., 290 F. Supp. 186 (D. Nev. 1968); Kaye, *Statistical Evidence of Discrimination in Jury Selection*, in Statistical Methods in Discrimination Litigation 13-32 (D.H. Kaye & M. Aickin eds. 1986).

[17] The theory has strong roots in the classical interpretation. Leading formulations of logical relation theory are found in John M. Keynes, *A Treatise on Probability* (1921), H. Jeffreys, *Theory of Probability* (3d ed. 1961), and R. Carnap, *supra* note 14.

[18] The writings of Leonard J. Savage and Bruno de Finetti have been influential in advocating and defending the personalist perspective. *See* Leonard J. Savage, *The Foundations of Statistics* (1954); Bruno de Finetti, *Theory of Probability*, Vol. 1 & 2 (1974, 1975). See also the pathbreaking essay, Frank P. Ramsey, Truth and Probability, in *The Foundations of Mathematics and Other Essays* 156 (R.B. Braithwaite ed. 1931), reprinted in H.E. Kyburg & H.E. Smokler, *supra* note 9, at 23. Less formal treatments of subjective probability date back to J. Bernoulli, Ars Conjectandi (Basel 1713).

[19] *See, e.g.*, I.J. Good, *supra* note 8, at 20 ("Some attacks and defenses of the Bayesian position assume that it is unique so it should be helpful to point out that there are at least 46656 different interpretations."). Thus, it may be valuable to distinguish

between a subjective theory that rejects the existence of objective chance and the personalist theory that views at least some personal probabilities as estimates of a more objective quantity. *See* D.H. Mellor, *The Matter of Chance 2* (1971).

[20] Actually, the "representation theorems" to this effect establish the existence of both a probability and a utility function such that an individual who is coherent acts to maximize expected utility. *See* Peter Fishburn, *The Axioms of Subjective Probability*, 1 Statistical Sci. 335 (1986) for a brief survey of this work. Because of the powerful link between subjective probability and cardinal utility and the mandate to maximize expected utility, the personalist interpretation also is called a normative theory of subjective expected utility. See, e.g., Glenn Shafer, *Savage Revisited*, 1 Statistical Sci. 463 (1986).

[21] For proofs, see the papers cited in Brian Skyrms, Dynamic Coherence, in *Foundations of Statistical Inference* 233 (I.B. MacNeill & G.J. Umphrey eds. 1987).

[22] L. Jonathan Cohen and R. Lea Brilmayer (with Lewis Kronhauser) have contributed the most powerful attacks. For an early reply to their views on the implausibility of the betting metaphor, see Kaye, *The Laws of Probability and the Law of the Land*, 47 U. Chi. L. Rev. 34 (1979).

[23] Cf. John L. Pollack, *The Paradox of the Preface*, 53 Phil. Sci. 246, 247 (1986) (using the phrase "nomic probability" for "the kind of probability involved in statistical laws of nature").

[24] D.H. Mellor, *supra* note 19, at xi.

[25] J.L. Mackie, *supra* note 6, at 179–87.

[26] L. Jonathan Cohen, *Subjective Probability and the Paradox of the Gatecrasher*, 1981 Ariz. St. L.J. 627, 633–34; D.H. Kaye, *Paradoxes, Gedanken Experiments and the Burden of Proof: A Response to Dr. Cohen's Reply*, 1981 Ariz. St. L.J. 635, 644–45.

[27] D.H. Mellor, *supra* note 19, at xii.

[28] Brian Skyrms, *supra* note 8, at 18–19. Expressing changes in partial beliefs resulting from the acquisition of new evidence and dealing with the problem of conditioning on uncertain evidence leads to the growing body of literature on probability kinematics and Jeffrey's rule. *See* Richard C. Jeffrey, *The Logic of Decision* (2d ed. 1983). I have seen no discussions of Jeffrey's rule in the legal literature.

[29] Brian Skyrms, *supra* note 8, at 214.

[30] Cf. Richard C. Jeffrey, *Dracula Meets Wolfman: Acceptance vs. Partial Belief*, in Induction, Accpetance and Rational Belief 157, 165, 177 (Marshall Swain ed. 1970) (coherence is a necessary but far from sufficient condition for rationality); Paul Horwich, *supra* note 3, at 33 (characterizing this view as " strong rationalism").

[31] Thus, for Shafer and Tversky the frequency, propensity and betting interpretations are merely separate manifestations of "Bayesian semantics." Glenn Shafer & Amos Tversky, *Languages and Designs for Probability Judgment*, 9 Cognitive Sci. 309, 316 (1985). Isaac Levi, who maintains that probability judgments ("credal states") often are indeterminate, uses a convex set of probability functions to model this indeterminacy. Isaac Levi, *The Enterprise of Knowledge: An Essay on Knowledge, Credal Probability, and Chance* (1980). The resulting theory has been called "broadly

Bayesian" although it "departs from Bayesian orthodoxy in several respects." Patrick Maher, Book Review, 51 Phil. Sci. 690, 691 (1984).

[32] Cf. Bartlett, *Discussion on Professor Pratt's Paper*, 27 J. Royal Statistical Soc'y (Series B) 192, 197 (1965) ("Bayesians should also take care to distinguish their various denominations of *Bayesian Epistemologist, Bayesian Orthodox*, and *Bayesian Savages*.").

[33] Likelihood 2 (1972).

[34] For brevity, I ignore less dominant but competing theories such as fiducial probability and likelihood. On the former, *see* Teddy Seidenfeld, *Philosophical Problems of Statistical Inference: Learning from R.A. Fisher* (1979). On the latter, *see* A.W.F. Edwards, *supra* note 33. I ignore also the Shafer–Dempster theory of belief functions that has been advanced as an alternative to conventional theories of inference. *See* Glenn Shafer, *A Mathematical Theory of Evidence* (1976); Glenn Shafer, *Lindley's Paradox*, 77 J. Am. Statistical Ass'n 325 (1982); A.P. Dempster, *A Generalization of Bayesian Inference*, 30 J. Royal Statistical Soc'y (Series B) 205 (1968); David H. Krantz & John Miyamoto, *Priors and Likelihood Ratios as Evidence*, 78 J. Am. Statistical Ass'n 418 (1983). This theory has been mentioned in the legal literature, but, perhaps owing to its complexity, no serious use of it has been made there. Cf. Stephen E. Fienberg & Mark J. Schervish, *The Relevance of Bayesian Inference for the Presentation of Statistical Evidence and Legal Decisionmaking*, 66 B.U. L. Rev. 771, 790–91 (1986) (questioning the usefulness of belief functions). For discussions of the various approaches for handling indeterminacy and imprecision in subjective probability, see the commentary following Fishburn, *supra* note 20, as well as Isaac Levi, *Imprecision and Indeterminacy in Probability Judgment*, 52 Phil. Sci. 390 (1985).

[35] Adapted from Vic Barnett, *supra* note 8, at 28–53.

[36] Where the batch is very large compared to n, the formula is approximately $f(x) = {_nC_x}\Theta_i^x(1-\Theta_i)^{n-x}$. The term ${_nC_x}$ represents the number of combinations of n things taken x at a time.

[37] Although this point has been made time and again in the statistical literature, it is not always appreciated fully in the legal literature. Kaye, Apples and Oranges: Confidence Coefficients Versus the Burden of Persuasion, __ Cornell L. Rev. __ (1987) (in press).

[38] For a survey of the pitfalls of classical hypothesis testing in litigation and some recommendations for improvements, *see* Kaye, *Is Proof of Statistical Significance Relevant?* 61 Wash. L. Rev. 1333 (1986).

[39] If we were limited to sample data on the prior batches, empirical Bayes' methods could be used.

[40] This Beta distribution is $(99!/19!79!)\Theta^{19}(1-\Theta)^{79}$. Vic Barnett, *supra* note 8, graphs it at page 48.

[41] From $\pi(\Theta)$, we also can see that the probability is .95 that Θ_i lies between .128 and .283. Thus, (.128, .283) is a Bayesian prior confidence interval for Θ_i. Likewise, computations based on $\pi(\Theta)$ show that the prior probability that $\Theta_i \leq .24$ is .835. Vic Barnett, *supra* note 8, at 48.

[42] In this case, the likelihood function is proportional to $\Theta^x(1-\Theta)^{n-x}$.

[43] *See, e.g.*, A. Birnbaum, *Likelihood*, in 1 Int'l Enc. Statistics 519, 520 (W. Kruskal & J. Tanur eds. 1978).

[44] The proportionality constant is the reciprocal of the prior probability of the observed sample proportion, $[\int l(\Theta|t_i)\pi(\Theta) \, d\Theta]^{-1}$.

[45] For recent discussions of the relationship between inference based on such posterior distributions and classical inference, *see* George Casella & Roger L. Berger, *Reconciling Bayesian and Frequentist Evidence in the One-Sided Testing Problem*, 82 J. Am. Statistical Ass'n 106 (1987); James O. Berger & Thomas Selke, *Testing a Point Null Hypothesis: The Irreconcilability of P Values and Evidence*, 82 J. Am. Statistical Ass'n 112 (1987).

[46] The Bayesian 95 per cent confidence interval moves from the prior region of (.128, .283) to the posterior region of (.220, .309), and the probability that $\Theta_i \le .24$ drops from the prior values of .835 to the posterior value of .147. That is, the sample data has made the probability distribution for Θ less diffuse and has shifted it toward higher values. Vic Barnett, *supra* note 8, at 48, 50.

[47] *Cf.* Ward Edwards, Harold Lindman and Leonard J. Savage, *Bayesian Statistical Inference for Psychological Research*, 70 Psych. Rev. 193, 194 (1963) ("Bayesian statistics is so named for the rather inadequate reason that it has many more occasions to apply Bayes' theorem than classical statistics has.").

[48] Thus, the WOOZYMAN problem was constructed so that Θ was a random variable whose value Θ_i for the ith batch we estimated from the sample observation t_i and the prior distribution. Interpreting the prior and posterior distributions of Θ as limiting relative frequency distributions is meaningful, but carrying this interpretation over to Θ_i is problematic. Since Θ_i is the fixed but unknown value for Θ in the current batch, it is not a random variable with a probability distribution that can be verified by repeated observations. Cf. Vic Barnett, *supra* note 8, at 51. For less equivocal examples of "objective" Bayesian inference, *see id.* at 193-94; George Box & George Tiao, *supra* note 5, at 12-13.

[49] But *cf.* W.C. Salmon, *supra* note 3, at 128-31 (using Bayesian inference with frequentist probabilities that must be translated into the "weight" of a scientific hypothesis to model scientific inference).

[50] Mikel Aickin, *How Should Statisticians Believe?* An Overview of the Shafer-Dempster Approach 1 (1982) (ms).

[51] And its generalization, Jeffrey's rule. *See supra* note 28.

[52] The proportionality constant is $[\sum_{j=1}^{k} \Pr(E|H_j) \Pr(H_j)]^{-1}$.

[53] Michael Finkelstein & William B. Fairley, *A Bayesian Approach to Identification Evidence*, 83 Harv. L. Rev. 489 (1970); Ellman & Kaye, *Probabilities and Proof: Can HLA and Blood Testing Prove Paternity?* 55 N.Y.U. L. Rev. 1131 (1979).

[54] *E.g.*, Richard Lempert, *Modeling Relevance*, 75 Mich. L. Rev. 1021 (1977).

[55] *E.g.*, David L. Faigman, *Bayes' Theorem in the Trial Process: Instructing Jurors on the Value of Probabilistic Evidence*, Law & Human Behavior (in press); William C. Thompson & Edward L. Schumann, *Interpretation of Statistical Evidence in Criminal Trials*, 11 Law & Human Behavior 167 (1987).

[56] Making Decisions 51 (2d ed. 1985).
[57] For a continuous random variable x,

$$R[\delta(x), \Theta_i] = \int L[\delta(x), \Theta_i] f(x|\Theta_i) \, dx,$$

where $f(x|\Theta_i)$ is the probability density of x for a given Θ_i.
[58] This reasoning gives rise to the following table:

	Θ_0 (20% defectives)	Θ_1 (30% defectives)
δ_1	0	B
δ_2	A Pr(x = 1$|\Theta_0$) = .2A	B Pr(x = 0$|\Theta_1$) = .7B
δ_3	A Pr(x = 0$|\Theta_0$) = .8A	B Pr(x = 1$|\Theta_1$) = .3B
δ_4	A	0

Table 2. Risk function $R[\delta(x), \Theta_i]$ for some decision rules $\delta(x)$ for WOOZYMAN

[59] For a survey of other strategies, *see* William J. Baumol, *Operations Research and Economic Theory* 460–475 (4th ed. 1977).

[60] Applying this analysis to the WOOZYMAN problem, and using p_i as an abbreviation for $Pr(\Theta_i)$, we arrive at the following expected posterior losses for the four decision rules:

$\delta(x)$	$r(\delta, \pi)$
δ_1	Bp_1
δ_2	A Pr(x = 1$|\Theta_0$)p_0 + B Pr(x = 0$|\Theta_1$)p_1 = .2Ap_0 + .7Bp_1
δ_3	A Pr(x = 0$|\Theta_0$)p_0 + B Pr(x = 1$|\Theta_1$)p_1 = .8Ap_0 + .3Bp_1
δ_4	Ap_0

Table 3. Expected posterior losses $r(\delta, \Theta_i)$ for some decision rules δ for WOOZYMAN

Ignoring other conceivable decision rules, the Bayes risk would be the value of r in Table 2 that turns out to be the smallest upon substituting applicable values for the losses A and B and the prior probabilities p_0 and p_1.

The term "Bayes risk" also has been applied to the expected posterior loss $r(\delta, \pi)$ for any δ rather than $\min_\delta r(\delta, \pi)$. G.P. Beaumont, *Intermediate Mathematical Statistics* 66 (1980); B.W. Lindgren, *The Elements of Decision Theory* 124 (1971).

[61] The expected posterior loss is

$$r(\delta, \pi) = \int R(\delta, \Theta) \pi(\Theta) \, d\Theta = \int \left\{ \int L[\delta(x), \Theta] f(x|\Theta) \, dx \right\} \pi(\Theta) \, d\Theta.$$

See supra note 51. Substituting the likelihood function $l(\Theta|x)$ for $f(x|\Theta)$ and rearranging, we have

$$r(\delta, \pi) \propto \int d\Theta \int L[\delta(x), \Theta)]l(\Theta|x)\pi(\Theta)\, dx.$$

But since Bayes rule (1) states that $l(\Theta|x)\pi(\Theta) \propto \pi(\Theta|x)$,

$$r(\delta, \pi) \propto \int d\Theta \int L[\delta(x), \Theta]\pi(\Theta|x)\, dx.$$

Reversing the order of integration, we conclude

$$r(\delta, \pi) \propto \int dx \int L[\delta(x), \Theta]\pi(\Theta|x)\, d\Theta.$$

For the same derivation in the case of discrete states and actions, *see* B.W. Lindgren, *supra* note 60, at 124.

The reformulation shows why the expectation of the risk $R(\delta, \Theta)$ over the prior distribution $\pi(\Theta)$ may be called the expected posterior loss. It is essentially the loss function averaged over the posterior distribution $\pi(\Theta|x)$, then averaged over the different data that might arise.

To find the rule that keeps $r(\delta, \pi)$ to a minimum, we need merely find the $\tilde{\delta}$ that minimizes the interior integral. This approach, which avoids the averaging over potential data, can be both computationally and conceptually advantageous. See Vic Barnett, supra note 8, at 255–56 (describing this "extensive form" of analysis that leads to the Bayes' decision rule in relation to the current data alone).

[62] *See supra* note 20. For discussions from the perspective of economic theory, *see* Jacob Marschak, *Decision Making: Economic Aspects*, 1 Int'l Enc. Statistics 116 (W.H. Kruskal & J.M. Tanur eds. 1978); William Baumol, *supra* note 59.

[63] *See, e.g.*, Terry Connolly, *Decision Theory, Reasonable Doubt, and the Utility of Erroneous Acquittals*, 11 Law & Human Behav. 101 (1987); Francis C. Dane, *In Search of Reasonable Doubt: A Systematic Examination of Selected Quantification Approaches*, 9 Law & Human Behav. 141 (1985). Anne W. Martin & David Schum, *Quantifying Burdens of Proof: A Likelihood Ratio Approach*, 27 Jurimetrics J. __ (1987) (in press), presents an interesting methodological critique of the early empirical work.

[64] For example, some commentators insist that decision theory is inapposite because it does not include a decision not to decide. While the binary decision problem restricts itself to only two decisions, BDT can accommodate decisions to gather more evidence. For instance, we could extend the decision space in the WOOZYMAN problem to include d_3 – a decision to sample and test more woozies before disposing of the batch. To my knowledge, the problem of naked statistical evidence has yet to be approached in this formal fashion.

David H. Kaye,
Professor of Law,
Arizona State University College of Law.

RONALD J. ALLEN*

A RECONCEPTUALIZATION OF CIVIL TRIALS†

The last two decades have seen an explosion of creative effort directed at the twin tasks of explicating the nature of the reasoning process employed by factfinders at trials and relating that process of reasoning to the rules governing the trial of disputes. Working primarily from the assumption that some form of conventional probability theory[1] must at least roughly describe the nature of the relevant reasoning process, theorists have created models of rationality for factfinders that rely heavily on ideas of conventional probability. This process commenced in earnestness among legal scholars approximately twenty years ago with Professor John Kaplan's demonstration that a decisionmaker could accommodate the ambiguity that permeates the normal trial of a civil dispute by employing the approach to evidentiary matters that is integral to Bayes's Theorem.[2] In addition, he demonstrated through the application of simple decision theory models how rules of decision could be derived that incorporate preferences and perceptions of utility of either the decisionmaker or of the larger society of which the decisionmaker is part.[3]

In the last two decades, a substantial literature has investigated the implications of conventional probability for the trial of civil disputes. For example, the implications of Bayesian theory have been explored in detail by scholars sophisticated in the relevant methodologies,[4] and a general although not unanimous consensus appears to have emerged that a careful use of Bayesian methodologies—one that is sensitive to its requirements and limitations—should prove useful in reducing errors resulting from trials.[5] Similarly, recent work has extended and deepened considerably the legal system's understanding of the relationship between soft, unquantifiable data and "naked statistical" evidence.[6] Perhaps most importantly, and certainly most impressively, researchers have recently turned their attention to substantive rules with quite interesting results. In particular, the consequences of differing types of remedial rules have been examined, disclosing a wealth of unanticipated consequences.[7] Indeed, even the courts may some day heed the message of this new knowledge, if *Sindell v. Abbott Laboratories*[8] is any indication.

Notwithstanding these impressive achievements, limits on the ability to reconcile conventional views of probability with conventional views of trials are being discovered. It is becoming increasingly obvious, for example, that Bayesian approaches can best be used heuristically as guides to rational

P. Tillers and E. D. Green (eds.), Probability and Inference in the Law of Evidence, 21–60.
© *1988 All Rights Reserved.*
Kluwer Academic Publishers.

thought and not as specific blueprints for forensic decisionmaking.[9] Similarly, there is apparently little disagreement that verdicts based solely on overtly probabilistic evidence[10] should be rare, indeed disfavored, occurrences due to the lack of context in such cases by which the overtly probabilistic evidence may be measured.[11] Moreover, in all but the area of remedial rules there have been no significant creative breakthroughs of late. Indeed, much of the present work seems either fixed at the level of already well-debated issues or focused on relatively insignificant issues that lend themselves to a probabilistic analysis.

There is, in short, a sense of a bit of malaise. Advancement has not proceeded rapidly, and for good reason if Professor Jonathan Cohen is correct. Professor Cohen has recently mounted a frontal assault on the theory that conventional probability is the proper paradigm for the trial of disputes.[12] He argues in great detail that conventional probability simply does not work very well to describe or guide the trial of disputes. What is needed, he argues, is a different conceptualization, one that traces its roots to Bacon and Mill rather than Pascal, and he proceeds to develop in even greater detail a rigorous Baconian theory of inductive probability.[13] According to Professor Cohen, this form of probability theory much better describes the actual operation of trials and thus ought to be a preferred conceptualization. He is unclear, however, as to whether any prescriptions can be drawn from his conceptualization.

As impressive as Professor Cohen's achievements are, and they are quite impressive, his work will not determine the final conceptualization of the trial of disputes. When Professor Cohen's work is interpreted for application to the trial of disputes, it possesses highly analogous attributes to the theory it was designed to replace. Upon reflection, this does not strike me as counter-intuitive. Both conventional and inductive probability theorists are attempting to bring rational thought to bear on identical phenomena. Unless one or the other contrasting camps has widely missed the mark in their perceptions of reality, or unless we exist at a widely irrational plane in the universe, the efforts of rational individuals to describe and explain similar observations should tend to converge.

Counter-intuitive or not, the convergence of these theories leaves a large problem: the theoretical framework for the trial of disputes appears seriously inadequate. Perhaps this is because our level of understanding is not yet sufficient to allow us adequately to explain the relevant phenomena. Perhaps, but I think not. Rather, I think that both conventional and inductive theorists have uncovered inadequacies in the manner in which we conduct trials. Thus, effort should not be directed toward reducing the range of dispute between various ways to conceive of the idea of probability, nor should it be directed toward efforts to create still another formal probability theory in the hopes that it will avoid the limitations of the two models to which I have been referring.[14] My suspicion is that any other theory in the end will describe the same phenomena in a manner highly analogous to our present conceptualizations. Indeed, a contrary belief seems almost a negation of the implications of rational thought. It is, in short, not our conceptualizations of probability that are in need of serious revision. It is instead our conceptualization of trials.

I intend to propose here a new manner of conceptualizing trials. I will begin first by briefly reviewing the limits of the conventional model of probability as it applies to the trial of civil disputes, and I will add a few new wrinkles to that worn garment. I will then demonstrate that similar limitations are inherent in Professor Cohen's model. My purpose in doing this is only to demonstrate that conventional views of probability and Cohen's views are highly analogous, and not radically different, when interpreted for application to the trial of civil disputes. It is from this similarity that I draw the conclusion that we observe two quite rigorous descriptions of identical phenomena. I will then proceed to sketch out a reconceptualization of civil trials that is designed to ameliorate many of the limitations inherent in our current conceptualization of trials and that is independent of any theory of probability that could be employed by it. The reconceptualization, in other words, works just as well no matter what theory of probability is employed to analyze the relevant factual inquiries at trial.

I. The Limits of Conventional Models of Probability Briefly Reviewed and Extended

Two types of difficulties are posed by the effort to reconcile conventional views of probability with conventional views of trials.[15] There are, first, what might somewhat loosely be called formal problems, in particular the problems of conjunction and negation. In addition, there is a disturbingly strained and unrealistic quality to much of the work of conventional probability theorists that engenders the suspicion that there is a serious problem lurking somewhere. This suspicion is intensified rather than assuaged by the efforts of these theorists to respond to their critics.

A. *Formal Limits of Pascalian Probability*

Conventional probability theory entails rules of negation and conjunction that are difficult to reconcile with the Anglo-American system of civil trials. The implications of negation are completely discussed in the literature and need only be briefly reiterated here.[16] An axiom of conventional probability is that the probability of any fact plus the probability of its negation must equal 1.0. Thus, if the probability of the plaintiff's case being factually true is .500001, the probability of the defendant's case being true, which is the negation of the plaintiff's, must equal .499999. In such a case, the normal burden of persuasion rule mandates a verdict for the plaintiff, yet this seems unfair to some commentators in light of the high probability that the defendant is not factually liable.[17] I will return to this problem below in analyzing Professor Kaye's response to it.

The second problem of conventional probability theory results from the rules of conjunction that specify that the probability of two independent events occurring is the product of their separate probabilities.[18] If the probability of getting a head is ½ on the flip of a coin, the probability of getting two in a row—or of getting two heads from two identical coins—is ½ × ½, or ¼. This is a serious problem for an account of civil trials in conventional probability terms. Jurors are generally instructed in civil cases that they must find each element to the level required by the relevant standard of proof.[19] That standard is normally a preponderance of the evidence, which is usually defined as "more probable than not" or as "50%+."[20]

Nonetheless, presumably a plaintiff deserves to win a civil trial only if all of the elements of his cause of action are true. We would say that an error was made if a plaintiff recovered for an intentional tort where the defendant caused the injury but did not intend it, or intended it but did not cause it. This understanding, however, is not consistent with the conjunction rule.*

Suppose a jury found the probability of intentionality to be .6 and that of causation to be .6 as well. Assuming that these elements are independent, the probability of their conjunction is .36. Thus, the probability of at least one of them not being true—which should result in a defendant's verdict—is $1 - .36$, or $.64$.[21]

The significance of this phenomenon is that there is a divergence between how we instruct juries and how we wish trials to come out. If these numbers were to be at all accurate assessments of the class of cases into which this individual case fell—that is to say that over the long run in about two-thirds of such cases at least one of the necessary elements of recovery is not true—then the system will be biased against defendants, and more errors favoring plaintiffs than defendants will be made.

These are well known problems and I will not belabor them here. I do wish to add to them certain new wrinkles that give further grounds for skepticism concerning either how consistent conventional theories of probability are with conventional theories of civil trials or the wisdom of one or the other of those theories. These wrinkles are generally derivative of the conjunction phenomenon and demonstrate a certain arbitrariness in decisionmaking.

One implication of the conjunction principle is that it injects a certain inequality of treatment into the trial of disputes that is a function of the number of elements of a cause of action. Compare two causes of action. Assume that the first has two elements and that the second has three. To see the inequality of treatment, assume that in both causes of action each element is established to a probability of .75, which more than satisfies the requirement of a preponderance of the evidence. But note the consequences of the conjunction principle. In the first case, the probability of both elements being true, again assuming independence, is .75 × .75, or .56. In that case, a verdict for the plaintiff is obviously justifiable notwithstanding the effect of conjunction. Now consider the result in the second case where the probability that all three elements are true is .75 × .75 × .75, or .42. Here we have an example where errors will favor plaintiffs over defendants if the jury is given the normal instruction to return a verdict for the plaintiff if it finds each element to be true by a preponderance of the evidence. In addition, we have another problem that emerges from comparing the results in the two cases. Defendants as a class are considerably worse off in the second case than in the first even if the jury finds each individual element in both cases by a preponderance of the evidence.

To generalize, a verdict may be returned for a plaintiff if there are two issues whenever there is a slightly lower probability than 1 − (.5 × .5), or .75, that a defendant should not be liable. When there are three issues, a verdict for a plaintiff may be returned whenever there is a slightly lower probability than 1 − (.5 × .5 × .5), or .875, that a defendant should not be liable. Moreover, inequality cannot be eliminated by requiring the product of the individual elements to exceed .5. The effect of doing that merely shifts the differential treatment from defendants to plaintiffs. Since each additional element will generally lower the probability of the conjunction, the more elements there are the higher is the probability to which, on average, each will have to be established in order to reach a specified level, regardless of whether it is .5 or something else. Thus, according to conventional probability theory as it would apply in this modified situation, plaintiffs' tasks will become more difficult as each new independent element is added. As a result, plaintiffs will be treated differentially based upon the fortuity[22] of the number of elements in a cause of action, whereas under our present rules defendants are treated differentially based on the number of elements. If conventional probability theory is applicable to the trial of disputes, one or the other disparity must exist, given our present conceptualization of trials.

There is yet another curiosity emanating from the interaction of the conjunction principle of conventional probability and the conventional view of trials. Compare two cases, each containing three elements. Assume that in the first case each element is established to a probability of .6, thus resulting in a verdict for the plaintiff. Assume that in the second case two elements are established to a probability of .9 and the third element is established to a probability of .4. Since a probability of .4 would not meet the preponderance of the evidence standard, the second case would result in a verdict for the defendant. However, if the three elements are independent, there is a $1 - (.6 \times .6 \times .6)$ or .78, probability that at least one of the elements is not true in the first case, and a $1 - (.9 \times .9 \times .4)$, or .68, probability that at least one element is not true in the second case. In other words, the probability is higher in the first case than in the second that the defendant is *not* factually liable, yet under the conventional view of trials a verdict will be returned for a plaintiff in the first case and the defendant in the second.

There are other formal problems implicit in, although not derived from, the conventional conceptualization of trials. I present them here because they are analogous to those just discussed, and their solution, developed in Part III, is the same. Consider from a somewhat different perspective the nature of erroneous judgments that will result if jurors are instructed to find for plaintiffs only if each necessary element is established to a specified probability. Assume that a cause of action has two elements, X and Y. There are four subsets that can be created of all such cases: 1) X and Y are both factually true; 2) X and Y are both factually false; 3) X is true and Y is false; and 4) X is false and Y is true. In subset 1, there are three types of errors that can be made: an error can be made on either element separately or on both of them. In each case, however, the result will favor defendants. Plaintiffs are entitled to a verdict in subset 1, and any error will result in a verdict for the defendant.

Now consider subset 2. Here, two out of the three possible types of errors will not result in an erroneous judgment. Only if the factfinder makes an error with respect to both issues will an erroneous judgment for plaintiffs result. Thus, defendants again seem to be relatively advantaged by this phenomenon, although the precise effect is a function of the actual distribution of errors that occurs.

Combining the analysis of these two subsets, one in which plaintiffs deserve a verdict and one in which defendants deserve a verdict, demonstrates that erroneous verdicts for defendants will be reached as a result of three types of errors whereas erroneous verdicts for plaintiffs will be reached as a result of only one type of error. If the objective at trials is in part to equalize errors among defendants and plaintiffs, this phenomenon should be troublesome.

A consideration of subsets 3 and 4 is not as startling, but it is of some interest. In both cases, defendants deserve verdicts. Thus, the only erroneous verdict that can result is a verdict for plaintiffs. Of the six kinds of errors that can be made, only two of them will result in erroneous verdicts for plaintiffs: erroneous verdicts for plaintiffs will result where an error is made with respect to the factually false issue in each subset. All other errors will still result in verdicts for defendants.

To some extent these consequences may be rationalized away by asserting that sometimes defendants are favored and sometimes plaintiffs are favored. That, however, is a weak explanation for what appears to be a crazy-quilt process. Although it is true that of the twelve kinds of errors that can be made, three result in erroneous verdicts for plaintiffs, three result in erroneous verdicts for defendants, while six leave the ultimate verdict unchanged, they are nonetheless distributed by an apparently nonsensical rule: when both elements are true or both false, defendants are favored; when one is true and one false, plaintiffs are favored.

The distribution of erroneous judgments, coupled with the implications of the conjunction principle, do not paint a picture of a perfectly rational system. Something is amiss. That something, as I elaborate in Part III, is the conventional view of trials. Before doing so, however, I will address the efforts of certain scholars to explain some anomalies that the conventional view of probability gives rise to in the trial of civil disputes. Curiously, these efforts tend to reinforce the conclusion that something is awry somewhere.

B. *The Reconciliation of Conventional Probability and the Conventional Conception of Civil Trials*

The efforts to elaborate the implications of conventional probability theory, as well as to respond to its critics, accentuate rather than ameliorate the concern that there is a poor fit between probability theory and the conventional view of civil trials. This is particularly evident in what is the best of this genre: the examinations of recovery rules by Professor David Kaye[23] and Professors Orloff and Stedinger.[24]

In his very interesting article, Professor Kaye hypothesizes a case where the litigated issue is causation. He then demonstrates that applying to that issue a burden of persuasion rule of a preponderance of the evidence conceptualized as a probability measure of greater than .5 will minimize the sum of defendants who are wrongfully required to compensate plaintiffs and plaintiffs who are wrongfully denied recovery. In addition, it will result in minimizing the total amount of money wrongfully paid by factually liable defendants and not obtained by factually deserving plaintiffs.[25] Professors Orloff and Stedinger extended Kaye's analysis to show that a different burden of persuasion rule is optimal if the policy is to reduce the incidence of large errors. If that is the policy, an expected value rule should be employed that provides for a plaintiff to recover an amount equal to the magnitude of the damages multiplied by the probability that the defendant is liable.[26]

Both of these efforts are impressive and are filled with insights. Furthermore, they seem superficially plausible in the sense that they engender the impression that they could in fact be discussing, and their prescriptions could be easily applied to, the trial of civil disputes. The problem is that this superficial plausibility melts away upon further examination. Both of these efforts make a simplifying assumption that in each case there is only a single litigated issue. Without that simplifying assumption, the prescriptions of these works begin to diverge radically from the present system of trials.

This is most evident in Kaye's work. By assuming that there is only a single issue, Kaye has assumed away the problem of conjunction. He assumes that all relevant issues other than the one under consideration are proven to a probability of one. If that assumption is relaxed, then his argument requires that the probability of the conjunction of all elements of the relevant cause of action be greater than .5. In other words, while he purports to be discussing the system of civil trials, his discussion entails a radical change in the system from one where jurors are instructed to apply the relevant burden of persuasion to each element to a system where they would be instructed to apply the burden of persuasion to the conjunction of all elements.

In fact, Professor Kaye's discussion entails an even more radical approach. When he expands the analysis to include the possibility that there is more than one person who could be liable, he concludes that liability should attach to that person, including the plaintiff, for whom the probability of liability is greatest, and without regard to whether the probability of the most probable culprit exceeds .5 or anything else for that matter.[27] Thus, upon generalization, Kaye is defending the normal preponderance rule only in a very narrowly constricted context. His analysis, rather than showing the consanguinity of conventional probability theory and the normal rules of civil trials, argues instead for a radical modification of civil trials.[28]

The work of Orloff and Stedinger has an analogous attribute. The example that they employ assumes that the probability of causation is the only litigated issue.[29] When that assumption is relaxed, liability becomes a function of the conjunction of all legally relevant elements, as it does in Kaye's work. Moreover, Orloff and Stedinger do not consider the possibility of multiple defendants. Had they done so, presumably their analysis would have led them to the conclusion that, if reduction of large errors is the goal, a plaintiff should be allowed to recover the "expected value" from all possible defendants.[30]

That may not seem so shocking when the number of possible defendants is limited and easily definable, as in *Sindell*, but it leads to breathtaking possibilities when that is not the case. The implication of their quite persuasive argument is that all persons potentially in the causal chain, or more precisely the legal chain, of a legal injury who can not show to a certainty that they should not be held liable, should be held liable, and thus become, in essence, insurers against social harm. That might be a very good result to reach, but it is quite a distance from our present system of civil dispute resolution.

While some of the works elaborating the implications of probability theory do not appear upon examination to be defending or explicating the present system of civil trials, other efforts that more directly attempt to demonstrate that there is no "fundamental dissonance between mathematical probability theory and forensic proof,"[31] somewhat paradoxically tend to demonstrate precisely that there is just such a dissonance. I will use as my example the best of this type of work, which is the dispute between Professors Kaye and Cohen over the now famous Paradox of the Gatecrasher.

To understand how odd the debate over the Gatecrasher hypothetical seems, one must bear in mind that the disputants on virtually all sides of the probability debates appear to have the reduction of errors at trial as an important value. Indeed, it appears to be the primary value of Professor Kaye.[32] In light of that, consider his response to that part of Professor Cohen's challenge to conventional probability implicit in his Gatecrasher hypothetical. Professor Kaye, essentially quoting from Cohen, presents the hypothetical thusly:

> Consider a case in which it is common ground that 499 people paid for admission to a rodeo, and that 1,000 are counted on the seats, of whom A is one. Suppose no tickets were issued and there can be no testimony as to whether A paid for admission or climbed over the fence. So there is a .501 probability, on the admitted facts, that he did not pay. The conventionally accepted theory of probability would apparently imply that in such circumstances the rodeo organizers are entitled to judgment against A for the admission money, since the balance of the probability would lie in their favor. But it seems manifestly unjust that A should lose when there is an agreed probability of as high as .499 that he in fact paid for admission.
>
> Indeed, if the organizers were really entitled to judgment against A, they would be entitled to judgment against each person in the same position as A. So they might conceivably be entitled to recover 1,000 admission prices, when it was admitted that 499 had actually paid. The absurd injustice of this suffices to show that there is something wrong somewhere. But where?[33]

Professor Kaye defends against these troubling implications of conventional probability theory for the trial of civil disputes in two ways. First, he asserts that although the "objective probability" of plaintiff's story being accurate is .501, "it may be appropriate to treat the subjective probability as less than one-half, and therefore insufficient to support a verdict for plaintiff, simply to create an incentive for plaintiffs to do more than establish the background statistics."[34] There are two problems with this explanation, however. First, it is obviously an attempt to explain away the troublesome aspects of the hypothetical by implicitly rejecting them. The only sensible way to understand the hypothetical is that it presents the question of what should be done when this is all the evidence there is. The answer Professor Kaye gives is to get more evidence. That may be a good idea, but it does not respond to this problem. In the context of this problem, not some other problem that Professor Kaye may prefer to talk about, the answer is that the plaintiff loses—even though Professor Kaye, in the same article, recognizes that the proper conclusion from the point of view of conventional probability is that the plaintiff should win.[35]

A second problem emerges if Professor Kaye's modification of the hypothetical is accepted. If the plaintiff can produce more evidence, then so can the defendant. If this statistical data is all that is presented, it is because that is all both parties wish to present. If one assumes that the classes of plaintiffs and defendants should be treated as equivalently as possible, then one class ought not to bear the costs of the defaults of both classes. Moreover, as I develop in greater detail in Part III, is it not clear to me why a court, or a court system, as a general rule should concern itself with the nature of the evidence produced by participants in a private dispute. If the parties are willing to let the dispute be resolved on this basis, as a general rule they should be allowed to make that choice. Professor Kaye's argument conflates appropriate rules of decision with appropriate discovery sanctions. The only time a court should follow Professor Kaye's advice in this context is if it believes a party (either party) is refusing to disclose relevant information in the discovery process.[36]

At any rate, Professor Kaye's first response sounds very much like an ad hoc rationalization rather than a convincing argument that there is not as much of a conflict as Professor Cohen asserts between the present system of civil litigation and conventional conceptualizations of probability. His second response is in the same vein. Relying on Bayes's Theorem, he argues:

> The very fact that the paradoxical plaintiffs, at the conclusion of their case, have failed to supply any particularized evidence about defendant is itself an important datum. Suppose a juror accepts the statistic about

the number of paying customers at face value. For him, the subjective probability, P(X), that defendant did not pay is .501. But, if he stops to reflect on the fact that this is all there is to the case, he should revise this probability in light of this new item of "evidence." Under the preponderance of the evidence standard, he should find for defendant if the revised subjective probability, P(X|E), is one-half or less. This will be the case only if the fraction f [in a Bayesian calculation] is more than P(X)/[1 − P(X)], or 1.004. Hence, if it is even slightly more likely that the rodeo organizers would have been able to come forward with more evidence about how the defendant A came onto the premises without paying if he had actually done so, then f could be taken to exceed this figure. Consequently, at the conclusion of plaintiff's case, this rational juror, following the dictates of the probability theory as it is conventionally understood, will think the probability that A is liable is one-half or less and find for A.[37]

As nice as this sounds, again it is not the problem. The problem is what should happen if the hypothesized evidence is all there is at the end of the entire case rather than the plaintiff's case-in-chief. At the end of the entire case, both the plaintiff and the defendant would have had the chance to produce more evidence, and both would have failed to do so. The result, one would think, would be that the inference Kaye discusses would arise on both sides and cancel each other out. Indeed, this rather sensible conclusion can be avoided only by another series of ad hoc moves, such as asserting that plaintiff's default gives rise to a stronger inference than defendant's (or vice versa).

Another curious aspect of Professor Kaye's argument is that it results in defeating the objective of minimizing error. If one understands Professor Cohen's hypothetical to include the fact that the statistical data is all that can be offered in each case, Professor Kaye would apparently deny recovery in each case. That would result in 499 correct decisions and 501 incorrect ones. If recovery were allowed in each case, by contrast, fewer mistakes would be made (499 instead of 501) and less money would be wrongfully paid (whatever the cost of the ticket is times the number of mistakes). This would result in a windfall to the plaintiff, but denying recovery results in a larger windfall, overall, to defendants. Professor Kaye recognizes this,[38] which makes his rationalization even more puzzling.[39]

The most distressing aspect of arguments similar to Kaye's response to the gatecrasher hypothetical is that they are ill-defined and appear to be internally inconsistent. Although neither point may directly further the argument that analyses of civil trials in terms of conventional probability do not work terribly well, both tend to corroborate the ad hoc nature of much of that genre.

Implicit in much of the literature on the implications of conventional probability theory for the trial of civil disputes is a dichotomy of evidence into that which is quantified and that which is not. This distinction often is articulated in terms of "statistical evidence" or some derivative thereof on the one hand, and evidence which "personalizes" the case on the other.[40] The former is usually categorical and definite ("501 out of a thousand did not buy tickets") and the latter is specific to an individual and complex ("I saw X take money out of his wallet and exchange it for what appeared to be tickets to the rodeo"). These categories do not exist as mutually exclusive sets, however, although they may reflect points on a spectrum with respect to certain variables.

Although the probabilists do not typically address in detail their epistemological views, they appear generally to hold the view that certainty about prior events is unobtainable.[41] If that is the case, then what distinguishes the quantified from the unquanitified is the clarity of the ambiguity of the evidence. Thus, testimony that 501 out of a thousand people in attendance did not pay for admission is not to make a cold and unassailable statement of a universal truth. It is instead to assert an inference drawn from a set of observations, or to assert the observations (or a summary thereof) themselves, any of which may contain error. Moreover, to make such assertions is to "personalize" data with respect to any person in the audience. Such a person has been culled out from all the rest of humanity and placed in a group of individuals with respect to whom there is some reason to believe that they have committed an actionable wrong. Such evidence may not be very personal, but the point remains that it is "personal" to some degree. What is omitted in the writings about statistical evidence and probability theory is any effort to specify why any particular degree of "personalness" should be treated differently from some other, where and why the line is to be drawn.

Look at the matter from the flip side of the coin. Suppose in the gatecrasher hypothetical that the operator of the rodeo testified that a particular defendant did not buy a ticket. He knows this, he asserts, because the defendant looks unusual to the operator, the operator sold all the tickets himself, and he would have remembered such an unusual character. This evidence would appear to meet the standards of admission to the set of unquantified data.[42] Now consider how a factfinder will analyze that data. If the factfinder will analyze it by reference to his own experience, and I see no other way for him to act apart from reliance on inspiration or intuition, he will have to compare that data, consciously or unconsciously, to his own perceptions of his own experiences and the inferences he drew from those experiences. Regardless whether that is done in a deductive, inductive or analogical manner, eventually a generalization will be formed (again, even if unconsciously) and the "evidence" compared to it.[43] A factfinder, in short, will convert "personalized" data into categorical data to analyze them. Thus, the distinction between quantified and unquantified evidence is again exposed as one of degree in its most crucial variable.

This perspective also demonstrates the curious inconsistency that meanders through arguments about quantified evidence. Viewed from the perspective developed here, the resistance to statistical evidence paradoxically amounts to favoring evidence whose limits are ambiguous over that whose limits are clearer, and indeed creates a bias against increased sophistication in the evidentiary process. As knowledge increases about some matter so that more and more confident statements can be made about it, greater skepticism is engendered about its admissibility as evidence and whether verdicts may rest upon it. The implications of probability theory are directly to the contrary, of course. Probability theory teaches that as ambiguity increases, reliance on the data should decrease.[44] Again there seems to be a dissonance, for which there is no good explanation, between the theory of civil trials and that of conventional probability.

II. The Inductivist Alternative to Conventional Probability

The lesson that at least one distinguished scholar has drawn from the uncomfortable fit between conventional probability and the trial of civil disputes is that trials should be conceptualized from a different perspective than that of conventional probability. According to Professor Cohen, a theory of inductive probability that permits ordinal statements to be made about the relative likelihood of events better captures the essence of civil trials than does conventional probability, which is a cardinal system that requires that the likelihood of an event be associated with a number between zero and one.[45]

Professor Cohen has developed his theory of inductive probability in quite a rigorous fashion.[46] To my knowledge, no aspect of his mathematics has been invalidated by a demonstration of contradictions. The primary criticisms directed toward his efforts have been less technical, focusing rather on important ambiguities contained in his interpretation of how his theory applies in the context of trials.[47] I intend to demonstrate here that not only are there ambiguities in his interpretation, but that even a generous interpretation demonstrates "paradoxes" closely analogous to those he attributes to conventional probability as it applies in the trial setting. Before discussing these paradoxes, a brief description of Professor Cohen's theory is in order. Dr. David A. Schum has provided just such a description that also places Professor Cohen's work into historical context:

> John Stuart Mill proposed a collection of specific methods for induction in his treatise *System of Logic*. Most present-day students of experimental design in various areas of behavioral, biological, and physical sciences study extensions of Mill's method without being aware of it. Mill is frequently not given appropriate credit for systematizing the design of empirical research. A variety of procedures exist for introducing various experimental "controls" so that one can isolate valid causes by removing the confounding effects of other possible alternative causes. Cohen tells us that the process of grading inductive support that one proposition can give another has a close affinity to three of Mill's methods for induction. The *method of agreement* establishes the co-presence of a cause and effect; the *method of difference* establishes the co-absence of a cause and effect; and the *method of concomitant variation* establishes the covariation of cause and effect.
>
> Suppose a situation in which we entertain a particular hypothesis Hj which explains a characteristic of some phenomenon of interest; how do we obtain inductive support for Hj? There may, of course, be other plausible hypotheses or explanations. Imagine now a series of tests which can discriminate among alternative hypotheses. Each test involves some relevant variable which can be manipulated independently of all others. The complexity of the test sequence increases as we proceed because at each stage a new relevant variable is added to those already present. As the test sequence proceeds, some hypotheses are falsified by test results and are eliminated from consideration. Suppose Hj survives the process of elimination. The degree or grade of support given by the test sequence to Hj depends upon the complexity level of the test that Hj attains. At some point we run out of relevant variables to manipulate or we run out of time or money and so we stop testing; the surviving hypothesis or hypotheses win the day in this process of eliminative induction. Mill's methods for induction provide the essential logic for the design of the test sequence.
>
> Formally, Cohen identifies a *support function*, s(H,E), which is read "The support for H, given test result E." Suppose there are n test levels 1, 2, 3, . . . , i, . . . n which represent increasingly complex tests. If H resists falsification or elimination up to test level i we can say that the grade of inductive support for H, given test result E, is s(H,E) = i/n. Test result E gives the ith grade of support where n is the highest grade possible. The support function value s(H,E) = i/n says that H has support up to level i and no higher; i.e., H was falsified at level i + 1.

Suppose the test sequence is replicable. If you object to the test result the person performing the test sequence says, "do it yourself." If you do and achieve the same result, replicability has provided a measure of confidence in the test result. A test sequence replicated enough times becomes, as the author says, a "solid evidential fact." Thus, if $s(H,E) \geq i/n$, on the basis of replicable or "solid" evidence we are entitled to conclude that $s(H) \geq i/n$; i.e., we can talk about the support for H without having to qualify it with a particular test result. Suppose $s(Hj,E) = 0$; Hj is falsified by the simplest test. Hypothesis H_k, however, passes test i but is falsified by test $i + 1$; $s(H_k,E)$ remains at level i/n and does not drop to zero because H_k obviously has more support than Hj which had none at all.

It is always possible, and the author cites examples of when it has happened, that a theory or proposition H may be true but fails to explain certain effects. Anomalies do occur and any theory of induction must be able to handle them. It would seem foolish to suppose that no theory could remain acceptable when confronted with counterevidence. Perhaps, with some slight modification, theory H can be rescued and resist being falsified at some level; one can buy support for H by revising it.

So far we have an inductive support function $s(H,E)$ which maps ordered pairs of propositions H, E into $n + 1$ fractions from 0 to n/n, where $s(H,E) = 0$ means that H is falsified by the simplest possible test and $s(H,E) = n/n$ means that H is not falsified or eliminated throughout the entire test sequence and has the highest level of support. The value of n may not be specifiable in any practical application. Technically, this presents no problem since it is apparent that the support function assigns values with only ordinal properties since there is no apparent equal unit of "test level difficulty" specifiable. Thus $s(H,E)$ only *ranks* evidential support; the numbers thus obtained are not additive nor can they be used to form ratios.[48]

As even this brief summary makes clear, Cohen has created a probability theory that differs greatly from conventional probability. Nevertheless, when it is interpreted for application to the trial of civil disputes, it has many implications that are quite similar to those of conventional probability.

A. *The Problems of Conjunction, Negation, and Ad Hoc Rationalization*

Unlike conventional probabilities, inductivist probabilities cannot be added or multiplied. Conjunction can occur, of course, but it occurs in a different manner, according to Professor Cohen:

> The inductivist analysis, however, has no difficulty in dealing with complex civil cases. Either the probabilities of the component elements

are incommensurable, in which case no probability-value can plausibly be assigned to their conjunction and separate assignments to each must suffice. Or the conjunction principle for inductive probability gives a quite satisfactory and paradox-free result. The conjunction of two or more propositions about the same category of subject-matter . . . has the same inductive probability on given evidence as each conjunct, if the conjuncts are equally probable on that evidence, or as the least probable of them, if they are not.[49]

Professor Cohen asserts that conventional conjunction problems are avoided by emphasizing the second of the two possibilities referred to in the quoted material—the conjunction of two or more propositions about the same category of subject matter. However, the conjunction problems emerge most clearly when independent events are assumed, and presumably independent events would be "incommensurable" under Cohen's analysis.[50] Accordingly, it is the first, not the second, of the two possibilities that can most usefully be considered. When that is done, a striking similarity is evidenced between inductivist and conventional approaches.

According to the inductivist approach to the trial of disputes, probability statements about the conjunction of incommensurable events cannot be made, and a plaintiff can win only by establishing each of the legally relevant and contested facts. To "establish" a fact means only that it is more probable than it is contradictory, not that it is certain. Therefore, mistakes will be made. Assume that in a certain set of cases there are two incommensurable, legally relevant elements to be established and that each is established in each case by a "balance of probability." In which of these cases—all verdicts for the plaintiff—should there have been a verdict for the defendant? The answer is obvious—in any case where an error on any element was made since in those cases at least one of the elements of the plaintiff's cause of action will not be true. Thus, under inductivist probability, errors will aggregate as a function of errors on discrete issues. In cases where plaintiffs should win there will be some wrongful defendants' verdicts because of an error on issue one, and others because of an error on issue two. In cases where the defendants should win, there will be erroneous plaintiffs' verdicts whenever an error is made on the element, or elements, that should have been resolved in favor of the defendant.

Compare these implications to those of the conjunction principle of conventional probability. To express the probability of the conjunction of two independent events as the product of their separate probabilities is simply to give a mathematical representation to the rate at which errors will aggregate as a function of errors on the separate issues, at least that is the meaning of the principle so far as it is of any relevance to the legal system.[51] To give a simple example, suppose two evenly balanced coins were to be flipped and a prediction was to be made as to their outcomes. If one predicted that two heads would result, the probability of that prediction being accurate would be ½ × ½, or ¼. What that means as applied to a series of similar events is that the prediction would be wrong approximately 75% of the time because either one coin or the other, or both, will "probably" come up tails about three of every four tosses of the two coins. Errors in the conjunct prediction will aggregate, in short, as a result of errors on discrete issues just as they would under an inductivist approach.

Thus, both inductivist and conventional theories have remarkably similar implications on the conjunction issue when the theories are interpreted for application to the trial of civil disputes. Indeed, each of the points made previously about conjunction will apply equally well to inductivist conjunction, including the curious inequality phenomena resulting from the number of elements in a cause of action. Accordingly, from the point of view of error allocation, the implications of inductivist theory, when interpreted for application to the trial of civil disputes, are highly analogous to and not radically different from those of conventional theory.

Even Professor Cohen's criticism of the negation principle is equally applicable to inductive probability. His primary criticism is that the negation principle operates to require a verdict for a plaintiff even if there is a significant chance (just barely under .5) that a defendant is not factually liable. He gives as an example of this problem the Gatecrasher hypothetical discussed earlier.[52] Again, though, when inductivist theory is applied to the trial of civil disputes, the same problem is present. In addition, Professor Cohen's discussion of the Gatecrasher hypothetical to show that the problem of negation is not present demonstrates yet another parallel between conventional and inductivist theory—the proponents of both engage in ad hoc rationalization in efforts to minimize the distance between the implications of their preferred theories and the conventional conception of trials.

The desired effect, and indeed the only relevant meaning, of applying the principle of negation to the trial of civil disputes so that a verdict is returned for plaintiffs whenever the relevant probability exceeds .5 is simply to decide the case in such a way as to hopefully reduce erroneous outcomes over time. It may be that in the set of all the cases in which the relevant probability for the plaintiff is .51, there will be a ratio of correct to incorrect results of 51/49. That is better than its inverse, however, and the only implication of the negation principle is to effectuate this desired outcome.

Now, take that same set of cases and analyze it from the point of view of inductive probability. Can it be that a set of plaintiffs do not prove their cases "on a balance of probability," when, over the long run, there will be fewer errors if plaintiff's verdicts are returned than if not? If that is so, I no longer know what the words mean that are being employed.

Professor Cohen asserts that it is so, however, and he argues that the proper result in the Gatecrasher hypothetical from the inductivist point of view is a verdict for the defendant. I think the real point of his argument, although obviously unintended, is to demonstrate that he—like his protagonists—is forced to engage in ad hoc rationalization to deflect the undesirable implications of his own theory. His argument runs thusly:

> Indeed on an inductivist interpretation there can be no case against the man at the rodeo in the circumstances described. If there is no evidence specifically against him, he cannot be brought under any inductively supported generalization from which it could be inferred that he did not pay for admission. Hence in order to elucidate why there can be no case against him we do not need to resort to some ad hoc stratagem. We do not need to postulate a legal rule ordaining some specific inadmissibility of evidence, such as the inadmissibility of statistical evidence in relation to voluntary acts. The heart of the matter is that there just is no inductive evidence against that particular man. So, if inductive probabilities are at issue, we can say quite simply that there is no evidence against him.[53]

Professor Cohen may wish to "say . . . that there is no evidence against" the defendant, but of course that assertion is false. In order to avoid the obvious implication that "on the balance of probability" the plaintiff in Professor Cohen's hypothetical has proven his case on the evidence hypothesized, Professor Cohen ironically relies on the unsupportable distinction between "naked statistical evidence" and evidence that is personalized, just as do conventional probability theorists. That distinction is no more plausible in Professor Cohen's hands than elsewhere. Moreover, there is nothing in Professor Cohen's mathematics—at least that I can detect—that makes the distinction upon which Professor Cohen relies. Thus, he is quite clearly engaging in ad hoc strategems, his protestations to the contrary notwithstanding.

There are additional interesting comparisons between inductivist and conventional theory, some of which suggest that an inductivist conceptualization of trials is not just analogous to a conventional one but is in fact inferior in certain respects. For example, the most insightful work to date of conventional probability theorists has been the efforts to explicate the nature of damages rules. Such efforts, by contrast, are impossible in certain circumstances under an inductivist interpretation and in others possess, once again, analogous features to those of conventional explications of civil trials.

The relative inferiority of inductivist theory is most clear with respect to the case of multiple possible tort-feasors. Because of the meaning ascribed to statements about the probability of responsibility for an outcome, it is coherent and perhaps sensible to allocate damages among the possible tort-feasors proportionate to their likely responsibility for the event in question.[54] This possibility is not obtainable through an inductivist approach, however, for as Professor Cohen says, "inductive probabilities about matters of fact are only rankable."[55] A fact may be said to be more likely than its contradictory or than some other fact, but one cannot allocate damages on that basis.

This points out an interesting "paradox" of inductivist theory. Suppose facts along the lines of *Sindell*, where it is clear that one of eight defendants wrongfully harmed the plaintiff. Assume further that the eight defendants are independent actors, but there is no evidence to distinguish among them so far as their respective relationship to the plaintiff is concerned. Applying inductivist theory to the case, it is quite clear that the plaintiff cannot recover from any of the eight since it is obvious that "the balance of probability" will never favor the plaintiff over any single defendant, for any single defendant will always be able to show a great likelihood that someone else is liable—one of the other seven.

Perhaps the response to this would be to hold all eight liable, but that response is not derivable from the theory Professor Cohen has developed. It would instead amount to a change in the trial system to improve its compatibility with inductivist theory. Further, it would be a change that leads to the same radical revision of our existing tort system as does the conventional probability analysis. If the plaintiff can recover here from all eight defendants because one of them is liable, any plaintiff in any case where it is clear that someone should be liable to him should be able to recover from every one possibly in the legal chain leading to the injury for just the same reason. That would be acceptable only if there is a way to make comparisons between defendants that will determine the likelihood of their individual liability. It is just that comparison that inductivist theory cannot make. Moreover, the inability to make that comparison has other secondary consequences, such as making comparative negligence impossible.

Inductivist theory may be able to provide an analogue for Professor Kaye's suggestion that the single most probable cause of an event be determined and liability ascribed to it. However, the manner in which that would be done leads to interesting pragmatic problems very similar to those afflicting the proposals to employ Bayesian analysis at trial.[56]

Suppose again a case with eight defendants. Although the matter is not entirely free from doubt due to Professor Cohen's ambiguous use of terms such as "incommensurable" and "same category of subject-matter," presumably inductivist theory would allow a statement of which defendant is the one who most probably caused a particular event. This would be done by having the factfinder compare the probability of each defendant to that of every other defendant. Only one of the eight should emerge as more likely than each of the others, for otherwise a contradiction would develop.

However, to make such a comparison would require a factfinder to engage in an analytical process as complicated as that which underlies the application of Bayes's Theorem to eight dependent elements, since each defendant's case would have to be compared to every other defendant's case. In addition, it should be noted that the present system does not operate in that fashion, which is not to say that it should not, of course. Still, this incongruence between the implications of inductivist theory and the actual operation of the trial system obviously undercuts Professor Cohen's assertion that inductivist approaches more accurately capture what in fact occurs at trial.

There is one other aspect of inductive probability that demonstrates both a theoretical and a practical problem in the use of an inductivist approach. The formal difficulty leads to the pragmatic one, and so I shall begin there.

Professor Cohen asserts that inductivist theory can explain burdens of persuasion other than the standard of preponderance of the evidence.[57] Unfortunately, this assertion is not explained. Presumably what Professor Cohen has in mind is that a higher level of inductive support emerges each time a hypothesis survives another attempt at falsification. In short, as the number of successful experiments approaches the limit n, the greater is the support for the hypothesis.

The difficulty with this conception is that it provides no consistency among cases and leads to ad hoc results. First of all, n will vary from case to case. Moreover, what it means to resist falsification at level 1, or 2, or whatever, will vary from case to case just because statements that are not about the same category of subject matter are not comparable. Thus, even if it were meaningful (a matter I have doubts about) to say that clear and convincing evidence means validation to a certain number of tests or a certain percentage of possible tests, the result would be to impose quite different evidentiary standards in each case. Secondly, even these suggestions cannot be followed, for n may be unspecifiable.[58] Obviously, if n is not specifiable, portions of n are not specifiable, at least not in a manner of any relevance to trials.

To be sure, the fact that n is not specifiable in a case does not mean that it is never specifiable. Thus, it might prove insightful to look to the process that underlies the specification of n to see what lessons one can learn. Remarkably, that process leads to the conclusion that Professor Cohen's efforts provide an alternative method of analyzing relative frequencies (and vice versa). Thus, it is not at all surprising that many of the implications of conventional probability are reflected in inductivist theory.

These implications are most evident in Professor Cohen's discussion of testing procedures that conform to an inductivist approach. He describes Karl von Frisch's investigation of the behavior of bees as such a case.[59] Von Frisch investigated how communication occurs among bees and whether they could discriminate between colors, odors, tastes, and shapes. He proceeded by constructing tests that permitted variables to be manipulated and the results observed. When a particular manipulation correlated with a behavioral change, and was replicated in subsequent tests, an inference of causality was drawn. In short, the relative frequency of the observed data led to speculations about causality.

Two points deserve to be made here. First, the von Frisch experiments may be an excellent paradigm for Professor Cohen's work, but they are not terribly useful paradigms for the trial of civil disputes. Trials do not proceed by the process of manipulating variables, observing the outcome and then replicating the experiment, although to be sure the beliefs of individual factfinders may emerge over their lifetime from some type of analogous process. Still, the crucial point is that Professor Cohen is conflating a careful and controlled process of experimentation containing planned manipulation of variables with employing much less carefully constructed beliefs to analyze evidence that is not subject to any sort of similar manipulation. Thus, the assertion that the von Frisch experiments capture the essence of the inductivist approach does not advance the proposition that the inductivist approach captures what occurs at trial.

Perhaps Professor Cohen's point is that although the von Frisch experiments do not capture precisely what does go on, nonetheless what does occur approximates von Frisch's approach. To some extent such claims would be true. Presumably factfinders do analyze the evidence before them from the perspective of their own beliefs about the nature of the relevant universe. Moreover, many of those beliefs undoubtedly come from the observation of variables interacting in various ways. For example, it is probably commonly observed that intoxicated people are more careless and less aware of their surroundings. Thus, a demonstration that an intoxicated person driving a car was involved in an accident will raise the probability that he or she caused it through inattention. Now, this can be conceptualized, as Professor Cohen would have it, as an example of inductivist probability where the prior observation of manipulated variables leads to certain conclusions. It can also be conceptualized as the application of a relative frequency approach. Prior observation leads to the conclusion that a certain set of events is usually divided up into certain subsets of a certain size. The size of these subsets indicates the likelihood that the event under investigation at trial falls into one or another subset, given the evidence.

I see very little difference between these conceptualizations so far as they are relevant to the trial of civil disputes, a conclusion that I do not find counter-intuitive at all. Both conventional and inductivist theorists are bringing rational thought to bear upon the same phenomenon. Although it is true that there are a multitude, perhaps an infinite number, of ways to describe any particular phenomenon, any rational approach will perforce have the capacity to overlap that of any other rational approach. When these opposing theories are applied to the trial of civil disputes, quite similar implications must result, for those implications will be generated primarily by the phenomenon under investigation. Take the conjunction principle, for example. If more than a single decision about a matter has to be made, and if mistakes will be made as those decisions are made, how could the total number of errors made not aggregate as a function of errors made in each category of decisions?

This is not to say, of course, that there are no differences between varying rational approaches to any particular question, nor is it to say that differing approaches do not or will not yield insights. It is simply to say that there will be broadly similar implications that result from bringing to bear differing conceptions of rational thought on any particular phenomenon. Thus, that there are certain differences in the implications of inductivist and conventional theory is not surprising, but neither is the fact that they demonstrate many similar implications.

Still, there is at least one major problem. A number of the implications of both theories are troublesome. There are formal limits, pragmatic difficulties and the literature is filled with ad hoc rationalizations of troublesome implications of various theories of probability. These do not, however, reflect problems with our theories of reasoning. Instead, they reflect problems with our theory of trials. What is being demonstrated is not that we need a new conceptualization of probability to apply to trials; rather, we need a new conceptualization of trials that responds to the difficulties posed by our understanding of the meaning of probability.

III. A Reconceptualization of Civil Trials

The unsettling implications of both conventional and inductivist analyses of civil trials are a function primarily of the fact that the conventional conception of civil trials requires comparing the probability of the plaintiff's elements to that of their negation.[60] It is just this conception that produces the problem of negation that is troubling to both Cohen and Kaye.[61] If the probability of the defendant's factual assertion being true is one minus the probability of the plaintiff's factual assertion being true, then whenever the probability of the plaintiff's assertion exceeds .5, a verdict should be returned for the plaintiff.[62] Moreover, the conclusion is the same if, as Cohen conceptualizes it, the plaintiff's burden is to establish the relevant facts as more probable than their negation. It is also this comparative process applied to more than a single factual issue that results in the "paradoxes"[63] of conjunction that Cohen finds in conventional theory, and that I find in his. If one determines the probability of the plaintiff's elements being true by reference to a conventionally conceived burden of persuasion rule, and then allows a verdict for the plaintiff when each element is established, rather than when the conjunction of them is established, the paradoxes do occur. These are not problems in either conception of probability, however. Rather, these are problems in our conceptualization of trials. Many of the unsettling implications of our understanding of probability can be eliminated or ameliorated by conceptualizing trials in a new way. I intend to propose and examine just such a conceptualization here.

My proposal has two parts to it. The first step is to conceive of trials as comparing the probability of the fully specified case of the plaintiff to the probability of the equally well specified case of the defendant. The second step is to structure trials in such a way that will permit the parties to determine how far they will push the particularity or singularity of the relevant facts.[64] These two propositions will be discussed in turn.

A. *Comparing Equally Well-Specified Cases*

Perhaps the single most troublesome implication of probability theory stems from the fact that erroneous judgements in cases will aggregate as a function of errors on discrete issues, as is represented by the conjunction and negation principles. The set of problems associated with or derived from the conjunction principle exists in the trial of disputes because trials are presently conceived of as a comparison between the probability of the plaintiff's assertions and their negations, and in conventional probability the probability of that negation is the probability associated with the plaintiff's elements subtracted from one (a concept formally absent but functionally present in inductivist theory). There is another way to conceptualize trials, however. Just as plaintiffs presently are required to specify with particularity at some point the nature of their claims and factual assertions,[65] defendants could also be required to respond with equally specific and affirmative allegations rather than with simple denials. The trier of fact could then compare its view of the likelihood of the plaintiff's case to that of the defendant's.

Such a conceptualization has numerous advantages over the present conceptualization of trials. First, it moves toward greater equality of treatment of the sets of plaintiffs and defendants. Second, the problem of conjunction is obviated in large measure because the probability of two series of allegations would be compared rather than a series of allegations with their negations. Third, this view requires a greater concentration on specific factual allegations on the part of the defendant, which may lead to a sharper focus on disputed factual matters. This in turn may lead to a commensurate reduction in the amount of extraneous material dealt with at trials, not only saving time and money but simplifying the fact-finding process as well.

Perhaps most importantly, this conceptualization may lead to fewer errors being made at trial. This can be demonstrated by freeing Professor Kaye's treatment of multiple possible sources of liability for an actionable wrong from its present artificial limitation of assuming only a single litigated issue. If one replaces in his efforts the probabilities of single elements with probabilities of the conjunction of all elements, one obtains a more general theory that liability should attach to the most likely sequence of events that explains the litigated event. The proposed conceptualization of trials would operationalize this more general theory by requiring the parties to assert what they believe are the most likely sequences of events leading to the event in question and then instructing the jury to choose between them. To be sure, this would allow a plaintiff to recover from a defendant when the jury concludes that the probability of the plaintiff's case is low, but that of the defendant's is lower. This would be inconsistent with minimizing errors only when there is yet another possible sequence of events leading to the event in question, such as some other person who is more probably liable than the defendant. In that case, however, the defendant would be allowed to implead that third party.

This proposal also can accommodate the Orloff and Stedinger view of allocation rules after it, too, is freed of the artificial constraint of assuming a single litigated issue. Orloff and Stedinger's work may be generalized in precisely the same manner as Kaye's. The question that would then remain is simply the political one of which allocation scheme is preferred.

A number of implications of this proposal deserve elaboration. The first is that it involves a dramatically different role for single elements than the current system of trials possesses. If a plaintiff and a defendant assert quite different factual allegations, with only a few common points, a factfinder could determine that the probability of a common element favors the defendant but that, taking each case as a whole, the probabilities favor the plaintiff. Nonetheless, the work of Kaye, Orloff and Stedinger shows upon generalization that to return a verdict for the defendant in this circumstance would lead to increased errors. Thus, the real lesson here is the counterproductive consequences of the present focus at trial on the individual elements of the plaintiff's case. If, however, there is a dispute over only one fact that has mutually exclusive possibilities (for example, was the light green or red when the car entered the intersection), then the probability of the respective cases will be determined by the appraisal of that single fact. In that case, the proposal would operate as the system does presently.

This conception of trials also eliminates the formal problems resulting from instructing the jury to find each, rather than all, elements to a specified level of probability. The jury will be comparing two fully specified versions of reality, rather than comparing discrete issues to their negations. As a result, the problems of conjunction do not create paradoxes where verdicts will be returned for plaintiffs even though there is an enormously high probability that at least one of the plaintiff's necessary elements is not true. Rather, the conjunction effect will be contained within *both* parties' evidentiary proffers. Similarly, the bizarre effect of distributing errors differentially over the parties based upon whether all or some of the elements are in fact true will not occur. Errors will occur, of course, but they will effect both parties in the same manner.

There is one obvious objection to this proposal. The proposal rests upon the distinction between a simple denial of the plaintiff's case and an affirmative allegation of the defendant's case. I am quite sure that no bright line separates these two concepts and that they are points on a spectrum. Nonetheless, the proposal made here is coherent so long as the parties are required to be fairly specific, although I cannot say what "fairly" means with any specificity.

B. *Exploring the Particularity of the Case*

Another indirect implication of treating plaintiffs and defendants in as equivalent a manner as possible is that the parties, not the state, should decide to what extent they wish to explore the particularities of the case. The primary concern of the state should be to provide the mechanisms by which the parties can explore the singularity of the relevant events to the extent that they choose to do so. This certainly means providing and enforcing liberal discovery rules. It may in addition mean permitting assistance to be given to the trier of fact to help it analyze the data provided by the parties.

Civil disputes in our culture are primarily disputes between parties that, so far as the law is concerned, are entitled to equal respect. The primary obligation of the state is to provide a disinterested forum that will assist in the rational resolution of the dispute. It is the parties' dispute, however. Thus, as a general matter the state has no serious ground to concern itself with the level of particularity of the evidence that the parties wish to provide the factfinder. While there may be a social interest in minimizing mistakes and I am willing to assume that mistakes are reduced as the level of specificity of the evidence increases, that does not seem to me adequate to impose upon the parties a case structure that they do not prefer. Rules governing the propriety of allowing verdicts to rest on quantitative evidence, or on the admissibility of such evidence, are in essence rules of relevancy. The only justification for such rules is that the evidence can not be understood, which includes evidence being put to an inappropriate purpose such as unfairly prejudicing a party. The state interest in minimizing errors is satisfied by creating the conditions whereby the parties are able to probe the uniqueness of the litigated event to the extent they choose to do so based on coherent evidentiary proffers. If they choose not to pursue particularity very extensively, and to rely instead on relatively crude statistical data, for example, then they should be competent to make that choice.[66]

The primary method that facilitates the parties in probing the uniqueness of the relevant events is the discovery process. Thus, broad and liberal discovery rules should be implemented and enforced. In fact, those cases where courts have not allowed verdicts based upon statistical evidence make much more sense if viewed as involving a sanction for discovery violations or an inference drawn from the failure to produce available evidence.[67] Suppose, for example, a case where a pedestrian is injured by a blue bus, and the plaintiff shows that 100 out of 102 blue buses are owned by the defendant. If both parties are willing to let the factfinder decide the case on that basis alone, I see no reason for the state to intervene to forbid that choice. Furthermore, if the state does wish to forbid it, there is no reason why that policy should redound solely to the detriment of plaintiffs, as it generally does.[68] However, it would be a different matter if a court is convinced that one party has evidence that it has not adduced or turned over to the opponent in discovery, thus forbidding the opponent to particularize the evidence to a greater degree.

There are other consequences that may result from this conceptualization. For example, the pursuit of rationality by the parties ought not to be stymied by the potential lack of understanding by the jury; and as cases become more complex, the difficulty of a rational analysis of the evidence most likely tends to increase. Accordingly, the parties should be allowed the creative use of experts to educate the factfinder on the problems presented by efforts to rationally evaluate the evidence. Although this point may appear to be indifferent to the level of particularity of evidence employed at trial, it nevertheless should result in encouraging issues to be joined at trial at the level of the most particularized evidence available. Presumably, the more particularized the evidence, the more persuasive in general it will tend to be if understood by the factfinder. Thus, greater latitude to explain complex matters encourages the parties to rely on more particularized evidence and to respond in kind to such proffers by opponents.

Another conceivable by-product of my general scheme may be an increase in the use of various types of experts. Expert testimony that would improve the factfinder's ability to rationally deliberate on the evidence would be encouraged. This, in turn, may disadvantage parties who are at a relative financial disadvantage. Accordingly, it may be necessary to institutionalize means to offset that advantage. If there is a general consensus that the state should provide a forum for the rational resolution of disputes, then a justification for providing the means by which that can be accomplished obviously exists.[69]

IV. A Final Paradox

I wish to address one last difficulty that may appear to beset the theory I have developed here. One of the implications of the theory is that if a party wished to do so, he should be permitted to advance alternative explanations of the relevant factual inquiry that are consistent with a verdict for that party, and that the factfinder should return a verdict for the party who has advanced the single most likely version of the facts.[70] An alternative, however, is that the decisionmaker should return a verdict in favor of the party whose factual propositions *collectively* are the most likely. For example, assume each party advances two competing versions of reality. Assume further that the jury would assess the likelihood of plaintiff's version one to be .05 and that of version two to be .3, while it would assess the likelihood of the defendant's version one to be .2 and that of version two to be .2. Assume further the radically unlikely but helpfully simplifying proposition that these four versions of reality are completely independent of each other.

Under my earlier argument, the jury should return a verdict for the plaintiff in this situation because the single most likely course of events is plaintiff's version two. Still, the probability is greater that one of the defendant's versions is correct than that one of the plaintiff's versions is correct. Thus, if the objective is to reduce errors, a verdict should be returned for the defendant rather than for the plaintiff. In short, a generalization of my theory is that a verdict should be returned for that party for whom it is most likely true that one of the party's factual versions is correct.

This creates a mild paradox, or at least a bit of irony. I have been developing my proposal as though it were somewhat radical, but this point, which I believe accurate, may appear to make my argument a rather convoluted justification for the status quo. The reason for this is that the plaintiff's burden will be to establish some version of the facts consistent with what the law requires for a verdict. Accordingly, the plaintiff will attempt to prove factual propositions that, if believed, will result in an inference of the necessary elements of his cause of action. The defendant, on the other hand, will attempt to prove factual assertions that will result in the inference that at least one of the necessary elements in the plaintiff's cause of action is not true. Thus, the final form of the inferential process under my theory may appear to be identical to that which occurs now. If the probability of the plaintiff's factual assertions concerning any single element is less than .5, the defendant can admit all other elements and defend solely with respect to this particular element. If the plaintiff can only establish an element to some probability less than .5, then the defendant has established that this element did not occur to more than .5, and the single most likely sequence or sequences of events favors the defendant.

I think all of this is right except the last point. A jury in evaluating evidence should conclude that the probability of the defendant's versions of reality is the probability of the plaintiff's versions subtracted from one only if it feels that it has before it all relevant versions of reality. Although it is an empirical matter, I doubt that this is often the case.

Take another oversimplified example that makes this point. Assume there is a cause of action that in the jury's view entails ten equally likely factual explanations.[71] Assume further that the plaintiff asserts four of them to be true and supports those assertions with credible evidence. In addition, assume that the defendant asserts three of the ten possibilities as true, and supports those assertions with credible evidence. Lastly, assume that the versions asserted by each party justify a verdict for that party.

In this hypothetical, under the conventional view of trials, presumably the defendant should win. Of the ten possibilities that the jury thinks may explain the relevant state of affairs, the plaintiff has provided evidence that only four of those may be true. If there are ten possible explanations of an event, four of which favor the plaintiff and thus six of which do not, certainly the plaintiff has not proven his case to a preponderance of the evidence. Indeed, this is true regardless of whether the defendant produces any evidence with respect to its three factual assertions.

The difficulty with this explanation, which highlights for me the single most troublesome aspect of our present conceptualization of trials, is that it results in resolving all ambiguity against the plaintiff. The "thus" in the penultimate sentence of the preceding paragraph is inaccurate, in other words. Presumably the jury does not know how to evaluate the three possibilities for which no evidence has been produced. In that case, a verdict for the defendant results in all possible inferences for which neither party has produced evidence being drawn against the plaintiff. A better view, I would suggest, is that the ambiguity in a case should be distributed over the parties. That is what my theory would do, and why on the facts of this hypothetical a verdict should be returned for the plaintiff.

V. The Continuing Process of Reconceptualization

The efforts to establish the superiority of conventional or inductivist approaches to the trial of civil disputes have not achieved their objective. Upon being interpreted for application to the process of civil trials, both generate quite similar implications, and it would be astonishing were this not so. The trial of civil disputes is a phenomenon with observable characteristics. Any rational thought process brought to bear upon it must reflect those characteristics, whether that process is inductivist or conventional in its nature, or one of the other conceptualizations of rationality and probability that have been developed recently, such as fuzzy set theory[72] or Shafer's idea of belief functions.[73] This is not to say that much has not been learned by these various efforts, or that they do not differ in many important respects. It is to say, however, that to the extent that these varying conceptions are employed to explain the phenomena of trials, there will be broad areas of similar implications. Moreover, many of those implications are troublesome. These implications are not to be avoided by changing the explanation of them; rather, they can be avoided only by changing the nature of the observed phenomenon.

I have sketched out one suggestion for just such a change. It is just a sketch, though, and it is obviously incomplete. The basic conception presented here will have to be tested by further work, and undoubtedly unanticipated implications will emerge. In addition, the implications of this approach for a whole range of trial-related issues must be examined. For example, what would be meant under my approach to impose a burden of production? Can burdens of persuasion other than by a preponderance be brought within the analysis? What would the standards be for preclusive motions such as directed verdicts and summary judgments? What happens to the idea of insufficiency of the evidence? What would be the role of inferences, presumptions and comments on the evidence? Will this conceptualization stimulate the filing of civil suits? These and other issues await development.[74]

Although an appraisal of the wisdom of my proposed conceptualization of trials will have to await its further explorations, I take some comfort in the fact that it has the curious feature of combining the wisdom of the common law with the essence of modern practice, thus being more theoretically than pragmatically radical. The common law system of pleading had as its objective the narrowing and clarifying of the range of factual dispute.[75] That, in essence, captures the crux of my proposal. Moreover, it is my understanding that the dominant view of contemporary litigation is that to convince a jury requires the presentation of an affirmative story, regardless of one's formal posture in the case.[76] Again, that is quite consistent with the theoretical approach presented here.[77]

There is one last implication of my proposal that deserves mention that is both troublesome and liberating. The approach suggested here could not be applied to criminal trials without a drastic reordering of the present procedures employed in the criminal process—so drastic as to be implausible. Perhaps that may weigh against my proposal by demonstrating a limited capacity for generalization. I think there is a more cogent proposition lurking here, though, and that is that the objectives and purposes of criminal trials may be so far different from those of civil trials as to make them alien to each other. Unlike civil trials, the purpose of criminal trials is not to provide a forum where parties essentially equal before the law can resolve their private disputes. Rather, a criminal trial involves unequal parties upon whom reciprocal burdens are not placed. The objective of a criminal trial is not to choose among stories of the parties. Rather, it is to determine whether or not the only plausible explanation of the event in question is that the defendant is guilty as charged. Moreover, as a check upon a potentially abusive government, the defendant is permitted to sit passively and to require the government to establish just that proposition. To do so requires that the government not only establish its own case but to negate any reasonable explanations of the relevant affairs consistent with innocence.[78]

If these assertions are true, it should not be surprising at all that a conceptualization of civil trials will not encompass criminal trials. Different tasks will call for different procedures. The mere fact that there are superficial similarities between entities does not guarantee fundamental similarity. Civil trials are designed to resolve disputes in an amicable fashion among parties who are indistinguishable before the law. Criminal trials pit an individual against the virtually inexorable power of the state. As a result, the concept of certainty assumes much greater importance in criminal than in civil trials. The implications of that difference deserve careful evaluation, but that is a task for another day.

NOTES

† © 1986 by Ronald J. Allen.

* Professor of Law, Northwestern University. I am greatly indebted to Professors Peter Tillers, of the New England School of Law, David Schum, of George Mason University, David Kaye, of the Arizona State University College of Law, Reid Hastie, of Northwestern University, and Michael Green and Serena Stier of the University of Iowa College of Law, for having read and commented on an earlier draft of this article. In addition, I had the pleasure of, and learned a considerable amount from, delivering this paper at Faculty Workshops at Northwestern University School of Law and Cornell Law School. In both instances, I received valuable suggestions from too many individuals to name.

[1] I refer to any form of axiomatic probability theory, such as Pascalian or classical probability, as "conventional." The points I intend to make here are applicable no matter which axiomatic theory is viewed as analytically superior or as better capturing what occurs in litigation. *See* L. COHEN, THE PROBABLE AND THE PROVABLE 116 (1977) ("The paradoxes . . . are common to all normal formulations of the Pascalian calculus.").

[2] *See* Kaplan, *Decision Theory and the Factfinding Process,* 20 STAN. L. REV. 1065 (1968).

[3] *Id.* at 1065-83.

[4] *See generally* 1A J. WIGMORE, EVIDENCE, § 37 (P. Tillers rev. ed. 1983).

[5] *See, e.g.,* Lempert, *Modeling Relevance,* 75 MICH. L. REV. 1021 (1977).

[6] *See, e.g.,* Kaye, *The Limits of the Preponderance of the Evidence Standard: Justifiably Naked Statistical Evidence and Multiple Causation,* 1982 AM. B. FOUND. RES. J. 487; Nesson, *Reasonable Doubt and Permissive Inferences: The Value of Complexity,* 92 HARV. L. REV. 1187 (1979).

[7] *See generally* Kaye, *supra* note 6; Orloff & Stedinger, *A Framework for Evaluating the Preponderance of the Evidence Standard,* 131 U. PA. L. REV. 1159 (1983).

[8] 26 Cal. 3d 588, 607 P.2d 924, 163 Cal. Rptr. 132 (articulating the market share liability theory and allocating the risk of loss to each of several manufacturers when precise manufacturer can not be identified), *cert. denied,* 449 U.S. 912 (1980).

⁹ *See* Callen, *Notes on a Grand Illusion: Some Limits on the Use of the Bayesian Theory in Evidence Law*, 57 IND. L.J. 1 (1982).

¹⁰ By "overtly probabilistic" I mean evidence that is largely the presentation of statistical analyses. I have some difficulty with the proposition that there is an important qualitative difference between "statistical" evidence and other kinds of evidence, for it seems to me that all evidence is "probabilistic" in a nontrivial sense. The primary variable in comparing evidentiary proffers is quality of the evidence. The ambiguity in some evidence is quite obvious and quantifiable; the ambiguity in other evidence is itself much more ambiguous. *See infra* notes 41-45 and accompanying text.

¹¹ As I elaborate in Part III, I am unpersuaded by this general view and think that generally it should be up to the parties to decide what type of evidence they will rely on. *See infra* Part IIIB.

¹² *See* L. COHEN, THE PROBABLE AND THE PROVABLE (1977).

¹³ *See, e.g.*, L. COHEN, THE IMPLICATIONS OF INDUCTION (1970).

¹⁴ *But see* G. SHAFER, A MATHEMATICAL THEORY OF EVIDENCE (1976); Comment, *Mathematics, Fuzzy Negligence, and the Logic of Res Ipsa Loquitur*, 75 NW. U.L. REV. 147 (1980).

¹⁵ My concern is with the use of probability theory to explicate the trial process. There are other issues posed by probability theory, however. One example is the dispute over whether jurors should be encouraged to employ mathematical modes of decisionmaking. *Compare* Finkelstein & Fairley, *A Bayesian Approach to Identification Evidence*, 83 HARV. L. REV. 489 (1970) (advocating explicit use of Bayes's Theorem to aid jury in assessing significance of identification evidence) *with* Tribe, *Trial By Mathematics: Precision and Ritual in the Legal Process*, 84 HARV. L. REV. 1329 (1971) (costs of attempting to integrate mathematical techniques into factfinding process at trial exceeds benefits). Another is the question whether individuals think in conventional probabilistic fashions. *See* D. KAHNEMAN, P. SLOVIC, & A. TVERSKY, JUDGMENT UNDER UNCERTAINTY: HEURISTICS AND BIASES (1982) [hereinafter KAHNEMAN].

¹⁶ *See, e.g.*, L. COHEN, *supra* note 1, at 74-81; R. EGGLESTON, EVIDENCE, PROOF, AND PROBABILITY 40-43 (2d ed. 1983); Williams, *The Mathematics of Proof*, 1979 CRIM. L. REV. 297.

¹⁷ *See* L. COHEN, *supra* note 1.

¹⁸ For the purposes of this article, I will generally assume that factual issues are independent, even though this is hardly ever the case. I make this assumption simply for purposes of simplification. In the event of dependence, the phenomena that I describe will still generally occur, but to a lesser degree.

¹⁹ *See, e.g.*, 11 E. DEVITT & C. BLACKMAR, FEDERAL JURY PRACTICE AND INSTRUCTIONS § 71.14 (3d ed. 1977 & Supp. 1986).

²⁰ MCCORMICK ON EVIDENCE § 339 (E. Cleary 3d ed. 1984).

* For an earlier discussion of the conjunction problem in this context, see Nesson, *The Evidence or the Event? On Judicial Proof and the Acceptability of Verdicts*, 98 HARV. L. REV. 1357, 1385-90 (1985) eds.

[21] I wish to make another assumption explicit at this point. The primary goal served by burden of persuasion rules is to allocate errors over plaintiffs and defendants. I will proceed for the moment on that assumption and will qualify it at points in this paper even though the economists would argue that this assumption may be unconvincing generally. I should point out that even if that assumption is rejected and some other value than, for example, equalizing errors over plaintiffs and defendants is substituted, the question remains of how to attempt to allocate factual error in light of the substituted policy. For example, recent tort scholarship under the influence of economic theory has argued that verdicts are returned for plaintiffs under various rules that achieve certain conceptions of efficiency. *See, e.g.,* Grady, *A New Positive Economic Theory of Negligence,* 92 YALE L.J. 799 (1983). Even if this is so, the determination of the factual setting of any particular litigated event will contain the same risk of factual error that is present in trials under more conventional views of dispute resolution. Thus, the question will arise how to allocate those risks of error given the desired social values to be maximized.

[22] "Fortuity" may not seem the right choice of word, but from the point of view of the analysis in the text, the number of elements in a cause of action undoubtedly is fortuitous. I doubt that legislators or common law courts gave any thought to the points made here in fashioning the elements of offenses (or affirmative defenses, for that matter).

[23] Kaye, *supra* note 6.
[24] Orloff & Stedinger, *supra* note 7.
[25] Kaye, *supra* note 6, at 494-503.
[26] Orloff & Stedinger, *supra* note 7, at 1166.
[27] Kaye, *supra* note 6, at 503-08.

[28] This entire genre, going all the way back to Kaplan, *supra* notes 2-3, assumes that the set of plaintiffs going to trial and the set of defendants going to trial will be approximately the same size. Absent knowledge to the contrary, that is a reasonable assumption to make. Nonetheless, the effect of any burden of persuasion rule will be in large measure a function of the size of those sets. Allen, *The Restoration of* In re Winship: *A Comment on Burdens of Persuasion in Criminal Cases after* Patterson v. New York, 76 MICH. L. REV. 30, 47 n.65 (1977). In criminal cases, this is a highly unrealistic assumption, since presumably the screening devices of the pretrial process remove most innocent people from the system.

[29] Orloff & Stedlinger, *supra* note 7, at 1162.
[30] *Id.* at 1160 n.6 (citing *Sindell* with apparent approval).
[31] Kaye, *Paradoxes, Gedanken Experiments and the Burden of Proof: A Response to Dr. Cohen's Reply,* 1981 ARIZ. ST. L.J. 635, 645.
[32] Kaye, *The Laws of Probability and the Law of the Land,* 47 U. CHI. L. REV. 35-36, 38 (1979). Cohen's views are less clear. He may simply be trying to describe in a more articulate fashion what he thinks is going on in the trial of disputes. His book, THE PROBABLE AND THE PROVABLE, contains no suggestion that errors will be reduced by embracing his analysis. Nor does it offer any suggestions for changing any currently employed procedure at or before trial. In his contribution to this

volume, he asserts that his thesis is "essentially a normative one, concerned with answering the question 'What is the legally correct way to judge proofs?,' not a factual one, concerned with answering the question 'What is the way proofs are actually judged?'" Cohen, *The Role of Evidential Weight in Criminal Proof*, infra at 113. Cohen must be referring here to analysts of the trial process, not to judges or juries, for once again no suggestions for change are offered.

[33] Kaye, *The Paradox of the Gatecrasher and Other Stories*, 1979 ARIZ. ST. L.J. 101.

[34] *Id.* at 106.

[35] *Id.* at 103.

[36] What I hope is undue caution militates in favor of pointing out that I am assuming that a mature discovery process is in place. If that assumption is wrong, then there is room for other forms of incentives to produce evidence. Again, however, they should apply to both plaintiffs and defendants.

[37] Kaye, *supra* note 33, at 107-08.

[38] *Id.* at 101.

[39] There may be other reasons to dislike this result, but none of those reasons are relevant to the present discussion. One might conclude, for example, that rodeo operators can and should build better fences so that this type of situation will not develop. In that case, denying recovery will encourage them to take such actions. Alternatively, one might conclude that it is not asking too much of patrons to keep their ticket stubs, just as any other receipt would normally be kept, and that failure to do so may result in having to pay once more for the privilege of attending the show. These economic considerations are independent of the error minimization concerns on which I am presently concentrating. Again, though, should a court or legislature decide to implement a recovery scheme based upon one of the ideas suggested in this note, the question of how to allocate the risk of factual error will still exist. What if, for example, a patron presents a badly faded ticket stub that may or may not be a stub from a ticket for this rodeo? What if an operator claims every reasonable precaution has been taken, but the defendant disputes the matter? A method of allocating the risk of error on such factual issues will have to be provided, so we will end up once more talking about errors.

[40] *See, e.g.,* Nesson, *supra* note 6; Tribe, *supra* note 15 at 1340-43.

[41] *See, e.g.,* Tribe, *supra* note 15 at 1330 n.2; Kaye, *supra* note 32 at 45 n.41.

[42] *See* Nesson, *supra* note 6.

[43] These are the "heuristics" that the psychologists, in particular, are fond of discussing. *See* KAHNEMAN, *supra* note 15.

[44] *See* D. BARNES, STATISTICS AS PROOF: FUNDAMENTALS OF QUANTITATIVE EVIDENCE 143-230 (1983).

[45] *See* T. FINE, THEORIES OF PROBABILITY: AN EXAMINATION OF FOUNDATIONS 65 (1973).

[46] *See, e.g.,* L. COHEN, *supra* note 13.

[47] Wagner, Book Review, 1979 DUKE L.J. 1071, 1077-81.

[48] Schum, *A Review of a Case Against Blaise Pascal and His Heirs*, 77 MICH. L. REV. 446, 458-60 (1979).

[49] L. COHEN, *supra* note 13, at 266.

[50] If that is wrong, I am at a loss to understand commensurability. The phenomenon that I am about to describe in the text is also present with "commensurable" events for which, according to Professor Cohen, the lowest probability is the crucial one for forensic purposes. That means that if the plaintiff fails to prove any issue, whether commensurable with some other issue or not, then the plaintiff loses. Accordingly, here again errors will aggregate as errors on those discrete issues, whether commensurable with other issues or not. If, however, "commensurable" means that proof of one issue entails proof of the other, then one is simply talking of issues that are in a one to one relationship (if X is true, Y must be true). In that case, nothing useful emerges from Cohen's analysis that I can see. It collapses to the proposition that the plaintiff must prove by a preponderance the one issue in the case.

[51] Moreover, this is true even if one emphasizes a subjective theory of probability. The speculation about subjective theories and degrees of belief is not unrelated to an effort to reduce errors at trial. At least I hope it is not.

[52] L. COHEN, *supra* note 12, at 74-76.

[53] *Id.* at 271.

[54] King, *Causation, Valuation, and Chance in Personal Injury Torts Involving Preexisting Conditions and Future Consequences*, 90 YALE L.J. 1353, 1396-97 (1981); *see supra* note 8.

[55] L. COHEN, *supra* note 13, at 226.

[56] *See* Callen, *supra* note 9, at 10-15.

[57] L. COHEN, *supra* note 13, at 225.

[58] *See* Schum, *supra* note 48, at 460.

[59] L. COHEN, *supra* note 13, at 129-33.

[60] I am ignoring affirmative defenses because they do not seem to affect the analysis.

[61] *See supra* notes 33-49, 52-53 and accompanying text.

[62] The real paradox here is why we hold the plaintiff to the conventionally conceived preponderance of the evidence standard, not why we only require that. *See infra* Part III.

[63] *See* L. COHEN, *supra* note 13, at 58-67.

[64] I also see a value in encouraging the parties to reduce the scope of factual disagreements. Accordingly, there are times that supplemental practices, such as allowing the jury to infer from its resolution of previous factual issues that subsequent factual issues should be resolved in favor of the same party, appear attractive to me. The point would be to discourage the propounding of highly dubious factual propositions that may serve only to "muddy the waters." I am not yet ready to say more about such matters, however.

[65] This is done through such devices as pleading, discovery, pre-trial conferences, and the like.

[66] I am not yet ready to provide a complete catalog of the exceptions to this view. The outlines of justifiable exceptions are obvious enough, however. Case structure should not be left to the choice of the parties when there is some public interest at stake that would suffer otherwise. In addition, there may be sets of cases

in which following the proscription in the text will lead to increasing rather than decreasing erroneous results due to the inability of defendants to generate evidence demonstrating their lack of liability. I must say, it is difficult to conceive of such cases. The blue bus hypotheticals certainly are not examples. If all you know is that a blue bus ran over the plaintiff, and company A runs 80% of the blue buses in town, and company B runs 20%, disallowing a verdict for plaintiffs in such cases is probably going to lead to more rather than fewer errors. Moreover, unless there is reason to believe that the discovery process is not functioning properly, not allowing verdicts in such a case makes plaintiffs bear the entire cost of the lack of evidence in both parties' possession. To be sure, allowing a verdict for plaintiffs in this context will make company A bear more than its share of the cost. The answer to that is to embrace a proportional damages rule rather than make plaintiffs bear all the cost of the lack of more particularized evidence. It is important to note, however, that this is true regardless of the nature of the evidence—whether it is in quantitative form or not.

The textual discussion raises a number of related issues. For example, to what extent is the state justified in imposing on the parties rules concerning such issues as cross-examination or impeachment? Similarly, what is the justification for order of proof rules such as the Best Evidence Rule? These matters deserve fuller treatment. In brief, I would point out that there is a justification for creating a usable form of dispute resolution, since that seems to be an important component of contemporary life. What I object to is the state going beyond that minimum in civil cases, as it does when it forbids private parties to rely on perfectly good evidence. Although there is a justification for a state-created case structure that is neutral in regards to how particular evidentiary proffers must be, parties to litigation by agreement should be permitted to alter that structure. If parties accept an alteration, I see no reason for the rest of us to forbid that choice. What I hope is undue caution again counsels that I point out that other collective values may at some point intrude to forbid parties from resolving their disputes in certain ways, such as fights to the death, for example.

[67] Support for the proposition that courts are reluctant to allow cases to be decided on the basis of "statistical evidence" is greatly exaggerated in the literature. For example, Nesson, *The Evidence or the Event? On Judicial Proof and the Acceptability of Verdicts,* 98 HARV. L. REV. 1357, 1380 (1985), asserts that "[p]laintiffs in such cases would almost certainly lose by directed verdict; the evidence would never reach the jury." The support for that proposition is Guenther v. Armstrong Rubber Co., 406 F.2d 1315 (3d Cir. 1969), where, in reversing a directed verdict *for the defendant* and remanding for a new trial, the court in passing referred to one "probabilistic" argument raised by the plaintiff with disapproval. That, however, was dictum and it was in the context of sending the case back for a new trial. Nesson also cites Smith v. Rapid Transit, Inc., 317 Mass. 469, 58 N.E.2d 754 (1945), in which a verdict for the defendant was sustained. *Smith* is difficult to view as a "statistical evidence" case, however. The plaintiff did not rely on any such evidence. She merely asserted that she was forced off the road by a bus and in addition proved that Rapid Transit, Inc. was the only bus company operating regularly on the road

where the accident occurred. In appraising the strength of the evidence, the court concluded that it was a matter of "conjecture" who owned the bus and that "[t]he most that can be said of the evidence in the instant case is that perhaps the mathematical chances somewhat favor the proposition that a bus of the defendant caused the accident. This was not enough." *Smith,* 317 Mass. at 470, 58 N.E.2d at 755. That is the language of a traditional sufficiency of the evidence decision. Nesson does not mention here the case that *Smith* relied on, Sargent v. Massachusetts Accident Co., 307 Mass. 246, 29 N.E.2d 825 (1940). The *Sargent* court did make the assertion that evidence is insufficient when "mathematically the chances somewhat favor a proposition to be proved . . . ." *Id.* at 250, 29 N.E.2d at 827. However, the decision of the court reversed a directed verdict for the defendant and entered a directed verdict for the plaintiff in a factual context that is easily as probabilistic as that in *Smith*.

An example of a court employing a directed verdict as a sanction for the evidentiary practices of the plaintiff may be Galbraith v. Busch, 267 N.Y. 230, 196 N.E. 36 (1935), where the court reversed a jury verdict for the plaintiff and remanded for a new trial where the plaintiff had failed to call the defendant, who could have considerably dispelled the ambiguity about the nature of the litigated events. *See* Rubinfeld, *Econometrics in the Court Room*, 85 COLUM. L. REV. 1048, 1048 (1985) ("The use of statistical methods for resolving disputes has found increasing acceptance within the adversary system.").

[68] I know of no cases forbidding "statistical" defenses, for example. *See* Lilley v. Dow Chemical Company, 611 F. Supp. 1267 (D.C.N.Y. 1985), resting a summary judgement for defendant in part on statistical evidence.

[69] In a recent article, Professor Nesson has constructed a justification for the status quo based upon his assertions that "[a] primary objective of the judicial process . . . is to project to society the legal rules that underlie judicial verdicts," Nesson, *supra* note 67, at 1357, and that the conventional proof rules "will produce the single most probable story." *Id.* at 1390. He demonstrates this latter assertion by arguing that if a cause of action has two elements, requiring the plaintiff to prove both of them by a preponderance will result in the probability of the conjunction of the two being more probable than the conjunction of any other combination of each element or its negation with the other element or its negation. *Id.* Although this argument has certain similarities to mine, I think it is problematic in its present form.

In the first place, the primary thesis confuses the effect of verdicts, which is to resolve disputes, with the effect of the underlying law relevant to any particular dispute. It is the law that projects substantive standards, not the verdicts in light of that law. The primary "affirming" that goes on at trials is that rules will be enforced if their violation is established, not what the rules are. Moreover, I know of no serious support for the proposition that anyone except the parties to litigation pays much attention to what goes on in the overwhelming proportion of courtrooms in this country. Yet, Professor Nesson asserts that a trial "is a drama that the public attends and from which it assimilates behavioral messages." His support for that proposition is a reference to speculation about the effect of court procedures in eighteenth-century England. *Id.* at 1360 n.14. My difficulties with Professor Nesson's central thesis are elaborated in greater detail in Allen, *Rationality, Mythology, and the "Acceptability of Verdicts" Thesis,* 66 B.U.L. REV. 541 (1986).

In addition to the main thesis being somewhat questionable, the discussion of proof rules is problematic. Professor Nesson asserts that negations are incoherent, implying that if A and B are the plaintiff's factual issues, not-A and not-B as an explanation of reality is not coherent. Nesson, *supra* note 67, at 1389. He proceeds to argue that if A and B are each proven to more than .5, the probability of the conjunction of A and B will be more probable than any other possibility, such as A and not-B, not-A and B, or not-A and not-B. But if these latter possibilities are "coherent" (and if they are not, why are they being compared to something that is?), that implies that their components are. That, of course, simply contradicts the earlier assertions about the incoherency of negations. It also causes another problem. If not-A and not-B is coherent, so is not-A *or* not-B, and not-A or not-B is more likely, given Professor Nesson's hypothetical, than A and B. Professor Nesson does not address why it is that only conjunctions matter. In part, this may be because his consideration of probability theory is inadequate. At one point, Professor Nesson asserts that conjunction problems do not arise with dependent elements. *Id.* at 1387. That is false, of course. The conjunction effect is reduced by the extent of the dependency, but it still occurs unless the two elements are completely dependent one on the other.

In any event, Professor Nesson's discussion of the proof rules is difficult to reconcile with his main thesis. If the concern of trials is to project the reasons for verdicts, it is certainly coherent to project a rule that defendants are not liable if the plaintiff fails to prove all (not "each") of the elements to a specified level of certainty. Again, this possibility is not considered. Thus, it is unclear why we should not be troubled by critiques such as Cohen's even if we accept Nesson's view.

There are other problems with Professor Nesson's argument, as well. For example, he purports to be defending the present proof rules, but his argument would work just as well so long as the plaintiff proved one element to more than .5 and all others to .5. That is not our system, of course.

In one respect, Professor Nesson's argument is similar to mine in its emphasis on stories. His error, in my view, is the attempt to emphasize stories in the context of the conventional conception of trials. His discussion of the persuasiveness of historical accounts is instructive in this regard. *Id.* at 1389. In deciding which historian has provided the most persuasive account, one would not compare an historian's account with its negation. Rather, one would compare one historian's account with that of another. That, of course, is the model I am advancing in this article.

[70] *See supra* text accompanying notes 61-67.

[71] The point about the jury thinking that there are ten possible explanations, although a crude over-simplification of reality, nonetheless presents an interesting problem. Much of the probability debate, indeed just about all of it, proceeds as though the concept of "evidence" were clear and coherent. I suggest it is neither. What counts as "evidence" is what sways a factfinder. That will be in part what is contained in the formal proffers at trial. In much larger part, I suggest it will be how what is proffered interacts with the intellectual tools that the decisionmaker brings to

bear on the problem. These tools will include the decisionmaker's understanding, knowledge, judgement and experience. It is the failure to make this distinction that allows Professor Cohen to criticize conventional probability theory in his contribution to this volume. He proceeds as though the only "evidence" a factfinder has is what is presented at trial, whereas a much more accurate perspective is that the evidence emerges from the process of judgement brought to bear on what is presented. Obviously, there are subtleties here that deserve extended exploration.

[72] Comment, *supra* note 14.

[73] G. SHAFER, *supra* note 14.

[74] Perhaps of overriding significance is the determination of the purposes served by various branches of the law. Whether, for example, the primary concern in tort law is accuracy in verdicts, and justice to the parties conceived of as some version of equal treatment, or whether tort law should pursue other values. Again, though, no matter how such issues are worked out, there will be the need for rules of decision to inform the jury how to decide when faced with uncertainty as to the historical facts.

[75] *See* Epstein, *Pleadings and Presumptions,* 40 U. CHI. L. REV. 556 (1973).

[76] D. BINDER & P. BERGMAN, FACT INVESTIGATION: FROM HYPOTHESIS TO PROOF 16 (1984).

[77] The intuitive sense of litigators that they must provide jurors with coherent presentations is beginning to receive support from the work of cognitive psychologists. Bennett and Feldman have provided an account of criminal trials as involving the construction of stories. W. BENNETT & M. FELDMAN, RECONSTRUCTING REALITY IN THE COURTROOM (1981). As they correctly point out, "[l]egal judgments must emerge from the juror's everyday cognitive repertoire." *Id.* at 64. These authors create a convincing argument that "[w]hether the concern is with how jurors apply legal statutes or how they process large bodies of information, stories provide the most obvious link between everyday analytical and communicational skills and the requirements of formal adjudication procedures." *Id.* at 10. Empirical support for analogous propositions is provided by Pennington & Hastie, *Evidence Evaluation in Complex Decision Making,* 51 J. PERSONALITY & SOC. PSYCHOLOGY 242 (1986), which found that trial evidence was represented in story form.

Both of these efforts involve criminal trials, however, and that is unfortunate, for the work is more easily applicable to civil trials. The problem is that neither can explain reasonable doubt. Pennington and Hastie do not even try, and Bennett and Feldman can only assert that the measure of doubt is "whether an inference is based on a set of connections that are internally consistent and that yield no other interpretations." W. BENNETT & M. FELDMAN, *supra*, at 64. That is not terribly helpful. Internal consistency should be a variable in all types of verdicts, and "yielding no other interpretations" either lacks a measure by which that conclusion is reached or simply refers to certainty, which is obviously somewhat problematic. The accounts of both pairs of authors, however, work well within the structure offered in this article.

[78] I believe that what I am suggesting here is similar to what Professor Cohen has discussed in his contribution to this volume. Unless I misunderstand his argument, one

way to understand my argument is that evidential weight of the evidence in criminal cases is of the utmost importance, and it must weigh heavily in favor of guilt. By contrast, in civil cases the decisionmaker's task is merely to compare the evidential weight provided by the parties for their respective factual assertions.

Although I am quite in agreement with what Professor Cohen has written on evidential weight, I must say that Professor Cohen's inability to say how heavy evidential weight must be in a criminal case to justify a conviction injects into his analysis a difficulty similar to that which he sees in the difficulty of going from a conditional to an unconditional probability assessment under conventional probability notions. In essence, his argument is that other evidence not before the decisionmaker may substantially qualify its willingness to rely on its assessment of the evidence before it. That, however, is also true if one conceives of the evidentiary process in Baconian terms. In either case, what is occurring is that a disinterested third party is analyzing the evidence in light of his or her own experience. If the decisionmaker believes that there is other evidence that should have been presented, its assessment of what has been presented will be influenced by that belief. Thus, if a statistic is provided to a criminal jury that, if true, shows a high probability of guilt, but the jury believes there is good reason to doubt its accuracy, the Pascalian probability it would assign to the evidence before it would be based on both the statistic and the reason it is of doubtful accuracy. *See supra* notes 41-45 and accompanying text concerning the meaning of "evidence." It is only by denying this possibility that Professor Cohen's demonstration of the inaptness of Pascalian accounts of criminal trials is made cogent, and there is no satisfactory reason for denying it.

Perhaps I should also point out an apparent tension in what I have argued here and my discussion of Professor Kaye's work, *see supra* notes 35-39 and accompanying text, where I argued against his suggestion that the failure of the plaintiff to produce evidence could count against him. I was there not denying that phenomenon but rather pointing out that it was a two-way street running against both plaintiffs and defendants.

Ronald J. Allen,
Professor of Law,
Northwestern University School of Law.

RICHARD LEMPERT*

THE NEW EVIDENCE SCHOLARSHIP: ANALYZING THE PROCESS OF PROOF†

I. Evidence Scholarship: From Rules to Proof

When I began teaching evidence seventeen years ago, the field was moribund. The great systematizers of the common law—Wigmore, Maguire, McCormick, Morgan and their ilk—had come and, if they had not all already gone, their work was largely finished. Not only was most of what passed for evidence scholarship barely worth the reading—the same, after all, could be said of many fields of law at most times—but disregarding student work, few scholars were writing regularly on evidentiary matters.

This situation changed with the proposal and adoption of the Federal Rules of Evidence. New talent was attracted to the field of evidence and lead articles on evidence proliferated in the law reviews. Too often, however, these articles followed the model "What's Wrong with the Twenty-Ninth Exception to the Hearsay Rule and How the Addition of Three Words Can Correct the Problem."[1] They were seldom interesting and if they had potential utility it was rarely realized, for the federal rules remain today largely as they were when enacted.[2] The work was, in short, a timid kind of deconstructionism with no overarching critical theory to give it life. But the interest in evidence inspired by the federal rules, even if rarely revealed in memorable work on that topic, was all to the good. Genuinely talented people have become excited about exploring evidentiary issues.

Today I think we are seeing the fruits of this burgeoning of interest and the talent it has attracted. Evidence is being transformed from a field concerned with the articulation of rules to a field concerned with the process of proof. Wigmore's other great work is being rediscovered,[3] and disciplines outside the law, like mathematics, psychology and philosophy, are being plumbed for the guidance they can give.[4]

This volume is a testimony to this third wave (in my teaching lifetime) of evidence scholarship; indeed, nowhere is the concern for proof more central than in that body of scholarship which seeks to build on or criticize mathematical models as modes of proof or as a means of understanding trial processes. What I propose to do in this paper is, first, discuss briefly the ends to which this body of work is directed. Then, I shall develop some themes, stimulated by a portion of Professor Allen's paper and by the larger literature, including other papers prepared for the Boston symposium, of which it is a part. Finally, since this paper originated in an invitation to comment on Professor Allen's contribution to this volume, I shall examine closely Professor Allen's suggestion that rather than rethink our mathematics we should reconceptualize trials.

II. The Use of Models

The literature that relates theories of probability to theories of proof is largely a debate between those who advocate some role for Bayesian modes of inference in understanding or applying the law and those who criticize or reject this position.[5] The language of the debate is often mathematics or essentially mathematical metaphors, but in this case the abstraction of numbers does not preclude considerable passion. The passion suggests that something important is at stake. In fact, there are a variety of stakes, and what is in issue has important implications for the conflicting positions. Too often, however, the stakes are not sorted out, and our ability to assess the relative merits of Bayesian and other approaches to the problem of proof suffers as a result. Thus, our first step is to ask why a particular conceptualization of the proof process is being used or advocated. As Professors Schum[6] and Tillers[7] have suggested in the papers they prepared for this volume, some conceptualizations will be better suited for certain purposes than others.

A. *A Prescription for Action*

The article which triggered widespread interest in the applicability of Bayesian reasoning to trial processes—and arguably still accounts for residual passion—was the comment of Finklestein and Fairley on *People v. Collins*,[8] in which they argued that the real problem in that case lay not in the prosecutor's attempt to use statistical reasoning but in his failure to offer the jury statistical information in the form best suited to its decision-making task—i.e., Bayes's Theorem.[9] Professor Laurence Tribe, in his justly celebrated article *Trial by Mathematics*,[10] took issue with Finklestein and Fairley on both counts.

Among legal academics it is generally agreed that Tribe won this particular debate. As Professor Allen writes in his contribution to this volume, "It is becoming increasingly obvious, for example, that Bayesian approaches can best be used heuristically as guides to rational thought and not as specific blueprints for forensic decisionmaking."[11] This conclusion, is however, premature.[12] Statistical evidence has figured in litigation for more than a century,[13] and in recent years has become increasingly common and complex.[14] For example, a recent LEXIS search of statistical terms done for the National Research Council reports:

> A search of published opinions in federal courts with a computer-based legal information retrieval system reveals the dramatic growth since 1960 in cases involving some form of statistical evidence. Between January 1960 and September 1979 the terms 'statistic(s)' or 'statistical' appeared in about 3,000 or 4% of 83,769 reported District Court opinions. In the Courts of Appeals, the same terms appeared in 1,671 reported opinions.[15]

These uses include not only statistical descriptions of samples and populations, but also uses of the kind Finklestein and Fairley proposed—as identification evidence.[16] With both sorts of uses problems arise because juries are presented with frequentist statistics in situations where Bayesian approaches may be more appropriate to the task at hand.[17]

I shall not dwell on this issue except to make one point. Many participants in the Boston symposium paint with a broad brush in rejecting any place for Bayesian models in trial processes.[18] Often their arguments, or portions of their arguments, read as if statistical evidence has no place at all in trials. Those who criticize Bayesian models of the legal process and the suggested application of Bayesian approaches at trial must confront the reality that statistical evidence is offered in trials every day.

I do not mean to argue that this reality cannot be accommodated by critics of Bayesian models or proposed applications. It is not difficult to imagine a place for statistical evidence in a theory that focuses, as do most non-Bayesian theories of rational proof, on the relative weight of conflicting evidence, the plausible generalizations that trial evidence allows, or the coherence of evidence with some larger plausible story. More difficult challenges for those who reject the Bayesian perspective are to explain why, if statistical evidence is presented at trials, frequentist approaches should be preferred to Bayesian ones,[19] and to reevaluate arguments used to reject Bayesian approaches to proof where they do not accommodate the reality of the regular use of statistical evidence. Thus, arguments from the intuition that the law will not allow verdicts to rest on naked statistical evidence must accommodate or condemn a world in which the only admissible evidence of discrimination is embodied in a statistical model, or where the only admissible evidence linking the defendant to a crime is a hair match.[20] If the answer is, as it appears implicitly to be in the case of fingerprint evidence, that the statistical probabilities are sufficiently high as to be unproblematic, an explanation is required of why one level of irreducible and undeniable uncertainty is tolerable and another is not.

B. *Normative Models*

A second use that may be made of Bayes's Theorem and of competing probabilistic schemes is as a normative model. This use is not completely distinct from the first use I have discussed, as those who advocate instructing factfinders on Bayes's Theorem do so on the assumption that Bayesian reasoning is normatively appropriate for legal factfinding. Attacks on Bayes's Theorem as a normative model may be similarly motivated by the perception of a link between the status of Bayes's Theorem as a normative model for trial factfinding and the appropriateness of furnishing jurors with the information needed to reason in a Bayesian fashion. In part to preclude any movement in the latter direction, critics of "trials by mathematics" deny the possibility that Bayes's Theorem may be a normative model of how trial factfinders *should* proceed.

The practical consequence of promoting Bayesian approaches at trials does not, however, follow from accepting Bayes's Theorem as a normative model for adjudicative factfinding. First, there may be competing norms. Bayes's Theorem may model only part of the reaction we expect jurors to have to information, and emphasizing the Bayesian part may have adverse consequences for other values. The right of juries to nullify, for example, suggests that jury factfinding is valued for reasons that extend beyond the rational weighing of evidence. Encouraging jurors to be more Bayesian may cause them to emphasize the rational, evidence-weighing aspect of their task to the detriment of other aspects we wish to preserve. Going further and providing jurors with aids to Bayesian information processing (imagine jurors of the future punching into a computer their prior probabilities and likelihood ratios after each item of evidence) or even failing to provide jurors with more than statistical evidence may symbolically—and intolerably—denigrate messages we wish to convey about trials or human judgment.[21]

A second reason why Bayesian normativity does not necessarily suggest the desirability of changing current modes of proof is that it may be impossible to provide jurors with the information they need to process information in a Bayesian rational fashion. In particular, we may not be able to inform jurors adequately concerning those prior probabilities that provide the basis for assimilating new information, or about the degree of dependency among various items of evidence. The move usually made by Bayesian advocates to deal with this problem—to allow subjective estimates of these hard (perhaps impossible) to quantify values—is not necessarily sound, for obvious "second-best" problems are raised. Strictly Bayesian decisions with inaccurate priors or mistaken estimates of evidentiary dependency may lead to worse decisions, by the criteria of full-information Bayesian rationality, than the decisions jurors would reach through ordinary unsystematic and non-mathematical processes.[22]

A different second-best problem arises if, as is clearly the case, people are not natural Bayesian decisionmakers.[23] While a factfinder's entire decision-making process might be more rational in the sense of yielding, on the average, more accurate verdicts if the entire case could be presented statistically with accurate probabilities amenable to Bayesian manipulation, it does not necessarily follow that verdicts on the average will be more accurate if only portions of cases are so presented. Professor Tribe,[24] in his critique of trials by mathematics, advanced one important reason why this is so when he suggested that numerical information might dwarf "softer" data. To this observation one might add that Bayesian ways of presenting information might have a greater impact than the ways that information is presented for softer information processing. In each case, of course, the softer data or the data better suited for softer processing may be more informative (diagnostic) than the information that exists or can be presented in a more systematic and quantitative form.

The argument about the "hard" dwarfing the "soft" might apply even if legal factfinders were natural Bayesian reasoners, for they might attend to information based on its compatibility with their natural mode of reasoning. If factfinders aren't natural Bayesian reasoners further problems can arise when attempts are made to encourage them to think about all or part of a case in Bayesian terms. Factfinders might be confused in ways they would not be if they received no advice about how to combine evidence, or they might give the quantitative evidence too little rather than too much weight because the Bayesian decision-making style was unfamiliar or felt onerous. The net result might be less accurate verdicts than the system would produce if all evidence were qualitative and no word of Bayes's Theorem were breathed.

For these reasons at least, the debate about whether Bayes's Theorem is an accurate normative model for all or part of the trial process has no necessary implications for the way we choose to conduct trials. But these potential problems do not mean that it is a mistake to instruct juries or judges about Bayesian reasoning or to introduce more quantitative methods at trials. It is an *empirical* question whether the problems I discuss above are so serious that, even if Bayes's Theorem is an appropriate normative model for all or part of the trial process, no attempt should be made to make legal factfinders better Bayesian reasoners.[25] Let us assume, however, that the problems I point to preclude practical applications of the Bayesian perspective at trial. Of what use, then, is Bayes's Theorem and why does the normativity of the Bayesian approach matter? I will suggest two reasons.

First, Bayes's Theorem may be useful as a heuristic device. This is the use I made of it in my article *Modeling Relevance*.[26] While I am clearly the wrong person to offer unbiased testimony, I use Bayes's Theorem every year when I teach relevance in my evidence course, and it works. Students have a better understanding of issues relating to relevance than they had before I used this approach, and the class develops a common vocabulary with which to discuss relevance-related problems throughout the course.[27] I also continue to defend the propositions that the basic rules of relevance, Federal Rules of Evidence 401 and 402,[28] are captured, as a normative matter, by Bayes's Theorem, and that the question whether probative value is outweighed by such Rule 403 considerations as prejudice and waste of time may be illuminated in a Bayesian framework.[29]

This use of Bayes's Theorem treats the law's rules and procedures as normative and attempts to model them. One might also argue that rationality of the type represented by Bayes's Theorem should characterize legal factfinding and that if the law's norms do not conform to the requisites of Bayesian models of rationality they should be changed so that they do. This is in essence the move that Professor Allen makes when he argues that we should respond to the law's shortcomings vis-à-vis normative models of rationality by changing the way we think about trials. The boldness of this move is self-evident, and Professor Allen's attempt to ground the suggestion is equally original. He argues that the trial format as it currently exists in civil cases is incompatible not only with Bayesian conceptions of rational decisionmaking but with Baconian and other conceptions of rationality as well. Assuming this incompatibility exists, as I shall for purposes of this paper, it poses starkly the questions of whether and when legal norms should conform to norms defined in other systems. Even if the legal system has a norm of formally rational factfinding at its core, it is not clear that this is or should be the master norm in civil or criminal cases. I shall express my doubts on these and other scores as I comment on Professor Allen's specific suggestions in the last portion of this paper.

C. *Descriptive Models*

A third use that may be made of Bayesian or other models of reasoning is descriptive. One might argue, for example, that Bayes's Theorem reflects the way people actually process information. Much of the writing antagonistic to the Bayesian model appears to attack it on descriptive grounds. Indeed, some of the non-lawyers[30] who have offered non-Bayesian models of rational reasoning appear to have been attracted to the law because they saw in legal rules and practices good evidence that the prescriptions of Bayes's

Theorem are inconsistent with one form of apparently rational decisionmaking.

The effort to discredit Bayes's Theorem on descriptive grounds is somewhat puzzling, for we know that Bayes's Theorem does not successfully model the decisionmaking of people who think that they are reasoning rationally or that they are expected to do so. A body of research reveals that people are at best inconsistent in their reasoning when judged by the standard of Bayes's Theorem; they have particular difficulty in appreciating the extremes that cumulated probabilities can reach, and in some respects and on some issues they do not reason in a Bayesian fashion at all.[31]

I believe that the explanation for the puzzling attack on what we know to be a descriptively inadequate model lies in a confusion between the "is and the ought" like that which Professor Tillers notes in the paper he prepared for this volume.[32] First, it appears that the critics of Bayes's Theorem seek to demonstrate its normative inadequacy by showing that it does not adequately model human decisionmaking. Second, those who have offered alternatives to Bayesian rationality seek to support the descriptive adequacy of their schemes by showing that they conform to modes of reasoning which the law appears to prescribe.

Ordinarily one cannot derive an "is" from an "ought," so the first argument appears suspect. However, when the issue is not the abstract issue of what is right but rather whether the legal system should demand a certain kind of reasoning, a good argument can be made that the "is" - "ought" distinction should be dissolved. If people cannot in fact reason consistently in a Bayesian fashion or if the practical problems with Bayesian interventions at trials are too severe, it may be that normative demands for Bayesian type rationality are unsuitable for the law in that they provide no realistic standard to guide decisions. But even if this position is rational, it is by no means self-evidently correct. It is possible that the best way to achieve the law's goal of rational decisionmaking is to impose *Bayesian rational*[33] rules of evidence and procedure on trials even if the law's factfinders are not completely or even largely Bayesian in the way they approach the evidence presented to them.

Since the reverse is also possible, in that the law may choose to follow rules of evidence and procedure that are inconsistent with Bayesian requirements even if factfinders largely follow a Bayesian logic, the law's rules and procedures as norms, to the extent they are inconsistent with Bayesian approaches, are not necessarily informative about the descriptive adequacy of competing models of rational decisionmaking. Reference to legal issues may be a helpful source of speculation, and even of thought experiments, but it cannot resolve the key issues. Empirical experimentation of the kind engaged in by Professors Schum and Martin[34] is needed.

III. Subjective Probabilities and Objective Paradoxes

In my view the key issues raised by Bayesian models of trial decisionmaking are normative. We might ask whether Bayes's Theorem is consistent with the law's expectations of the fact-finding process, and we might ask whether the fact-finding process should conform with the requisites of Bayesian rationality. Some would argue that inconsistencies between Bayesian approaches and the norms of legal decisionmaking are as obvious as the inconsistency between the prescriptions of Bayes's Theorem and the way that people actually process information. Except with respect to the normative characteristics of Federal Rules 401 and 402,[35] I am agnostic on the issue of whether Bayes's Theorem adequately captures the law's expectations about how trials should proceed. I do, however, think that some of the arguments that are used to suggest the inapplicability of Bayes's Theorem as a model of actual trial processes are not as convincing as those who advance them and many of those who resist them apparently assume.

Two such arguments take the form of paradoxes. They juxtapose the apparent requirements of a Bayesian approach with the results we see in law, and argue from the apparent inconsistency between them that Bayes's Theorem is an inadequate model of the law's decision-making expectations. A third argument is that a Bayesian approach to trials does not take account of the weight of evidence and so suggests results—such as finding for plaintiffs in civil cases whenever the ultimate Bayesian probability is above .50, even though the evidence in support of the plaintiff is in some sense unconvincing—that are inconsistent with the results that are and should be reached at trials. Professor Allen relies on two "paradoxes," the conjunction paradox and the gatecrasher's paradox, in arguing that the Bayesian model is inconsistent with the rules that currently govern trials, and he notes the argument regarding the weight of the evidence as well. The latter argument is emphasized by Professors Brilmayer,[36] Cohen[37] and Shafer[38] in the papers they prepared for this volume.

One reason why paradoxes arise when results apparently mandated by Bayes's Theorem are compared with the result that seems intuitively right for a legal trial is that the paradoxes are based on the manipulation of objective probabilities while factfinding is based on subjective probabilities. While it is true, as Professor Savage[39] and others have shown, that subjective probabilities may be manipulated according to the same probability

A. The Conjunction Paradox

Consider first the conjunction paradox. It is well known that in a Pascalian system the probability of two independent events, A and B, equals the probability of A times the probability of B.[40] Professor Allen points out one apparent implication of this fact. If there are two independent elements that the plaintiff must prove to make a case, each representing a state of affairs the existence of which has no implications for the existence of the other and each of which exists with a probability of .75, their conjoint probability is .56, and the plaintiff should recover in a civil suit since the preponderance of the evidence standard is thought to mandate a verdict for the plaintiff whenever his case as a whole is more likely than not (i.e., has a greater than .50 chance) to be true. On the other hand, if there is a third element to the plaintiff's case that is similarly independent of the other two and with the same probability of existence, the probability that the plaintiff, who must prove the simultaneous existence of all three elements, has met his burden in all respects is .42, and the defendant should prevail. Yet the law apparently requires the plaintiff to prove each element of his case by only slightly more than a .50 probability which, given the preponderance standard, appears inconsistent with the requisites of Bayesian rationality and likely to lead to a plethora of wrongful plaintiff's verdicts. Not only does this recovery rule suggest that the law does not respect Bayesian decision-making processes, but the law seems to many to be right in this view, and it is difficult to spot the plethora of wrongful plaintiff's verdicts which a Bayesian approach would avoid. Thus, the law does not and should not respect the Bayesian version of wisdom. Decisions that are incorrect from the perspective of Bayes's Theorem are correct as a matter of both jurisprudence and justice.

There are a number of defenses that one who believes the law is Bayesian rational (or at least is not proved Bayesian irrational by this example) may make. Most defenses avoid the problem, but are no less likely to be correct on this account. One might, for example, argue that the probabilities favoring the existence of the different elements of a plaintiff's case are generally highly dependent, and that successful plaintiffs generally prove the separate elements of their cases by far more than a mere preponderance of the evidence. If these empirical assumptions are tenable, then in most cases in which plaintiffs recover the conjoint probability of the necessary elements is likely to above .50 as well.[41]

A second and more intriguing defense is that the law may be Bayesian rational without mandating a fact-finding process that combines elements in conformity with Bayes's Theorem or its underlying axioms.[42] If one were designing a legal system and attempting to maximize the number of correct verdicts in civil cases, the Bayesian rational strategy would not necessarily be to insist that factfinders assess the conjoint probabilities of separate elements in a Bayesian fashion and render a verdict only when the conjoint probability was in excess of .50. This decision rule is optimal only if the estimates of the probabilities of the various elements are objectively correct. But factfinders are limited to subjective estimates, while the law's goal is objective truth.

A Bayesian rational design for legal decisionmaking should take the subjectivity of factfinders into account. To deal only with the marginal case, we are interested in the contrast between the probability that a plaintiff actually deserves to recover, given that the factfinder *believes* that all elements necessary to the plaintiff's case have a slightly better than .50 chance of being true, and the probability that the plaintiff does not deserve to recover in these circumstances. It may be that total errors are minimized by allowing rather than denying recovery in this situation. In other words, the imperfections of human information processing and the fact that jurors will not have all the information relevant to the issues they must decide may mean that a rule that allows a recovery on conjoint subjective probabilities of less than .50, as long as the existence of each element is thought to be more likely than not, is from a systemic point of view Bayesian rational. More than any other easily applied rule, it may accurately separate those cases in which plaintiffs *objectively* deserve to recover from those cases in which they do not.[43]

B. *The Gatecrasher Paradox*

The paradox of the gatecrashers presents a different problem. Here 499 people have paid admission to a rodeo, but 1000 have attended; the other 501 people sneaked in without paying. Our intuition is that if the rodeo owner sued all the attendees and introduced no evidence other than the preceding statistic he could not recover from anyone, although in each case the probability that the defendant was a gatecrasher was above .50. Assuming that the preponderance of the evidence standard means a probability of liability above .50, this result seems inconsistent with the commands of Bayes's

Theorem. Thus, this paradox also implies that Bayes's Theorem yields results that are inconsistent with the results that are and should be reached at trials.[44]

Professor David Kaye, in a paper cited by Professor Allen, tries to deal with the paradox.[45] He recognizes that subjective probabilities may vary from apparently objective ones, and argues that it may be appropriate to hold as a matter of law that the factfinder's subjective probability in the gatecrasher hypothetical must be less than .50, and that a verdict should be directed for the defendant. This rule, Kaye suggests, gives plaintiffs an incentive to offer more than background statistics. Indeed, it is arguably unreasonable in this case to allow a juror to reach a subjective probability of liability above .50 since the plaintiff's failure to offer more information, when the rule is that he must, itself counts as evidence against the plaintiff. Professor Allen argues that Kaye's incentive rationale avoids the hypothetical, and the claim that is even stronger with respect to the spoliation argument.

If Professor Kaye's argument avoids the hypothetical, it at least suggests a good policy reason why the state should place the burden of acquiring more evidence on plaintiffs even in cases where it is impossible for plaintiffs to do so. The reason is that in the real world we will never be sure if the conditions of the hypothetical are met; much more often than not they won't be met, and more information will be available to the plaintiff. Thus, if we seek a rule which over the run of actual cases will minimize errors, a rule imposing a burden of producing more than background statistics on plaintiffs will do so. The fact that the law rejects the apparent commands of Bayesian rationality and the preponderance of the evidence standard in specific cases in order to minimize error in the long run does not mean that the law is, from a systemic standpoint, adopting a Bayesian irrational rule.[46]

Now let us put aside any policy reasons for directing a verdict against the plaintiffs and give a fully Bayesian argument that the plaintiff should not be able to get to the jury. The argument, following Professor Kaye,[47] is an argument from spoliation. It is that the plaintiff's failure to provide additional information is itself informative. Since the hypothetical provides that the plaintiff has offered evidence which on its face barely exceeds the more-probable-than-not threshold, any spoliation inference that should rationally be drawn against the plaintiff will entitle the defendant to a directed verdict.

Let us consider the objections that may be made against this position. The first is Professor Allen's argument that it may be as easy for the defendant to offer non-statistical evidence as it is for the plaintiff to do so. Indeed, in the gatecrasher situation the defendants are likely to be in a better position than the plaintiff to offer non-statistical evidence since if tickets were issued some innocent defendants might have kept their ticket stubs and all defendants could testify under oath that they had in fact paid. But despite these possibilities, Professor Allen's critique misses an important point. This point is that the law currently imposes a burden on plaintiffs—but not on defendants—to go forward with evidence. The question is whether at the close of the plaintiff's case, at a time when the defendant has had no chance to present evidence, a reasonable jury could find that the plaintiff's case is more probable than not. The fact that the plaintiff's case might be strengthened when the defendant presents or fails to present a case does not at this time enter into the court's decision. So long as this is the preexisting rule and plaintiffs have strong incentives to present accessible probative evidence, granting each defendant a directed verdict in the gatecrasher situation will be consistent with a Bayesian evaluation of the probabilities whenever it appears that other evidence supporting the plaintiff's claim would have been available had the claim been meritorious.

In the gatecrasher case, for example, if the plaintiff really thought that defendant A was a gatecrasher, why didn't he call A to the stand, put him under oath, and ask him whether he had paid for his admission? Since the notion that one "vouches" for the credibility of his witnesses has been generally discarded and "hostile" witnesses can be impeached, there is little to be lost. One cannot assume that A would lie and, even if A had lied, cross-examination might have revealed A's deception. A possible reason for plaintiff's decision not to call A is that the plaintiff, perhaps based on A's demeanor during a discovery deposition, had reason to believe that A was more likely than not telling the truth when he claimed he had payed. Thus the failure to call A may be informative, suggesting that it is more likely than not that A paid.[48]

There are at least two objections that might be made to this attempt to reconcile the gatecrasher paradox with Bayesian rationality. The first is Professor Allen's claim, directed against Kaye, that it avoids the hypothetical; the second is that we would decide the same way if 80% of the attendees at the rodeo had crashed the gate. In the latter circumstance, despite the spoliation inference, a reasonable jury at the end of the plaintiff's case might believe that it was more likely than not that a particular defendant had failed to pay. I shall deal with this latter objection first.

This possibility is embodied in another familiar problem case which is commonly regarded as inconsistent with the view that the law contemplates Bayesian rational factfinding. This case is usually expressed not in terms of gatecrashers but by reference to red and blue taxi cabs or blue and green buses, one of which has hit the plaintiff in a situation where both negligence and the impossibility of any more specific identification is clear. The argument proceeds on the assumption that a court would properly grant a directed verdict for the defendant in this situation. In my view the assumption is wrong, or if it is correct it is correct for a reason which suggests no necessary inconsistency between the application of a Bayesian model (in a world where the preponderance of the evidence standard is conceptualized as a greater than .50 probability that the defendant is liable) and the rules governing proof at trials.

First, consider the matter at the level of intuition. Assume we are confident that the plaintiff has actually been injured by a negligent bus driver, that the bus belongs to one of two companies, and that it is impossible for the plaintiff to offer any evidence about which company's bus was responsible. Is there anything offensive about denying the defendant's motion for a directed verdict at the close of the plaintiff's case? I see no great cost in allowing such a case to survive a directed verdict and forcing the larger bus company to present evidence—e.g., testimony as to its bus schedules, testimony from its drivers that on the night in question they had injured no one—or risk liability. After all, statistical evidence is introduced in trials all the time. In a sex discrimination suit, for example, the plaintiff's case may rest on an equation which is not only acknowledged to be an imperfect representation of the defendant's employment policies but, if perfect, indicates some probability (e.g., one chance in twenty) that the data would be equally or more suggestive of discrimination although the defendant had done nothing illegal.[49]

One troublesome observation that some people have made in discussing the blue bus hypothetical is that if such suits are allowed to proceed the larger bus company will end up paying for all the bus accidents in town. Putting aside the question of whether this is more troublesome than generating a string of injured and deserving plaintiffs who never recover and the fact that in most future bus cases plaintiffs either will be unable to explain their lack of non-statistical evidence or defendants will be able to rebut the plaintiffs' claims, let me simply note that the objection if well taken does not threaten a Bayesian characterization of trial fact-finding norms. Bayes's Theorem as a normative model only tells us how information should be processed. It doesn't tell us when other values should outweigh a probabilistic judgment that favors the plaintiff. Thus, a decision to direct a verdict for defendants because it is unfair that the larger bus company always pays does not call into question the claim that the Bayesian approach captures the essential norms of trial factfinding any more than does the exclusion of privileged information, which prevents a fully informed assessment of all the facts.

Finally, it might be argued that the law of the blue bus hypothetical is inconsistent with my characterization of what a sensitive intuition suggests. The case most often cited on this point is *Smith v. Rapid Transit Company*.[50] *Smith*, however, is only one case, and it is not the hypothetical case of the colored buses. Few consciences would have been shocked had *Smith* been decided the other way. *Smith* did not involve naked statistical evidence. There was other evidence, relating to bus schedules and the like, suggesting that if Mrs. Smith's story about her injury and her inability to offer more detailed information was believed, then the bus company was indeed responsible for her accident. While these additional facts might argue against my position if the decision in *Smith* were clearly correct, I think they undercut the power of *Smith* to persuade. Certainly those who make the case against allowing verdicts based on naked statistical evidence do not claim to be able to extend their arguments to situations where there is additional information of the kind that existed in *Smith*. Indeed, in somewhat analogous situations such as those in *Summers v. Tice*[51] and *Sindell v. Abbott Laboratories*[52] courts have been praised for creative solutions allowing recoveries to plaintiffs who because of the nature of their injuries cannot prove the responsibility of particular defendants by a preponderance of the evidence.

This brings me to the charge that I, like Kaye, have ducked the hypothetical. I shall respond by confession and avoidance, but the avoidance shall precede the confession. The problem of the gatecrasher may be acknowledged and avoided by pointing to the uncertainty—given the subjectivity of factfinding—that will exist about whether the conditions of the gatecrasher hypothetical are met. These conditions are that the only evidence against the defendants is that they are members of a group, 50.1% of whom owe the plaintiff money, and that the plaintiff cannot possibly bring any more specific evidence of violation against any of the defendants. These are stringent conditions and only the plaintiff will know fully—and then only if some effort is expended—if they pertain. If plaintiffs are allowed to get to the jury by showing that naked statistics are the only evidence available, they will have an incentive to falsely create the appearance that this is the case. In particular, they will seek to avoid the expense of ascertaining whether more particular evidence is available and the possibility that they will uncover evidence that exonerates innocent defendants who could not exonerate themselves at trial.[53] The result is that a general rule directing verdicts for defendants in gatecrasher situations is likely, over the run of cases, to lead to more accurate factfinding by the law's preponderance of the evidence standard. A rule, in other words, works better than subjective case-by-case determinations. A decision rule which appears not to be Bayesian rational in a particular case is consistent with a Bayesian rational system.

But suppose we are not allowed to appeal either to Kaye's value of providing plaintiffs with incentives to uncover specific information or to my value of achieving more correct decisions by the law's more-probable-than-not criterion over the run of cases. Then we must confront the hypothetical head on. In these circumstances, where the plaintiff cannot present non-statistical evidence and we cannot refer to other values, I think the intuitions that the plaintiff should not recover and that the law would not allow recovery are incorrect. They are incorrect because both our intuitions and the law's apparent rules are based on situations actually encountered or on the contemplation of situations which implicitly contain other ingredients, such as the possibility of spoliation inferences. If we encountered pure naked statistical inference cases like the gatecrasher hypothetical with any frequency, both our intuitions and the law's rules would change. Indeed, in cases like *Sindell* or in cases where statistical evidence is often the best if not the only mode of proof, both our intuitions and the law accept judgments that rest largely on the weight of impersonal, quantified probabilities which on their face make clear an irreducible probability of error.

If the pure gatecrasher case arose with any frequency, we would begin to feel sorry for starving rodeo operators (or we would strongly object to the hefty ticket prices that honest people were charged), and the law would accommodate itself to the situation. The simplest accommodation would be to allow the statistical evidence to place on defendants the burden of going forward, but this would only work if it were possible for defendants to present exonerative evidence. In the pure hypothetical, this possibility is ruled out. A second response would be to stick with the traditional rule to encourage rodeo operators to take steps to prevent gatecrashing. If this possibility is also ruled out, the hypothetical world in which gatecrasher cases continue to arise becomes increasingly distant from the world we live in. If we may nevertheless generalize from our legal world, we see, in the movement toward comparative negligence and, in cases like *Sindell* joint responsibility, the likelihood that strictly legal solutions, probably involving some form of proportionate recovery and liability, would arise. What is unlikely is that a rule directing verdicts against plaintiffs who can present only naked statistical evidence would endure. This seems right to us only because in this world our intuitions and the law have been conditioned by superficially similar cases from which careless generalization is easy.

Consider the following hypothetical case. 501 people pay to attend a rodeo. They pay their money and are allowed in but receive no ticket stubs. Then 499 people crash the gate. The incensed manager calls off the production before it begins and does not refund any money. A suit is brought by X to recover his admission fee. X can offer only the statistical evidence and—although it is difficult to imagine the situation—his failure to offer other evidence has no implication for the likelihood that he paid his way in. Will the rodeo manager be allowed to deny recovery to X and everyone like him?[54] Why should the resolution of the gatecrasher's paradox turn in the pure case on the identity of the moving party? Should the law allow this situation to exist where no other values, like the stability of expectations, exist? I think not.[55]

C. *The Weight of the Evidence*

Finally, let me turn to the weight of the evidence. The most fundamental and interesting challenge to Bayesian models of legal decisionmaking is that they do not describe how factfinders process evidence and cannot in principle do so since factfinders respond to a sense of the weight of the evidence which is incommensurate with Bayesian probabilities. The argument is that this *in principle* disjunction between what Bayes's Theorem requires and how legal factfinding proceeds means that Bayes's Theorem is both descriptively and normatively an inadequate model of what the law expects of its factfinders since it leads to results which would not and should not be reached by the legal system. For example, situations may exist in which on the balance of probabilities it appears more likely than not that the defendant is liable to the plaintiff but the plaintiff's evidence is so slight that he will not and should not be allowed to recover. If so, Bayes's Theorem neither captures the law's norms nor, where it differs from the law's norms, makes a normative claim that we should respect.[56]

I do not wish to assert that there is no *in principle* distinction between Bayesian models and weight-of-the-evidence models of decisionmaking, for coherent and inconsistent logical systems have been generated on the basis of these different perspectives. However, with respect to the law, I do not believe that a convincing case has been made that the Bayesian model yields results that are inconsistent with the results that arise when a court or jury evaluates the weight of the evidence in accordance with legal norms. The intuitive argument that it does generally rests on the proposition that one can imagine civil cases in which the probability that the defendant is liable appears from a Bayesian perspective to be above .50, but such scanty evidence has been offered in support of the plaintiff's case that a jury is properly not convinced that the defendant's liability has been shown by a preponderance of the evidence.

I question whether cases like this hypothetical in fact exist. I think the examples generally offered to support this proposition implicitly assume a Bayesian prior of 1:1, or something close to it, and then assume that whenever the likelihood ratio for the scanty evidence that is offered favors the plaintiff, applying Bayes's Theorem yields a posterior probability above .50. From this it is assumed that the plaintiff has, according to the Bayesian model, proved his case by a preponderance of the evidence although the case is unconvincing.

The mistake here is in implicitly or explicitly setting the Bayesian prior at 1:1. Although a civil jury should not favor one side over another, this is, in terms of my model of relevance, a matter captured by a regret matrix and not by Bayes's Theorem.[57] Bayes's Theorem refers solely to the processing of information—that is, to an empirical process. The law's norms cannot set the Bayesian prior; the "ought" of not favoring either side cannot determine the "is."[58] What then is the proper prior? This is, of course, the sticking point in all attempts to apply a Bayesian approach to real world decisionmaking. Although any subjective estimate of the prior odds is often tolerable when we are discussing the *in principle* applicability of Bayes's Theorem as a normative model of legal reasoning, a similar looseness is not tolerable when we are concerned with the posterior probabilities at which a Bayesian factfinder should arrive.

I suggest that in all cases, civil and criminal, the prior probability of the matter in question be set at an odds of one to one less than the number of actors in the world. In other words, at the outset of the fact-finding process the defendant should be thought to be no more likely to be guilty or liable than he would be if responsibility were allocated by a lottery in which every actor in the world had the same chance of being selected.[59] With absolutely no information, I see no other empirically justified starting point.[60] To pursue the argument, imagine a Bayesian information processing machine that starts with this suggested prior. The Bayesian machine may of necessity receive evidence sequentially, but it evaluates information together. Thus, the first likelihood ratio it calculates is:

$$\frac{P(A|H)}{P(A|not\text{-}H)}$$

where A is an item of evidence and H is an hypothesis—e.g., that the defendant acted negligently—at issue in the case. The second likelihood ratio the machine calculates is:

$$\frac{P(A\&B|H)}{P(A\&B|\text{not-}H)}$$

and so on until:

$$\frac{P(A\&B\&\ldots N|H)}{P(A\&B\&\ldots N|\text{not-}H)}$$

is calculated. This procedure allows the dependency of information to be taken fully into account, and it provides for the anomalous situation in which the direction in which evidence cuts is reversed upon the receipt of new evidence.[61] If after receiving each item of evidence the Bayesian machine spewed out a tentative posterior probability, we would see that early evidence drastically reduced the odds against the hypothesis. For example, in an auto accident negligence suit where a hit-and-run defendant denied involvement in the accident, the evidence that the defendant drove a car would reduce the odds of the defendant's responsibility from one to whatever number of people inhabit the world to one to whatever number of people drive cars. Evidence that the defendant usually drove in the vicinity of the accident would reduce the odds against the defendant's involvement by a further substantial amount. Ultimately, however, to prove that it was more likely than not that the defendant was responsible, more specific evidence would be required. I will assert—and can only assert, for this example must be left at the level of thought experiment—that whenever the final probability spewed out by the Bayesian machine proceeding in this fashion exceeded .50 on all the evidence presented, including permissible spoliation inferences, the evidence would be sufficiently weighty for the law to allow the verdict to stand. If so, there is no *in principle* distinction between decisions apparently justified by a Bayesian approach in a world where the preponderance of the evidence standard is conceptualized as a greater than .50 probability that the plaintiff's claim is correct, and those decisions that the law's norms require.[62]

I think that the description of the Bayesian machine takes care of the suggestion that Bayesian conclusions are in principle different from and less in accord with the law's norms about trial decisionmaking than the decisions reached by the non-Bayesian weighing of conflicting evidence. While the defense of Bayes's Theorem might thus stop here, there is one final point I would like to make, a point which involves accepting the implicit premise that the appropriate Bayesian prior in a civil case is an odds ratio of 1:1. I do this not because I think the assumption is correct but because it helps illuminate another problem caused by deriving objective paradoxes from subjective probabilities.

The claim is basically that if Bayes's Theorem is a model of trials the factfinder applying a preponderance of the evidence standard must always find for the plaintiff when the hypothesis the plaintiff seeks to establish has a better than .50 chance of being true, and must find for the defendant in all other instances. This argument, in Professor Allen's terms, is the argument of negation. Bayesian probabilities on H and not-H must sum to one. Thus there appears no way to withhold judgment and ask for more information.

The validity of this argument depends on how the law's standard is phrased. If the law commands factfinders to decide for the plaintiff whenever the posterior Bayesian probability exceeds .50 and for the defendant otherwise, the law is treating subjective probabilities or degrees of belief as objective ones and there is no third option. However, if the law recognizes the factfinder's subjectivity and commands factfinders to find for the plaintiff when they are *satisfied* that the posterior probability is above .50 and for the defendant otherwise, factfinders have three options. They should find for the plaintiff when they are satisfied that there is more than a .50 chance that the plaintiff deserves to recover. They should find for the defendant when they are satisfied that there is a .50 chance or less that the plaintiff should recover, and they should find for the defendant when they see gaps or other flaws in the plaintiff's case that makes them feel uneasy about giving the plaintiff the verdict even though, if forced to guess, they would think it more likely than not that the plaintiff should prevail.

If this third option seems unfamiliar in civil cases, it is not unfamiliar in the law. In criminal cases a "reasonable doubt" is commonly defined for jurors not in probabilistic terms but as the kind of doubt that would cause them to hesitate in a matter of great personal import. It is not irrational or inconsistent with the expectation that factfinders will reason in a Bayesian fashion to place on plaintiffs the burden of removing the factfinder's uncertainty about whether a probability that apparently exceeds .50 actually does so.[63] Factfinders are not Bayesian machines, and it may be that fact-finding accuracy will in the long run be enhanced rather than hindered by respecting difficult to identify sources of discomfort.[64]

Moreover, there is a common law tradition that jurors should not reach verdicts based largely on speculaton. Speculation may yield a subjective probability or degree of belief in the plaintiff's case that is above .50, but the probability estimate may be based on so little evidence that it is insufficient as a matter of law to justify a plaintiff's verdict no matter how high it is. Thus, the law may expect factfinders to reason in a Bayesian fashion, but it must also recognize that this expectation will at best be imperfectly achieved. In the resulting second-best world, it may make sense, even if the only concern is accurate factfinding, not to require the rigid transformation of uncertain Bayesian posterior probabilities above .50 into plaintiffs' verdicts. Legal factfinders, unlike our ideal Bayesian machine, may start with unduly high prior probabilities, or they may overvalue redundant evidence because it is difficult to appreciate the implications of dependency. Indeed, it may be that errors like these lead to difficult to articulate sources of unease with what on a conscious balance of the probabilities appears to be the appropriate final verdict.[65]

IV. Reconceptualizing Trials

If, as I have suggested in the preceding section, the paradoxes that Professor Allen and others draw on to show an inescapable disjunction between a Bayesian rational fact-finding process and the fact-finding process apparently prescribed by legal norms do not exist or do not suggest such a fundamental disjunction, it is unnecessary to proceed as Professor Allen does to "reconceptualize" trials. Thus, on my own view of the situation there may be nothing more to say.[66] However, since I have been invited to comment on Professor Allen's arguments and have hardly discharged this obligation to this point, I shall assume the disjunction Professor Allen articulates and focus on his suggested solution.

Before I turn to the specifics of Professor Allen's proposal I wish to emphasize the boldness of his scheme. It is bold in two ways. First, Professor Allen does not follow the lead of most critics of Bayesian rational models and argue that if real decisionmaking does not conform to our abstract models then the models should be changed. Rather he accepts the models, or at least the premises of rationality built into them, as normative and argues that if trials are not rational in the same way the models are, it is the rules of trial proof that should be changed. Second, although Professor Allen speaks about *reconceptualizing* trials, he in fact argues for *reformulating* trials. That is, he is not just suggesting a new way of conceiving of trials or thinking about them. Rather, he is prescribing a new set of legal rules. These rules have implications for the admissibility of evidence, the instructions given jurors, and the issuance of directed verdicts.

While I genuinely admire Professor Allen's boldness and creativity, I have considerable difficulty both with the specific suggestions he makes and the more general position that underlies them. Professor Allen's analysis begins with the observation that in civil cases the law should be indifferent between errors that favor plaintiffs and errors that favor defendants and should aim at minimizing the overall error rate in the long run. In other words, for Professor Allen the rules of proof should be neutral between the parties. This position is facially consistent with the preponderance of the evidence standard in civil cases, and so is firmly rooted in a legal norm. Professor Allen goes on to imply that rational decisionmaking, of a Pascalian or Baconian kind, will tend to bring about this state of affairs. Since he sees inconsistencies between the decisions likely to be reached using either of these two general approaches and the decisions that the current rules of trial practice are likely to yield, he argues that the rules of trial practice should be changed.

This approach, for all its originality, leaves important questions unanswered. The first is why we should define a rational system of proof for civil trials solely in terms of the minimization of equally weighted plaintiff and defense errors. To say that this decision respects the law's norm or definition of rationality as articulated in the civil burden of proof—the answer Professor Allen's discussion implies—will not do, for according to Professor Allen the pursuit of this norm requires changes in other norms of trial proof which on their face have as much claim to be respected as that implied by the civil burden. Moreover, it is clearly not irrational for a system of trial proof to pursue ends other than or in addition to the minimization of verdict errors, for trials serve values other than the dispensation of justice by the preponderance of the evidence standard.

Professor Allen acknowledges this point,[67] but his argument at times comes perilously close to denying it. For example, the scheme he suggests relies more than the current system on effective pretrial discovery. This suggests the likelihood of increased disputing over discovery-related matters. The potential social costs of discovery disputes are obvious, and the legal system's rules might well recognize them. The situation is similar where values reflect concerns less easily measured. For example, Professor Allen writes:

> The only justification for such rules is that some evidence cannot be understood, which includes being put to an inappropriate purpose such as unfairly prejudicing a party.[68]

This assertion is not self-evidently valid. With respect to the admissibility of evidence, privileges provide a clear counter-example, for they exclude evidence for reasons other than those Professor Allen asserts. With respect to the sufficiency question, it is similarly possible for values other than rational factfinding to prevail. For example, we may not want to allow a case like that of the gatecrashers to rest on statistical proof because we regard this as intolerably dehumanizing. The defendant is being found liable not for who he is personally but for a category to which he belongs. Just as we would not allow a defendant in a job discrimination case to offer statistical evidence showing that on the average blacks or women perform less well than whites or men, so we may not wish to allow civil suits to proceed based solely on evidence that the defendant belongs to some broadly defined class more than half of whose members owe the plaintiff money. The symbolic nature of allocating responsibility on the basis of the statistical characteristics of members of a broadly defined class may be especially and properly disvalued by the law when stigma attaches to the finding of responsibility, as it does when a person has been shown to have acted carelessly or, even worse, to have sneaked into a rodeo without paying.

Finally, it is not clear what it means to be neutral between parties. Professor Allen's proposal, for example, will largely foreclose directed verdicts for defendants at the close of plaintiff's cases, for the jury under his scheme will be charged not with weighing the general plausibility of the plaintiff's case, but with weighing the plausibility of the plaintiff's case against that of the case which the defendant offers. This allows plaintiffs to impose greater costs on defendants than they can under the current system, and may affect the propensity of plaintiffs to bring marginal or "strike" suits. Also, those cases that under the current system do not get to the jury will, by virtue of their weakness, almost always result in judgments for the defendant after the defense case is presented. Consequently, unnecessary costs, which are deadweight social losses, will be imposed on defendants.

A related question that Professor Allen leaves unanswered is why either a Pascalian or Baconian approach is an appropriate model of rationality for the trial system. I will not discuss this issue in detail, except to point out that a strength of Professor Allen's general argument becomes a weakness when the discussion switches to specific ways to reformulate trials. If Professor Allen is correct in his claim that both Pascalian and Baconian models do not fit the rules of proof that govern trials, trial processes appear irrational when judged by either of the predominant formal approaches to the question of human rationality. This perspective is original and is a strength of Professor Allen's argument, since it suggests that whatever our conception of rational decisionmaking, if that is our goal we should be troubled by the procedures that have emerged to govern trials.

This strength becomes a weakness, however, when specific reformulations of trial practices are suggested. Pascalian and Baconian approaches suggest different ways of evaluating the rationality of trial decisionmaking and of minimizing error by appropriate criteria of rationality. Having proposed a reformulation of trial procedures, Professor Allen must choose a model of rational decisionmaking against which his proposals can be judged. He does not. Even if Professor Allen's proposed revisions will eliminate certain features of trials that are paradoxical by reference to both Pascalian and Baconian models, it does not follow that the revised system will yield rational judgments by reference to the criteria of either.

In short, when I read in the introduction to Professor Allen's paper that he intended to offer a reconceptualization of trials, I thought he had in mind a project like that which Professor Nesson[69] and Professor Tribe[70] before him undertook. That is, I expected him to elaborate a model of the trial which attended to values other than rationality but—and this would have been the advance over Nesson and Tribe—had a clear place in it for rational information processing of a Baconian or a Pascalian kind. Instead, more than any other recent writer on this topic, Professor Allen is willing to let some formal model of rational decisionmaking dominate our views of how trials should proceed. In his scheme the normativity of such models lies not in the extent to which they are consistent with the law's rules about how decisionmaking should proceed but in their postulated compliance with a barely argued first premise that the civil law elevates rational, neutral decisionmaking above all other values. But there is a circle here, for it is not at all clear how *rational* decisionmaking proceeds. In suggesting that Pascalian and Baconian models, at least where their implications overlap, set a standard of rationality by which the appropriateness of legal procedures may be judged, Professor Allen elevates formal mathematical norms above legal ones. I, for one, am troubled in a way that I am not by efforts to formally model the rationality that legal rules require.[71]

Turning from Professor Allen's idea of what it means to reconceptualize trials to his specific proposals for changes in how we should proceed, I find yet more that troubles me. I alluded to some of these matters when I discussed some values that Professor Allen's perspective ignores—for example, the social costs of discovery disputes and increased defense expenditures in cases that now end in directed verdicts at the close of the plaintiff's case. Some of my other problems with Professor Allen's approach are even more fundamental. For example, Professor Allen's suggestion that the trial factfinder compare the probability of a fully specified plaintiff's case with that of a fully specified defendant's case assumes that neither party has an advantage in developing the facts of what occurred, and that the party who can tell the most probable story deserves to win. This suggestion seems to me to be either a prescription for what now occurs, including the possibility of a negation defense, or a move likely to lead to substantial injustice.

Suppose a defendant moves for a directed verdict at the close of the plaintiff's case. He is asking the court to find that under no reasonable reading of the evidence could a jury conclude that the plaintiff's claim is more likely than not to be true. If the plaintiff's case is as weak as the defendant claims, what can be gained from requiring the defendant to present a more plausible story than that implicit in his motion for a directed verdict; i.e., the event had to have happened in some other way? Nothing, I would argue.

If Professor Allen's proposal means that defendants would be required to offer particularized counter-stories in such circumstances, costs would be incurred. One, to which I have alluded, is the cost of forcing a defendant to mount an otherwise unnecessary defense. A second is the possibility that an unjust decision will result because the defendant does not know how to explain the phenomenon that the plaintiff seeks to hold him responsible for and so cannot tell a particularized counter-story.

Consider, for example, a malpractice action brought by a plaintiff of age ten against the doctor who delivered him. The claim is that a symptom—let us say a tremor that began affecting the plaintiff's left arm at age nine—was attributable to the way the doctor delivered him ten years before. The doctor may have no recollection of the delivery, for the plaintiff's birth may have been one of thousands of then unremarkable births the doctor aided, and the doctor may have no clue as to what in fact causes the plaintiff's tremor. Thus, all the doctor can say, unless perhaps he pays the costs of fully diagnosing the plaintiff's condition, is that whatever the cause of the tremor that emerged at age nine, it does not lie in an earlier unremarkable delivery. Surely we do not wish to impose the costs of discovering the cause of the plaintiff's condition on the defendant doctor. Not only may the cause be ultimately undiscoverable but, even if it is potentially discoverable, it may be unascertainable without the full cooperation of the plaintiff and his family.

Professor Allen suggests that in some circumstances his preferred mode of proceeding resembles the status quo. He writes, "A jury in evaluating evidence should conclude that the probability of the defendant's versions of reality is the probability of the plaintiff's versions subtracted from one only if it feels that it has before it all relevant versions of reality."[72] To say that an event happened in some way—that is in any possible way—other than as the plaintiff claimed is to offer the jury all relevant versions of reality since the plaintiff's claim and the defendant's negation exhaust the possible ways in which the accident might have occurred. If this sort of negation is allowed, then trials can proceed as they do now. If Professor Allen does not consider general demurrers of this sort exhaustive of possible realities, injustice may result when the plaintiff's claim is improbable, as in the malpractice example, but an innocent defendant cannot be expected to know much about what really happened.

Now consider the situation where the plaintiff's case-in-chief is under the current legal standard sufficient to survive a motion for a directed verdict. Here the defendant fails to offer a counter-story at his peril.[73] Thus it is not surprising that civil defendants, unlike criminal defendants who sometimes are content to present no evidence and argue that the prosecution has not proved its case beyond a reasonable doubt, invariably offer a defense when confronted with a plaintiff's case sufficient to get to the jury. While we have no systematic descriptions of what defense cases in civil trials typically consist of, it is likely that they present story lines that conflict with plaintiffs' stories.[74] The conflicting story may be as simple as "the plaintiff's witnesses cannot be believed" or as complex as is necessary to explain the plaintiff's injury while negating any implication that the defendant is responsible. Where a counter-story is offered, it is likely that factfinders decide by evaluating the relative plausibility of the competing versions. This is precisely what Professor Allen would have them do.

But Professor Allen writes as if deciding in this way is inconsistent with a system in which a "preponderance of the evidence" means "more likely than not" in the sense conveyed when we say that to prevail the plaintiff must show that there is at least a slightly better than .50 chance that his version of reality is correct. Thus in Professor Allen's proposed scheme a plaintiff who offered a story or stories that had a .40 chance of being true would prevail over a defendant whose version or versions of what happened had only a .30 chance of being right.

Again, however, I think we are misled because objective probabilities are being used to analyze a situation in which the factfinder employs subjective estimates. The factfinder is not asked to evaluate the true probability of the plaintiff's case. Rather he or she must decide *on the basis of the evidence in the case* whether the plaintiff's case is more likely than not—that is whether it has a more than .50 chance—to be true. In most cases when a factfinder is presented with a plaintiff's story that has a .40 chance of being true and a defendant's story that has a .30 chance of being true, it can conclude that, on the basis of the evidence before it, the plaintiff's story is more likely than not—not simply more likely than the defendant's story—to be true. This is in part because if none of the evidence presented suggests other explanations for the incident which objectively has a .30 chance of being true, probabilities evaluated on the basis of the extant evidence will *subjectively* sum to more than .70—perhaps even to 1.0. The factfinder will not be conscious of other possibilities and so will not accord them some probability.

Where the factfinder *is* aware of other possibilities a spoliation inference may close an otherwise apparent gap. Assume, for example, that a plaintiff presents a story that strikes the factfinder as having a .40 chance of accurately recounting what has happened. The defendant presents an exonerative story that the factfinder estimates as having a .30 chance of describing what occurred. An additional exonerative possibility—also with an apparent .30 probability of describing what has occurred—is obvious to the factfinder. May the factfinder find that the plaintiff has proved his case by a preponderance of the evidence if this requires a rational conclusion that there is somewhat more than a .50 chance that the plaintiff's story is true? I think so. The second exonerative possibility that was obvious to the factfinder should have been obvious to the defendant as well. The defendant's failure to offer evidence in support of this possibility suggests that the defendant, and perhaps the plaintiff as well, knows from evidence not presented that this apparently plausible theory is not supported by the facts. Thus, the factfinder may rationally rely on the spoliation inference and treat the case as if there is a four in seven chance that the plaintiff's version is correct.

Where the spoliation inference is unreasonable given the evidence, and the situation is as described above,[75] I believe that the preponderance of the evidence standard should ordinarily and usually does require a verdict for the defendant even though the plaintiff's story is more plausible than the account the defendant explicitly offers. It is, however, difficult to imagine such a situation, for in rebutting the spoliation inference, the defendant is, in effect, advancing the second exonerative story. He is seeking to persuade the factfinder that his failure to present more specific evidence in support of a plausible exonerative story does not suggest that the story is untrue. Specific supporting evidence is unavailable[76] for reasons that are neither conditioned on the truth of the story nor within the defendant's control.[77]

A related situation exists when the defendant's strategy is to destroy the plaintiff's case through cross-examination, and the defendant offers little or nothing in the way of a case-in-chief. By Professor Allen's scheme it might appear that in this situation the plaintiff should prevail, as even a highly questionable story appears to outweigh no story at all. But our instincts tell us this cannot be right. It is not right because in relying on cross-examination the defendant is telling a story, and if the cross-examination is devastating the defendant's story is more plausible than the plaintiff's. The defendant's story is that the sequence of events resulting in injury did not occur in the way it must have for legal liability to attach to the defendant. The sequence has this appearance only because the plaintiff has found witnesses who are poor observers, forgetful, or deceptive.

That the defendant cannot explain how the injury occurred does not preclude a defense verdict. On the evidence before it the factfinder cannot conclude that it is more likely than not that the injury occurred in a way that justifies a decision for the plaintiff. In attacking the plaintiff's case through cross-examination, the defendant is suggesting numerous ways in which the facts, if relayed by credible witnesses, might exonerate him. When cross-examination has been artful, there may be no need for a defense case-in-chief.

Thus, I disagree with Professor Allen's suggestion that defendants be required to respond to plaintiffs assertions with "equally specific and affirmative allegations rather than with simple denials."[78] I do not think Professor Allen's proposed reformulation of the rules of proof, a proposal rooted in his concern with the "negation problem," is workable in practice. It imposes unnecessary costs on defendants and may lead to injustice when either the factual situation or the rules of evidence are such that true defense stories are difficult to establish. Nor do I find the suggestion conceptually compelling. For the reasons I have just outlined, I don't think a negation problem of the kind Professor Allen outlines exists in most cases. This is because the factfinder must decide the case on the evidence before it, which will ordinarily involve comparing plaintiff and defense stories in such a way that the subjective probabilities of the plaintiff's and defendant's versions of events, taking into account spoliation inferences, will add up to one. Where this is not the case, I think it is appropriate to require the plaintiff to show that there is a better then .50 chance that he should prevail.[79] I see no good reason why the defendant should not be able to take advantage both of weaknesses in the plaintiff's case and of the factfinder's common-sense ability to spot theories specifically offered by neither party but rationally based on the evidence.[80] In the long run I think that justice in the sense of accurate verdicts will be enhanced. This may in part reflect my feeling that the paradoxes that Professor Allen and others pose are not as serious as some think, and my belief that even if they pose conceptual problems this does not mean that the rules currently regulating trial proof make the trial an irrational fact-finding institution.

In sum, I do not think that Professor Allen offers us a theoretically or practically useful reconceptualization of trials. Nevertheless, I admire Professor Allen's article for the creativity that underlies some of his most controversial claims. His novel positions make his effort an interesting and stimulating piece of work. In this respect Professor Allen's article illustrates the virtues of the new evidence scholarship, for it suggests the controversy and intellectual excitement that are likely to accompany the move from a concern for rules to a concern for proof.

NOTES

† © 1986 by Richard Lempert.

* Professor of Law & Sociology, University of Michigan Law School. J.D. 1968 University of Michigan Law School; Ph.D. 1971 University of Michigan. I am grateful to the following people for reading and commenting on a draft version of this paper: Stephen Burbank, Stephen Fienberg, David Kaye, Donald Regan, David Schum, Peter Tillers and Judith Thomson. The usual disclaimers of commentor responsibility apply; in some instances with more than the usual degree of force.

[1] This title not only suggests the focus of the type of scholarship I am thinking about but also typifies it in length and lack of grace.

[2] Some of this work may have influenced state codifications or judicial constructions of ambiguous language. Such an impact—if it occurred—is a genuine contribution that I do not denigrate. It reminds us that scholarship may be valuable without being interesting.

[3] J. WIGMORE, THE SCIENCE OF JUDICIAL PROOF AS GIVEN BY LOGIC, PSYCHOLOGY, AND GENERAL EXPERIENCE, AND ILLUSTRATED IN JUDICIAL TRIALS (3d ed. 1937).

[4] I realize that broad-brush painting of the kind I have done thus far calls out for detailed support by citation in the law review tradition. However, as an invited commentator in a symposium, I think the essayist tradition which does not demand detailed footnoting of what most people know is not out of place and has much to commend it. Citation to the works of the great common law critics and synthesizers is unnecessary for anyone who knows anything about evidence, and I would not cite authors of "Twenty-Ninth Exception" type articles in any event, for there is no virtue in embarrassing people or in creating unnecessary enemies. I shall discuss a major body of "process and proof" type work in what follows. Among the work I could cite under this head let me simply note the work on presumptions by Ronald Allen, *see, e.g.*, Allen, *Structuring Jury Decisionmaking in Criminal Cases: A Unified Constitutional Approach to Evidentiary Devices*, 94 HARV. L. REV. 321 (1980), and Charles Nesson, *see, e.g.*, Nesson, *Reasonable Doubt and Permissive Inferences: The Value of Complexity*, 92 HARV. L. REV. 1187 (1979), to emphasize that transforming the focus of evidence scholarship from rules to proof does not exclude familiar modes of doctrinal legal scholarship.

[5] Frequentists—the Bayesians' usual antagonists in discussions of the utility of the Bayesian approach to statistics—are generally non-participants in the discussion. *But see* Cohen, *Confidence in Probability: Burdens of Persuasion in a World of Imperfect Knowledge*, 60 N.Y.U. L. REV. 385 (1986).

[6] Schum, *Probability and the Processes of Discovery, Proof, and Choice, infra* at 213.

[7] Tillers, *Mapping Inferential Domains, infra* at 277.

[8] 68 Cal. 2d 319, 438 P.2d 33, 66 Cal. Rptr. 497 (1968) (en banc).

[9] Finkelstein & Fairley, *A Bayesian Approach to Identification Evidence*, 83 HARV. L. REV. 489 (1970). This was not the first article in the legal literature to

suggest the applicability of Bayes's Theorem to trial proof. *See* Cullison, *Probability Analysis of Judicial Fact-Finding: A Preliminary Outline of the Subjective Approach*, 1 TOL. L. REV. 538 (1969); Kaplan, *Decision Theory and the Factfinding Process*, 20 STAN. L. REV. 1065 (1968). *See also* Ball, *Moment of Truth: Probability Theory and Standards of Proof*, 14 VAND. L. REV. 807 (1961). However, these earlier discussions did not benefit from either the interest in probability theory engendered by *People v. Collins* or from making claims so bold and questionable as to arouse the interest of an antagonist as formidable as Laurence Tribe.

There is another more autobiographical sense in which the Finklestein and Fairley article together with Tribe's response helped generate some of the debate that followed. In the winter of 1972 I taught a seminar on problems in trials and proof to five students at the Yale Law School. We spent several weeks on the Bayesian debate, teaching ourselves Bayes's Theorem largely from the footnotes to the two articles. One conclusion we reached was that while Tribe had won the debate on the advisability of instructing jurors in Bayes's Theorem, the theorem had substantial potential as a model of legal standards and processes of proof. Half the members of that seminar—if I may consider myself a member—went on to publish articles reflecting this belief. Daniel Kornstein authored an article entitled *A Bayesian Model of Harmless Error*, 5 J. LEGAL STUD. 121 (1976), and I wrote *Modeling Relevance*, 75 MICH. L. REV. 1021 (1977). By far the most important member of this group was David Kaye, whose numerous contributions figure prominently in discussions of Bayes's Theorem, statistics, and trial processes, as can be seen from the many citations to his work in this volume.

[10] Tribe, *Trial by Mathematics: Precision and Ritual in the Legal Process*, 84 HARV. L. REV. 1329 (1971).

[11] Allen, *A Reconceptualization of Civil Trials*, *infra* at 21–22 (citing Callen, *Notes on a Grand Illusion: Some Limits on the Use of Bayesian Theory in Evidence Law*, 57 IND. L.J. 1 (1982)).

[12] It is, I should note, a view I once shared. *See* Lempert, *supra* note 9, at 1021.

[13] *See, e.g.*, Meier & Zabell, *Benjamin Peirce and the Howland Will*, 75 J. AM. STATISTICAL A., 497-506 (1980) (discussing Robinson v. Mandell, 20 Fed. Cas. 1027 (C.C.D. Mass. 1868)).

[14] *See, e.g.*, Fienberg, *The Increasing Sophistication of Statistical Assessments as Evidence in Discrimination Litigation*, 77 J. AM. STATISTICAL A. 784 (1982).

[15] THE EVOLVING ROLE OF STATISTICAL ASSESSMENTS AS EVIDENCE IN THE COURTS 11 (S. Fienberg ed. 1986) (Draft report of the Panel on Statistical Assessments as Evidence, National Research Council).

[16] Examples include efforts to identify the accused as the criminal through blood traces or hair samples. (The statistical basis of hair sample identifications is, I should note, suspect. *See id.* at 79-82.) Other kinds of identification evidence like fingerprints also depend on statistical inferences but the reliability of these tests is thought to be so high that their statistical base may be neglected.

[17] Lempert, *Statistics in the Courtroom: Building on Rubinfeld*, 85 COLUM. L. REV. 1098 (1985); *see supra* note 15 and accompanying text.

[18] *See, e.g.*, Brilmayer & Kornhauser, *Review, Quantitative Methods and Legal Decisions*, 46 U. CHI. L. REV. 116 (1978); L. COHEN, THE PROBABLE AND THE PROVABLE *passim* (1977).

[19] Even this may not be essential for those who reject Bayesian models of the proof process. Some role for a Bayesian presentation of evidence may be nested within a more encompassing non-Bayesian theory.

[20] Ordinarily the latter type of evidence will not be naked because some of the evidence that led, for example, to the decision to take a hair sample from the suspect will be admissible. But such statistical evidence may be naked when the evidence that leads to the suspect's arrest is not admissible or when, as in the case of fingerprints, "trace" evidence, the weight of which can only be evaluated statistically, is all that allows the suspect to be identified. Moreover, even where the statistical evidence is not naked, it may so far outweigh all other evidence in the case that it is clearly both crucial and dispositive. Where this is clear, one must confront the fact that the trier is relying on statistical evidence despite the irreducible and undeniable uncertainty that such evidence entails, thus posing the core problem of naked statistical evidence.

[21] *Cf.* Nesson, *The Evidence or the Event? On Judicial Proof and the Acceptability of Verdicts*, 98 HARV. L. REV. 1375 (1985).

[22] In this connection it is important to note that a juror's information processing is not in an important sense fully *personalistic*. Thus, in deciding how to instruct the jury we do not function, for example, as decision theorists trying to aid a client with an investment decision. The theorist can tell the client that if he estimates probabilities in a particular way, a certain decision is a wise one. Since the client is deciding for himself, it is appropriate for him to live with his estimates; he can hardly be heard to complain that he would have done better had he followed an apparently less rational way of combining information because his probability estimates were not accurate. Given what he believed at the time, his action was the best thing for him to do. A juror, however, is deciding for someone else and society must live with the juror's errors. Thus the juror's probability estimates may be subjective but they are not personal in the sense that the resulting decision is the best thing for him to do. The law is not interested in what is most satisfying or wisest for the juror, but only in accurate verdicts, leavened perhaps by other value considerations. While it may be appropriate to hold individuals deciding for themselves to the perverse consequences that sometimes attend second-best solutions, the law in deciding whether a particular kind of reasoning should be promoted cannot escape the need to evaluate fully the consequences of introducing an irreducibly subjective element.

A committed Bayesian subjectivist might argue that there is no such thing as an objective probability or that objective probabilities exist only in those limited circumstances where a known mechanism randomly generates outcomes from a larger set. It is clear, however, that for some purposes some probability estimates are better than others. A person who bets on outcomes of tosses of a fair coin will, for example, lose money if he gives odds based on the prior expectation that the coin has a 2/3 chance of coming up heads. Indeed, I would argue that working from a "realistic" prior, whether held initially or derived retrospectively by induction, is in law and most other settings an important ingredient of rational thought. For example, a juror who believes at the outset of litigation that the odds are 100:1 that a tort plaintiff injured himself through his own negligence would not be deciding the subsequent case rationally *within the meaning of the law* even if he used a perfectly Bayesian approach to evaluate all further items of evidence. The factfinder would from the law's standpoint be similarly irrational if in applying Bayes's Theorem the evidentiary value accorded the information entering the likelihood ratio were untenable. Rationality in law, I am arguing, involves a "sensible" view of the weight to be accorded information and not just the process of combining subjectively weighted evidence according to a scheme that meets the criteria of some normative model—like Bayes's Theorem—of rational decisionmaking. When I speak of a "realistic" view of a prior probability or a "sensible" view of the weight to be accorded information, I mean a view which, if information is rationally combined, will lead to a verdict that given the evidence presented is likely (i.e., is a good bet) to describe the situation the factfinder is charged with reconstructing.

[23] Schum & Martin, *Formal and Empirical Research on Cascaded Inference in Jurisprudence*, 17 LAW & SOC'Y REV. 105 (1982). *See also* D. KAHNEMAN, P. SLOVIC & A. TVERSKY, JUDGMENT UNDER UNCERTAINTY: HEURISTICS AND BIASES (1982); R. NISBETT & L. ROSS, HUMAN INFERENCE: STRATEGIES AND SHORTCOMINGS OF SOCIAL JUDGMENT (1980).

[24] Tribe, *supra* note 10.

[25] By the same token, Bayes's Theorem may be an inappropriate normative model for trials but it may be appropriate to instruct jurors in Bayesian approaches to information processing. The case may be particularly strong where the issue is not whether jurors are going to be given statistical evidence, but how that evidence is to be presented. *See supra* notes 13-20.

[26] Lempert, *supra* note 9.

[27] It is important in this connection to note that the classroom model includes not only Bayes's Theorem but also a utility matrix (called a "regret matrix") which allows basic value choices and concepts like prejudice to be captured.

[28] FED. R. EVID. 401. Definition of "relevant evidence":

"Relevant evidence" means evidence having any tendency to make the existence of any fact that is of consequence to the determination of the action more probable or less probable than it would be without the evidence.

FED. R. EVID. 402. Relevant evidence generally admissible; Irrelevant evidence inadmissible.

All relevant evidence is admissible, except as otherwise provided by the Constitution of the United States, by Act of Congress, by these rules, or by other rules prescribed by the Supreme Court pursuant to statutory authority. Evidence which is not relevant is not admissible.

[29] In his paper, Professor Shafer gently criticizes my earlier work and suggests that even heuristically Bayes's Theorem is inadequate because it ignores the fact that

evidence may be relevant on different issues. Shafer, *The Construction of Probability Arguments, infra* at 185. I don't disagree with Professor Shafer's point but I think his criticism of my use of a Bayesian model misunderstands the legal process. While the point is not articulated except when controversies arise, evidence is always offered on specific issues, and relevance is judged with respect to the issue on which it is offered. Thus, the model applies with respect to specific issues even if the evidence might have been admissible had the proponent argued for admissibility based on its tendency to prove some other point.

[30] L. COHEN, *supra* note 18; *see generally* Shafer, *supra* note 29.

[31] *See supra* note 23. Note, however, that the failure of Bayes's Theorem to consistently model actual behavior does not mean that it has no descriptive utility. It might still capture central tendencies in the way people reason about some or most issues. By analogy, the economic model of human behavior fails to capture much that is important about human decisionmaking; but at the aggregate level and even for certain purposes at the micro level it may offer an adequate or even a good model of the way people will react when confronted with particular incentives. Similarly, Bayes's Theorem may in some situations predict well particular or average effects that certain types of information are likely to have on human decisionmaking.

[32] Tillers, *infra* at 314.

[33] I use the term *Bayesian rational* to refer to a set of rules, a system, or a process that is "rational" in several respects. First, if evidence is to be combined with some inferential end in view, the process by which it is combined follows the postulates of Bayes's Theorem. Second, prior probabilities are taken seriously not just as is implied by the preceding criterion but also in that in the absence of additional evidence, the prior probability is treated as a posterior probability. Third, posterior probabilities are taken seriously in that they are regarded as the decisionmaker's best estimate of the probability that a state of affairs exists given the evidence—including base rate evidence—relating to the likely existence of that state of affairs. Fourth, the process, system, or rules value accurate outcomes where accuracy is defined as correctly evaluating a true state of affairs. This last requirement does not mean that it is Bayesian irrational to place different values on different possible decisions thus leading to decisions which presume states of affairs that are less likely than their alternatives. Nor does it mean that the system cannot regard values other than accuracy as important or even dispositive. Bayesian rationality does, however, mean that value issues must be confronted directly and cannot be ducked by fudging the likelihood that a state of affairs will be mischaracterized. For example, it is not Bayesian irrational to have a system which mandates acquittals when there is an .80 chance that the defendant committed the crime charged even if the acquittal is taken to mean that the defendant did not commit that crime. It would be irrational even if there are good value reasons to acquit to combine evidence which justifies a posterior probability of guilt of .80 so as to arrive at a posterior probability below .50 or to assume that an acquittal must reflect such a probability. Finally, what is Bayesian rational depends on the decision-making process's intended outcome. A Bayesian rational decision at the system level may not require factfinding in the individual case

which follows Bayes's Theorem. For example, assuming factfinders could process case-specific information in a Bayesian fashion, it might be Bayesian rational not to allow this. First, other values might suffer from instructing factfinders to apply Bayes's Theorem. Second, factfinders might systematically err in evaluating the evidence that is manipulated according to Bayes's Theorem and this error might be such that fewer long run errors are made by requiring factfinders to reason in some other way than that required by Bayes's Theorem. These kinds of decisions are Bayesian rational if the designer of the system evaluated in a Bayesian rational fashion the possibility that the specific value or fact-finding goals are more likely to be achieved under non-Bayesian than under Bayesian procedures. In other words, a committed Bayesian might conclude from empirical evidence or the implications of non-Bayesian value considerations that attempting to impose Bayesian procedures is not a good way to go about a specific decision-making task.

[34] Schum & Martin, *supra* note 23.

[35] Lempert, *supra* note 9.

[36] Brilmayer, *Second-Order Evidence and Bayesian Logic, infra* at 156-59.

[37] *See generally* Cohen, *On the Role of Evidential Weight in Criminal Proof, infra* at 113.

[38] *See generally* Shafer, *supra* note 29.

[39] L. SAVAGE, THE FOUNDATIONS OF STATISTICS (1954).

[40] This is one manifestation of the product rule. If there are two events A and B, $P(A \& B) = P(A) \times P(B|A)$; if $P(B|) = P(B)$ then events A and B (elements of a case in the following discussion) are independent. Note that unless $P(B|A) = 1$, $P(A \& B)$ will always be less than both $P(A)$ and $P(B)$. The non-independence of events does not vitiate the product rule. It simply means that $P(A \& B)$ is larger than it would be if $P(A)$ and $P(B)$ were independent. But, as I suggest below, the implications of this fact for the challenge posed by the conjunction paradox are by no means trivial.

[41] I am aware that the empirical assumptions that support this conclusion, like the assumption of perfect or substantial independence made by those who advance the conjunction paradox, have not been systematically tested. One obvious source of dependency in the evidence the factfinder receives is that evidence bearing on different elements may come from the same witness. Thus the credibility accorded to testimony supporting one element of a case is likely to be similar to the credibility accorded to testimony supporting another element of a case. This means that even if the elements of a case are conceptually independent the proof of these elements and the factfinder's subjective probabilities of their existence need not be. In short, the case in which a conjoint subjective probability favoring the plaintiff is .50 or less when the subjective probabilities relating to the elements of the plaintiff's case each exceed .50 may be empirically rare despite the mathematical plausibility of this situation.

[42] One might also argue that the conjunction paradox has nothing to do with the Bayesian rationality of trials. The argument is that Bayes's Theorem prescribes a way of combining *evidence*. Once the legally separate elements of a case have been

evaluated, there is no more evidence left to be combined. The factfinder has, for example, already evaluated the evidence bearing on issue A and reached a conclusion about its probable existence, and the same is true of issues B, C, etc. The implications of this series of findings is a legal question that Bayes's Theorem does not and cannot address. There is, on the other hand, no conjunction paradox with respect to evidence that tends to prove a legally required element since factfinders are expected to follow the conjunction rule in evaluating the evidence that bears on each legally separate element.

While I find this argument intriguing, I do not rest the case against the challenge of the conjunction paradox on it. The law's goal is, presumably, to allow recovery when it is more likely than not that the plaintiff deserves to recover and to deny recovery when this is not the case. Recovery is deserved at law—I assume—when the conjunction of all the elements that the plaintiff must prove to establish his case is more likely than not to exist. While Bayes's Theorem cannot be applied to evaluate this conjunction since it is not a conjunction of evidence, a basic rule of the probability calculus and one of the axioms on which Bayes's Theorem rests is that $P(A \& B) = P(A) \times P(B|A)$ or, where $P(B|A) = P(B)$, i.e., in the case of independent events, $P(A \& B) = P(A) \times P(B)$. Treating the elements of the case as events (e.g., the defendant was negligent, or the plaintiff suffered an injury) I believe that the challenge of the conjunction paradox is a fair one, for in a Bayesian rational system as I use the term, the product rule must be applied where the probability of a conjunction of events is in issue unless some other value or a Bayesian evaluation from a more encompassing perspective precludes it.

[43] Whether this is the case is, of course, an empirical question. I am simply arguing that the Bayesian rationality of the trial process is not *necessarily* disproved by the conjunction paradox.

[44] One may argue, as Anne Martin did at the Boston symposium that gave rise to this issue, that the gatecrasher's paradox has no implications for the appropriateness of Bayesian models of the legal system because Bayes's Theorem commands no result inconsistent with our intuition but instead suggests that our intuition, in the form of whatever prior hypothesis we held, should be followed. This is because under one reading of the problem there is no evidence to be combined. The argument is that the likelihood ratio, which is:

$$\frac{P(E|H)}{P(E|\text{not-}H)}$$

where E is the evidence and H is some hypothesis of interest, in this case that the defendant crashed the gate, is 1. This is true if we regard the evidence as the fact that 501 people crashed the gate and 499 did not. This evidence is equally likely to be received whether or not the defendant crashed the gate and so does nothing to change our estimation of the likelihood that the defendant is a gatecrasher from the likelihood we estimated before receiving this information. There is nothing remarkable about this. If evidence that is fully evaluated in the prior is again presented to the decisionmaker, it will be redundant and will not change the judgment that has been tentatively reached. What makes the gatecrasher case special is that it appears from this argument that there is no information available to the factfinder except information that can be used to establish a prior. Thus Bayesian information processing cannot proceed. Even if I accepted this argument, however, I could not avoid the challenge posed by the gatecrasher paradox, for as I stated in note 33, *supra*, taking prior probabilities seriously is a requisite of what I call Bayesian rationality. In the "pure" gatecrasher hypothetical—that is a situation with no other evidence bearing on the defendant's guilt and no permissible spoliation inferences—the prior probability of the defendant's gatecrashing is .501 and a plaintiff's verdict seem appropriate. Moreover, viewing the evidence and prior probabilities in other plausible lights, there is evidence that may be manipulated in a Bayesian fashion which yields a posterior probability of .501 that the defendant crashed the gate.

Treating the base rate information as all that need be shown to make out the plaintiff's case assumes that the defendant was at the rodeo and that the number of ticket buyers and gatecrashers is also given. Thus, there is no information apart from the base rate for a factfinder to assess. A different approach, which David Kaye and I took with some consultation, has the virtue of not assuming that the defendant was present at the rodeo, leaving this to be proved by the evidence, as would be the case at a trial. Professor Kaye's argument proceeds along the following lines.

Let R be the event that the defendant was at the rodeo; N_T the number of ticket buyers (499); N_R the number of people at the rodeo (1000); and N the number in the relevant population to which ticket buyers and gatecrashers belong. The posterior odds are the likelihood ratio times the prior odds:

$$\frac{P(\bar{T}|R)}{P(T|R)} = \frac{P(R|\bar{T})\ P(\bar{T})}{P(R|T)\ P(T)}. \tag{1}$$

First, consider the likelihood ratio $P(R|\bar{T})/P(R|T)$. If everyone who paid for a ticket went to the rodeo, then $P(R|T) = 1$. If everyone in the relevant population who did not pay for a ticket is equally likely to have gone to the rodeo, then $P(R|\bar{T})$, which is the probability that someone went to the rodeo given that he did not buy a ticket, is simply the number of people who went to the rodeo without buying a ticket, $N_{R\bar{T}} = N_R - N_T = 501$, divided by the number of people in the relevant population who did not buy tickets, $N_{\bar{T}} = N - N_T = N - 499$. Thus,

$$\frac{P(R|\bar{T})}{P(R|T)} = \frac{N_{R\bar{T}}/N_{\bar{T}}}{1} = \frac{501}{N - 499}. \tag{2}$$

Next, continuing to treat everybody in the relevant population as initially identical, the prior odds are the proportion of non-ticket buyers in the relevant population, $P(\bar{T}) = N_{\bar{T}}/N = (N - N_T)/N = (N - 499)/N$, divided by the proportion of ticket buyers, $P(T) = N_T/N = 499/N$. Hence,

$$\frac{P(\bar{T})}{P(T)} = \frac{(N - 499)/N}{N/499} = \frac{N - 499}{499}. \tag{3}$$

Substituting (2) and (3) into (1) gives the odds that a person who is shown to have been at the rodeo did not pay for a ticket:

$$\frac{P(\bar{T}|R)}{P(T|R)} = \frac{501}{N-499} \times \frac{N-499}{499} = \frac{501}{499}.$$

My own attempt to grapple with the problem reaches the same result more simply by directly estimating the prior odds according to the formula (in which GC stands for being a gatecrasher and R for being at the rodeo):

$$O(GC|R) = \frac{P(R|GC)}{P(R|\overline{GC})} \times O(GC).$$

This solution has the virtue of emphasizing that the factfinder may begin with the assumption that the defendant is no more likely than anyone else in the relevant population (of size N) to have been a gatecrasher. Thus, the prior odds are consistent both with one tenable definition of the presumption of innocence (albeit a matter not in issue in the example since we are supposing a civil suit) as well as with the situation of a factfinder who starts a trial with no knowledge other than the fact that an actionable offense, or in the gatecrasher case some number of offenses, has occurred. The fact that the defendant was at the rodeo is evidence and the number of ticket buyers at the rodeo is information needed to evaluate the implications of that evidence. The only necessary assumptions are that all gatecrashers and ticket buyers attended—and may be shown to have attended—the rodeo, and that everyone in the relevant population has an apparently equal propensity to crash the gate. The prior odds, $O(GC)$, are $501:(N-501)$, since knowing nothing but the number of gatecrashers the odds of the defendant's being a gatecrasher is the ratio of that number (501) to the number of people in the population who are not gatecrashers $(N-501)$.

Alternatively

$$O(GC) = \frac{P(GC)}{P(\overline{GC})} = \frac{501/N}{(N-501)/N} = \frac{501}{N-501}.$$

Since it is certain that the defendant will be at the rodeo if he is a gatecrasher, and the probability that he will be at the rodeo if he is not a gatecrasher equals the number of ticket buyers (499) divided by the total population minus the number of gatecrashers, the formula for the posterior odds becomes:

$$\frac{1/499}{(N-501)} \times \frac{501}{(N-501)} = \frac{501}{499}.$$

[45] Kaye, *The Paradox of the Gatecrasher and Other Stories*, 1979 ARIZ. ST. L.J.

101. *See also* Fienberg, *Gatecrashers, Blue Buses and the Bayesian Representation of Legal Evidence*, 66 B.U.L. REV. 693 (1986).

[46] Responding to Professor Kaye, Professor Allen also argues that if the plaintiff can produce more evidence so can the defendant. This does not follow, for the ability to produce more evidence is likely to be conditional on the existence of liability and on the issues the plaintiff chooses to raise. See the discussion in the text following note 67 *infra*.

[47] Kaye, *supra* note 45. *See also* Lempert, *supra* note 9; Tribe, *supra* note 10.

[48] There is, I will admit, only the weakest of inferences to be drawn from the plaintiff's failure to call A in the context of the gatecrasher hypothetical, but as the hypothetical is posed even a slight spoliation inference will change the balance of probabilities so that the plaintiff is no longer favored. In actual cases, or even in factually richer hypothetical cases, stronger inferences are likely to exist.

[49] Consider also antitrust actions, where the damages sought rest on statistical estimates of the harm done. The jury is not required to limit the plaintiff's damage award to the lower bound of a 95% confidence interval around a damage estimate. If, for example, the jury settles on the point estimate of damages, there is a substantial chance that the plaintiff has received more than he deserved, but we are not troubled by this. If damages had to be set at the lower bound of the 95% interval, plaintiffs would ordinarily receive less than they deserved.

[50] 317 Mass. 469, 58 N.E.2d 754 (1945). In *Smith* the Massachusetts Supreme Court held that a directed verdict was properly issued for the defendant when the only evidence suggesting that the defendant owned a bus allegedly responsible for the plaintiff's accident was that the defendant was the only bus company that had a franchise to operate a bus on the street where the accident occurred, and that the defendant company's bus schedule was perhaps consistent with the involvement of one of its buses in the accident. The court quoted from an earlier Massachusetts case: "[It is] not enough that mathematically the chances somewhat favor a proposition to be proved; for example, the fact that colored automobiles made in the current year outnumber black ones would not warrant a finding that an undescribed automobile of the current year is colored and not black . . . ." Sargent v. Massachusetts Accident Co., 307 Mass. 246, 250, 29 N.E.2d 825, 827 (1940).

[51] 33 Cal. 2d 80, 199 P.2d 1 (1948).

[52] 26 Cal. 3d 588, 607 P.2d 924, 163 Cal. Rptr. 132, *cert. denied*, 449 U.S. 912 (1980).

[53] In the extreme case where the plaintiff in the gatecrasher situation could identify specifically all 501 gatecrashers, the 499 others who might, on a balance of probabilities, have been found responsible will have their cases dismissed. Indeed, once three defendants are *specifically* identified as gatecrashers, in each subsequent case the revised naked statistics (499 ticket buyers; 498 unidentified gatecrashers) mean that it is more probable than not that a given defendant purchased a ticket. Thus, if naked statistical evidence were sufficient to establish a prima facie case, the plaintiff in the gatecrasher hypothetical would have a strong incentive to suppress better, more particular evidence that might be available in a few cases.

⁵⁴ With respect to the evidence available to the court, X's situation is like that of all those at the rodeo, paying customers and gatecrashers alike.

⁵⁵ Of course, I acknowledge that the law has tie-breaking rules that typically disadvantage those challenging the status quo. I am arguing that our intuitions about the right result in the gatecrasher case do not turn on the need for such rules and that we better appreciate the questionable basis of the intuitions that many people express regarding the pure case when we ponder the reverse situation.

⁵⁶ I am just discussing the weight of the evidence issue in respect to the law, but it is this issue which is at the heart of the broader debate about how we should think about evidence and the nature of rational reasoning. *See*, *e.g.*, L. COHEN, *supra* note 18; G. SHAFER, A MATHEMATICAL THEORY OF EVIDENCE (1976).

⁵⁷ Lempert, *supra* note 9, at 1032. In other words, no legal norm suggests that a factfinder should, before the parties present their cases, hold the tenative position that it is as likely as not that the plaintiff deserves to recover. Rather, the factfinder should throughout the litigation hold the belief that a mistaken decision for the plaintiff is as regretable as a mistaken decision for the defendant.

⁵⁸ The presumption of innocence may appear to do this, but I think it is largely concerned with an attitude the jurors should take in evaluating evidence rather than with specifying a prior probability to be revised in the light of the evidence in the case. It does, however, operate in a negative way with respect to the setting of priors, for it instructs jurors that certain information they have at the beginning of a case, such as the fact that the defendant has been arrested, should not enter into their prior probabilities. Moreover, the attitude captured by the presumption of innocence may be more consistent with some prior probabilities than with others.

⁵⁹ An exception exists in cases like the gatecrasher case where we know that more than one person has engaged in an activity. Here the appropriate prior is the ratio of the number of actors known to have engaged in that activity to the number of actors in the world minus the number known to have engaged in the activity. (Actors need not be human beings, but may be other entities such as corporations.)

Legal factfinders, of course, face issues other than the identity issue I am discussing in the text. With respect to issues like motive, it is more difficult to specify an objective referent that, even if only in theory, may be used to indicate appropriate prior odds. Nevertheless, I think the same principle does apply; for example, in the case of motive, the prior odds can be set in principle at an estimate of one to the number of all possible other motives that might explain a behavior at issue. That is, except insofar as a behavior itself suggests that some motives are more likely than others, a factfinder without further evidence should not, before receiving other evidence, regard one motive as more likely to explain that behavior than other possible motives.

⁶⁰ Some have suggested that 1:1 is the appropriate beginning odds for someone who is ignorant of a true state of affairs. I believe that Professor Shafer has shown the problems this can cause. G. SHAFER, *supra* note 56. With respect to my suggestion, I am not saying that this is the prior the factfinder should have before hearing the first item of evidence. Rather this is the prior the factfinder should proceed from as he or

she begins to think about the case. Much that is learned or assumed before any evidence is given will properly affect the prior a factfinder holds when the first item of evidence is presented. For example, it is reasonable for a factfinder to assume that only those in some local population of which the defendant is a member could have engaged in the charged behavior (e.g., the population of people who could have been in a particular location at a particular time). This will dramatically reduce the prior and is a harmless leap so long as evidence of the defendant's ability to get to the location is presented at the trial and is not "double counted."

[61] For example, evidence that the defendant in a murder case received a ticket for running a red light at a point from which it was difficult but not impossible to reach the scene of the crime by the time of the murder suggests the defendant is innocent, as it appears less likely than it would have without the evidence that the defendant was at the scene of the crime at the crucial time. However, if it is also revealed that the defendant had a full view of a police officer when he ran the red light, the evidence of the defendant's location is more probative of guilt than innocence; it now appears that the defendant wanted to get ticketed, a desire consistent with seeking to establish an alibi in contemplation of committing the crime.

[62] Note that with a Bayesian machine and objective information about probabilities the conjunction rule should apply where more than one element had to be proved to make out a case. That is, if the probability of each of three independent elements that the plaintiff had to prove to make his case was .75, the plaintiff would lose because there was less than a .50 chance that events had occurred in such a way that the plaintiff deserved to recover. I believe that if the accuracy and objectivity of my hypothetical machine could be guaranteed, the law would allow the conjunction rule, for the law's concern would be with the probability that the plaintiff was more likely than not entitled to recover on the case taken as a whole.

[63] *See also* Cohen, *supra* note 5. I think that Professor Cohen's concern with "weight of evidence" issues is well placed, but I do not mean to endorse all the arguments he makes for his position.

[64] Note that if the source of discomfort is, as it may often be, a spoliation inference—e.g., if defendant were really liable why didn't the plaintiff present evidence of X—the discomfort fits nicely into the Bayesian model and would, if the source were appreciated, lower the final probability judgment. As Arthur Conan Doyle showed us and as others have documented, *see* R. NISBETT & L. ROSS, *supra* note 23, people often do not appreciate the relevance of "negative" evidence. Nevertheless we may perhaps feel less confident of analyses because of dogs that never barked.

[65] In addition, other values that have nothing to do with the style of information processing expected of legal factfinders may justify the "not persuaded" alternative. We might, for example, want to induce the parties, especially the plaintiff, to offer more or better quality evidence, or we might not want factfinders to leave trials dissatisfied with the verdicts they felt compelled to return. If these are our motives, the legal system's choices do not prefer non-Bayesian to Bayesian modes of proof.

They simply elevate other values above a conception of rational factfinding that, however defined, has no place for such values in it.

⁶⁶ The same is true if one finds the models offered by Professors Cohen, Shafer and others to be models of rational decisionmaking and accepts the claim that the processes such models describe are generally similar to those that characterize trials. Professor Allen, of course, attempts to show that such models give rise to some of the same paradoxes as the Bayesian model and so, as models of the trial process, they share many of the Bayesian model's alleged deficiencies. I have some difficulty with the arguments that Professor Allen advances to make this point, but since the proponents of these models are, unlike the Reverend Bayes, still active scholars, I shall leave to them the task of answering that portion of Professor Allen's argument that bears on their theories.

⁶⁷ See Allen, *supra* at 46 & n.66.

⁶⁸ *Id.* at 46.

⁶⁹ Nesson, *supra* note 21.

⁷⁰ Tribe, *supra* note 10.

⁷¹ The difference lies in the implications of disjunctions between legal rules and formal models. If the model is independently normative, inconsistent legal rules are "wrong" and should be changed. If the law is normative, inconsistencies between the law and a model reflect the model's inadequacy. If Professor Allen could show that the law embodies some formal model of rationality as its *primary* norm, requiring secondary procedural norms to conform to the model would not be problematic, for it would be requiring them to conform to some higher legal norm. Professor Allen may believe that this is the situation he is describing, but he has not established it convincingly.

⁷² Allen, supra at 84.

⁷³ Note that the counter-story need not involve presenting a case or even a different version of the events. It may be simply a story, conveyed through cross-examination, that the plaintiff's version of events is not to be believed.

⁷⁴ *Cf.* W. BENNET & M. FELDMAN, RECONSTRUCTING REALITY IN THE COURTROOM: JUSTICE AND JUDGMENT IN AMERICAN CULTURE; O'Barr & Conley, *Litigant Satisfaction Versus Legal Adequacy In Small Claims Court Narratives*, 19 LAW & SOC'Y REV. 661 (1985).

⁷⁵ That is, the plaintiff presents an inculpatory story with a .40 chance of being true; the defendant presents an exonerative story with a .30 chance of being true; and there is an obvious second exonerative possibility with a .30 chance of being true that is not rendered less plausible by the defendant's failure to support it with evidence.

⁷⁶ If specific evidence supporting a third possibility or possibilities could be offered, the chances of the possibility should rise above the .30 probability that Professor Allen hypothesizes. This is so because the law's factfinder is required to decide cases on the evidence presented, evaluated in the light of common sense. If a factfinder believes that there is an exonerative possibility not directly advanced and only indirectly supported (i.e., not ruled out by the evidence offered) which has a .30

chance of being true, it is almost certain to accord the possibility a greater than .30 chance of being true if specific supporting evidence is offered.

[77] An exception to the defendant's ability to take advantage of obvious exonerative possiblities not belied by the spoliation inference exists when the spoliation inference is untenable because some legal rule, e.g., a rule of privilege, prevents the defendant from supporting the possibility. Here, even though the defendant's failure to offer evidence does not give rise to the inference that the possibility does not in fact exist, the factfinder ordinarily should not be able to consider the possibility in evaluating the strength of the plaintiff's case. This is because the policy reasons that preclude the defendant from offering the evidence usually preclude verdicts based on assumptions—however plausible—of what the evidence would have contained. Where they do not, as in jurisdictions that allow the invocation of privileges to be mentioned in civil cases, the situation reverts to that described in the text. Where policy reasons do preclude such assumptions, the state has decided to sacrifice accurate verdicts to some officially more important value.

[78] Allen, *supra* at 43.

[79] I also believe it appropriate to place on plaintiffs the burden of showing that there is a better than .50 chance that all legally necessary elements exist. While Professor Allen's suggestion that juries compare two fully specified versions of reality may eliminate the conjunction paradox, *id.* at 44, to the extent this is a problem, I believe the suggested solution's costs are too high. For example, the suggestion would apparently allow a tort plaintiff to escape a directed verdict at the close of his case even though he presents *no* evidence tending to establish the defendant's fault. Moreover, if the defendant presents no evidence that he was not at fault Professor Allen would apparently let the case get to the jury at the close of all the evidence, for his view appears to be that if the parties do not wish to explore some feature of the case in great depth a court should not force them to do so. *Id.* at 44–45. This suggestion is troublesome for a variety of reasons. The first, to which I have alluded, is the cost of forcing the defendant to mount a defense when the plaintiff has no case. A second is that the argument assumes that proof is equally accessible to both parties. It may be harder for the defendant to prove a negative—absence of fault—than it is for a plaintiff to prove a positive violation of some duty of care. Third, to the extent that the rule leads to recovery where defendants are not at fault, social resources would, at least in theory, be misallocated as certain potentially careless activities bear more than their share of the costs they generate. Fourth, to change the rule as Professor Allen proposes would negate the evidentiary value of the spoliation inference and perhaps reduce confidence in judgments. Currently, if a plaintiff presents no evidence of fault in his case-in-chief, we can be confident that a directed verdict is just because the burden placed on plaintiffs means that if evidence of the defendant's fault were available, the plaintiff would surely have presented it. Under Professor Allen's proposed scheme, the plaintiff's (or the defendant's) failure to present evidence on a hitherto legally necessary element of the plaintiff's case would justify only a lesser inference of nonexistence since the evidence might be withheld—perhaps knowing the other side could offer nothing on the point as well—

for tactical reasons. Finally, Professor Allen's scheme neglects certain symbolic and other hard to quantify values. The adversary system is not the law's highest value. We may not wish our legal system to allow parties to recover in negligence without evidence of defendant's fault even if both parties are willing to let the case go to the jury in this posture. The system does not exist to take cognizance of all disputes or to resolve all cases on the grounds the parties prefer.

The example of parties who in developing their specific stories choose not to present evidence of fault in a negligence case is an extreme example chosen to make a point. But I think many of the same arguments apply in other cases, such as Professor Allen's example of the case where the parties both choose to rely on relatively crude statistical data. *Id.* The issue is not just whether the parties should be allowed to choose what evidence to present as Professor Allen suggests. Rather, given a variety of policy reasons like those sketched above, the issue is what should be the consequence of a party's choice when the other side moves for a directed verdict. I don't see the adoption of Professor Allen's suggestions as improving the current rules.

[80] Note that this latter possibility is symmetrical. Common-sense theories not specifically advanced may fill what would otherwise be gaps in the plaintiff's case.

Richard Lempert,
Professor of Law and Sociology,
University of Michigan Law School.

RONALD J. ALLEN*

ANALYZING THE PROCESS OF PROOF: A BRIEF REJOINDER

In his article, *The New Evidence Scholarship: Analyzing the Process of Proof*,[1] Professor Lempert offers three criticisms of my work. He criticizes me for not attempting certain tasks that he would like to see undertaken; he doubts whether the gap between conventional probability theory and the conventional theory of trials is as large as I suggest; and he asserts that either my reconceptualization is dangerous or that it is merely a restatement of the status quo. I will respond briefly to each point.

Lempert expresses disappointment that I failed to produce "a model of the trial which attended to values other than rationality but . . . had a clear place in it for information processing of a Baconian or a Pascalian [or—I would add—any other rational] kind."[2] He also criticizes me for ignoring anterior questions, such as whether "some formal model of rationality" is or ought to be the law's "primary norm,"[3] and for not advancing in certain instances beyond a "barely argued first premise."[4]

It is standard academic practice to criticize articles for what they do not do rather than for what they do. I did not attempt what Professor Lempert had in mind but sought instead to demonstrate that certain rigorous ways of thinking about inference have strikingly similar implications when applied to the trial of civil disputes. Some of those implications are troublesome if error reduction is an important goal of the civil justice system, and I proposed a reconceptualization that may obviate some of those implications. Thus, I did not undertake to create the general theory that Professor Lempert desires, although I, too, would like to see it produced. I must say, though, that the efforts of legal theorists such as myself to produce a comprehensive theory would be greatly aided if the proponents of various forms of probability theory, of which Lempert is one, would also work toward a reconciliation of their competing views. In short, I am as disappointed in Lempert's failure to even attempt such an effort—and his concentration instead on yet another defense of Bayesianism—as he is in my failure to accomplish it.

Professor Lempert also objects to my "barely arguing" first premises. He is right about that; in fact, he understates the case. Not only did I not "barely argue" first premises, I did not argue them at all. Indeed, I do not even know what it means to "barely argue first premises." If a first premise could be argued, it would not be a first premise. There would be some other first premise anterior to it. Had I argued the first premises referred to by Lempert, I would then have been criticized for not arguing the "second set of first premises" upon which such an argument would rest, and we would descend into the abyss of an infinite regress. Ironically, I could just as easily criticize Lempert for his failure to defend first premises or indeed to even provide them clearly. In an effort to discourage such criticisms, he avoids taking a normative position, thus creating the appearance that only the internal logic of his views can be examined, when in fact even the most sterile sort of logic rests upon first premises.[5] Thus, Professor Lempert's criticisms can be turned back upon him, as they could with respect to any "logical" argument.

Rather than pursue that path, I will instead explore the logic of Lempert's views. I demonstrated in my article that there is a strained and ad hoc quality to the efforts of Bayesian proponents to respond to their critics. I did this in order to corroborate the implication that there is a dissonance between the conventional views of probability and trials. My belief was that if proponents of a particular view get into even more difficulty by their efforts to extricate themselves from some perceived problem, then that problem might just be real. The efforts of Bayesian defenders demonstrate this point quite clearly, and Professor Lempert's efforts are no different. Professor Lempert is a committed Bayesian, and the disutility of yet another defense of Bayesianism is evident in his attempt to show that the disquieting implications of conventional probability theory are really not so troubling after all.

Professor Lempert responds to the conjunction problem by advancing the "intriguing argument" that "the law may be Bayesian rational without mandating a fact-finding process that combines elements in conformity with Bayes's Theorem."[6] He explains this proposition in the following manner:

> [T]he imperfections of human information processing and the fact that jurors will not have all the information relevant to the issues they must decide may mean that a rule that allows a recovery on conjoint subjective probabilities of less than .50, as long as the existence of each element is thought to be more likely than not, is from a systemic point of view Bayesian rational in that more than any other easily applied rule it accurately separates those cases in which plaintiffs objectively deserve to recover from those cases in which they do not.[7]

In other words, jurors who are instructed to find each element by a preponderance of the evidence instead may find the conjunction of all the elements by a preponderance. This, according to Lempert, shows that "the Bayesian rationality of the trial process is not *necessarily* disproved by the conjunction paradox"[8] since after all it is "an empirical question."[9]

What is one supposed to say in response to that? We may not know how jury verdicts correlate with reality, but we do know how they would correlate if juries were capable of doing what we instruct them to do and followed those instructions. In the absence of such information, one can always speculate that things are hunky-dory, or one can examine the system of civil trials by assuming that jurors basically do what we ask them to do.

Of course, even more information about how jurors behave might not be sufficient, for Lempert's response is of the "no-lose" variety. We will always lack perfect knowledge. There will always be unanswered questions. In an epistemology in which it is a satisfactory response to an analytical effort to assert an unproven empirical hypothesis, analytical efforts lose their meaning and Bayesian analysis can never be "necessarily disproven." Professor Lempert, however, overlooks that his response is also of the "no-win" variety, for under his analysis Bayesian analysis can never be proven either. Tomorrow we may discover something that will make it clear that the Bayes's Theorem is gibberish.

An argument that rests entirely upon unconfirmed speculation is not very interesting. It tells us nothing that we do not already know, even if it does remind us that omniscience is not an attribute of human existence. That such an argument would even be advanced demonstrates precisely my point: the Bayesians are hard pressed to respond to the incompatibility of conventional probability theory and the conventional view of trials.

Consider also Professor Lempert's effort to defend the "Bayesian" view of the gatecrasher hypothetical. He asserts that we can justify not allowing the plaintiff to get to the jury as an incentive to produce more evidence.[10] According to Lempert, if the plaintiff really believed that defendant A was a gatecrasher, he would have called him to the stand and asked him whether he had paid for his admission: "One cannot assume that A would lie, and even if A had lied, cross-examination might have revealed A's deception."[11] By contrast, my suggestion that the defendant should take the stand if he wants to tell a story is, according to Lempert, unconvincing:

> Professor Allen's critique misses an important point. This point is that the law currently imposes a burden on plaintiffs—but not on defendants—to go forward with evidence. The question is whether at the close of plaintiff's case, at a time when the defendant has had no chance to present evidence, a reasonable jury could find that the plaintiff's case is more probable than not.[12]

In the first place, one can reasonably wonder who "missed the point." Professor Lempert wants to rely on prevailing law as it is at times, but only at highly selective times, an inconsistency that again demonstrates the ad hoc nature of his defense. For example, Lempert is wrong about the law of directed verdicts, at least as it is in practice. Cases are rare that direct verdicts against a plaintiff before the defendant presents a case if there is any basis at all for a plaintiff's verdict, and a "probabilistic" basis is normally sufficient.[13] All of the cases normally cited as antagonistic to "probabilistic evidence" appear to involve verdicts after both sides have rested.[14] Accordingly, the plaintiff in the gatecrasher hypothetical has met the requirement of presenting enough evidence so that a reasonable person could find in his favor, and a directed verdict for the defendant will not be entered.

Professor Lempert's willingness to rely on the "law" when it is convenient and to overlook it when its implications are contrary to his position is also evident in his "spoliation" argument. He argues that if the plaintiff in the gatecrasher case produces only statistical evidence, his failure to produce more counts as evidence against him, thus permitting a directed verdict. In response to my point that if the plaintiff can produce more evidence, so too can the defendant, Lempert argues that the law requires that the decision be made at the close of the plaintiff's case.[15] However, a spoliation inference is permitted only when a party is shown to have been in possession or control of relevant information and fails to produce it at trial.[16] That requires the defendant to demonstrate the necessary conditions, which is precisely what I want him to do in any event.

Moreover, although it is possible that a plaintiff who produced only statistical data in the gatecrasher situation could have produced more, it is also possible that he could not have done so. Accordingly, a reasonable person could find that the plaintiff had established his case by more than a 50% probability, even if a reasonable person could also find to the contrary. Once that conclusion is reached, a court's decision is simple. The case goes forward.

The curiousness of this whole chain of argument is fully revealed if Lempert's assertions are unpacked. He argues that in the gatecrasher scenario the law must worry that the plaintiff will hide evidence that exonerates defendants. Yet, how likely is it that the plaintiff will have access to exonerating evidence that defendants do not also possess? If there is any exonerating evidence that would demonstrate that a particular defendant had in fact paid, the most likely locus of the evidence is the defendant himself. The chance of the plaintiff suppressing exonerating evidence is too slim to justify a special rule, in any event. On the other hand, the chance of the defendant suppressing evidence is considerably greater. Indeed, the gatecrasher hypothetical begins to look like a res ipsa loquitur case. The lesson of such cases runs parallel to my argument: make the defendant explain himself.

There are other ways in which Professor Lempert's argument reflects a failure to apply his reasoning evenhandedly. Lempert tells us that "one cannot assume that A would lie." Defendants, in short, are of high moral fiber; plaintiffs as a class are somewhat less respectable. If plaintiffs are allowed to get to juries on evidence of the sort in the gatecrasher hypothetical, they "will have an incentive to falsely create the appearance" that the only evidence that exists is the "probabilistic" evidence.[17] Yet, why can defendants be counted on to tell the truth or have their lies exposed, but plaintiffs cannot be counted on to tell the truth or have their lies exposed? That question is unanswered. Lempert also overlooks the fact that directing verdicts against plaintiffs in this context will create the same incentives for defendants to hide evidence as allowing plaintiffs to proceed will create to encourage plaintiffs to hide evidence.

Moreover, although Lempert wants to create an incentive for plaintiffs to produce more evidence, he does not tell us, as he could not, under what conditions plaintiffs will have produced enough. Would it be sufficient if everyone but one person were a gatecrasher? If so, why is that different from 501 out of 1000? What is the principle that is to replace the principle of error minimization? As he says in a different context, if a distinction is going to be made on the basis of the strength of the underlying inference, "an explanation is required of why one level of irreducible and undeniable uncertainty is tolerable and another is not."[18] He does not give us such an explanation, although he is correct that he should have.

The ad hoc nature of Professor Lempert's defense of Bayesian approaches is also demonstrated by his comparison of the rodeo hypothetical to a standard blue bus hypothetical.[19] According to Lempert:

[If] we are confident that the plaintiff has actually been injured by a negligent bus driver, that the bus belongs to one of two companies, and that it is impossible for the plaintiff to offer any evidence about which company's bus was responsible . . . I see no great cost in allowing such a case to survive a directed verdict and forcing the [defendant] to present evidence.[20]

There are two problems with Lempert's argument. First, Lempert is once again playing fast and loose with the legal standards. The test for directed verdicts and summary judgments is whether there is an issue of material fact with respect to which reasonable people could disagree.[21] It is not whether we are "confident" that the plaintiff has established his elements, that negligence is clear, or that it is "impossible" for plaintiffs to produce more evidence. Such a standard would skew the trial system dramatically in favor of defendants by imposing extraordinarily high burdens on the plaintiff to avoid directed verdicts.

The second problem with Professor Lempert's argument is that it lacks meaningful criteria. What does it mean to be "confident" that an element "actually" is true or to know when it is "impossible" to present more evidence? Lempert makes no effort to define these terms for the obvious reason that he could not do so satisfactorily.

Put these two points together. Lempert relies on fictional legal standards, yet he criticizes me for not carefully attending to the law. He provides standards that are virtually meaningless, yet he criticizes me for "leaving important questions unanswered." The real message, I think, is that the ad hoc nature of the efforts of Bayesian proponents to demonstrate that the "paradoxes" emerging from the interaction of conventional views of probability and conventional views of trials are not really so serious after all.

Lempert's defense of conventional probability indirectly confirms that there is an uneasy fit between the conventional view of trials and the conventional view of probability.[22] Moreover, as I demonstrated in my article, a similar incompatibility is present with inductivist probability theory. I concluded from these observations that we should reconsider how we construct trials, if we are concerned about error reduction in the manner I assumed.

I agree with some of the things Professor Lempert has to say about this aspect of my argument. My reconceptualization does need to be studied further in an effort to carefully isolate its costs, benefits, and implications. I doubt, however, that great costs will be imposed and great injustice will be done to defendants by making them defend.[23]

The largest cost will be incurred in preparation before trial. Lempert is simply wrong in asserting that attorneys bank on cases being dismissed without full preparation, as he implies by his hypothetical.[24] The only additional cost involves examining whatever witnesses the defendant wishes to call. Of course, if nothing were to be gained from it, even this cost ought to be avoided, but it may turn out that the defendant's case is discernibly less persuasive than the plaintiff's, even though the plaintiff's was not very persuasive. Lempert asserts that it would be "unjust" to return verdicts for plaintiffs in this context. Yet, if the level of persuasiveness has any relationship to the likelihood that the parties' stories are true, then errors will be reduced if verdicts are returned for plaintiffs even if the plaintiff's story is only marginally convincing. Thus, the "injustice" that Lempert is referring to amounts to reducing the amount of error in the system. That, I must say, does not sound "unjust" at all; rather, it sounds like a good idea.

I take it that it must sound like a good idea to Professor Lempert, too, for notwithstanding his efforts to criticize my central thesis, in the end he embraces it. He informs us that my idea is not all that original, since Bayesian analysis requires that inferences be drawn based on the evidence, and the evidence may not include all the relevant possibilities. According to Lempert, if there is a .4 probability of plaintiff's case being true, and a .3 probability of defendant's case being true, and .3 probability unaccounted for, the plaintiff should win because the jurors will not be aware of the unaccounted for probability. Thus, "on the evidence" the probabilities will be more than .5 that the plaintiff should win.[25] That is precisely the reconceptualization that I have advanced, and it represents a very different characterization than Professor Lempert previously gave to burdens of proof.

In his book on evidence, Professor Lempert states that "one can show algebraically that regret is minimized by deciding for P whenever the probability of negligence is greater than .5 and deciding for D whenever the probability of negligence is less than .5. . . . This is what is meant by a burden of proof by the preponderance of the evidence."[26] There is no discussion of comparing stories, nor of the possibility of alternative explanations for a litigated event for which no evidence is produced. Nor is there such a discussion in Lempert's other major work concerning Bayes's Theorem.[27] In addition, this argument contradicts his argument about injustice because he now seems to be conceding, as he must, that errors will be reduced by deciding for whichever party presents the most likely explanation of what occurred. In any event, notwithstanding Professor Lempert's critical tone, I am pleased that he embraces the central thesis of my article.[28]

NOTES

[1] *Supra* at 61.
[2] *Id.* at 82–83.
[3] *Id.* at 83 n.1.
[4] *Id.* at 83.
[5] *Id.* at 82–83. After a lengthy discussion of the relationship between conventional probability, in particular Bayes's Theorem, and conventional conceptions of trials, Professor Lempert says of my analysis that "there is a circle here, for it is not at all clear how *rational* decision making proceeds." *Id.* He criticizes me for not showing that a formal model of rationality is "independently normative" (whatever that means) or that the law embraces "rationality as its *primary* norm." *Id.* at 83 n.1. Lempert appears to believe that coherent discourse can occur without assumptions as starting points. He is wrong about that. His failure to recognize that error raises

serious questions about his own work. If he does not have a conception of rationality and the goals of the civil justice system, then his discussion of Bayes's Theorem is of utterly no consequence.

⁶ *Id.* at 70.
⁷ *Id.*
⁸ *Id.* at 70 n.43 (emphasis in original).
⁹ *Id.*
¹⁰ *Id.* at 71–72.
¹¹ *Id.* at 72.
¹² *Id.* at 72.
¹³ *See, e.g.*, Kramer v. Weedhopper of Utah, Inc., 141 Ill. App. 3d 217, 222, 490 N.E.2d 104, 107 (1986).
¹⁴ *See* Allen, *A Reconceptualization of Civil Trials, infra* at 19, 46 n.67.
¹⁵ Lempert, *supra* at 71.
¹⁶ *See generally* II J. WIGMORE, EVIDENCE §§ 285-91 (J. Chadbourn rev. ed. 1979).
¹⁷ *Id.* at 74 & n.53.
¹⁸ *Id.* at 67.
¹⁹ *Id.* at 72.
²⁰ *Id.* at 72-73.
²¹ Anderson v. Liberty Lobby, Inc., 106 S. Ct. 2505, 2510 (1986); Celotex Corp. v. Catrett, 106 S. Ct. 2548, 2552 (1986).
²² Lempert recognizes this where he says that if the gatecrasher hypothetical is faced "head on" (that is, freed from all the rationalizing efforts that change the facts of the hypothetical), then the plaintiff should be allowed to recover. Lempert, *supra* note 1, at 461. I do not know why the conventional probabilists take so long to get to that obvious point. I suspect the reason is that there is discomfort in allowing a recovery where the odds of liability are so evenly balanced. That, however, is an incident of their theory, and it will be true no matter what the burden of persuasion is. If, for example, the plaintiff's burden is raised from .5 to .6, what happens when 601 out of 1000 people are gatecrashers? The spirited efforts of the conventional probabilists are mainly directed at attempting to show that this case will never arise, which in fact demonstrates their discomfort with the implications of their theory when the hard case does arise.
²³ *Id.* at 83-87.
²⁴ *Id.* at 84.
²⁵ The reader may wish to compare this aspect of Lempert's discussion to Allen, *supra* note 14, at Part IV, where I develop this point at length.
²⁶ R. LEMPERT & S. SALTZBURG, A MODERN APPROACH TO EVIDENCE, 163, 163 n.25 (2d ed. 1983).
²⁷ Lempert, *Modeling Relevance*, 75 MICH. L. REV. 1021 (1977).
²⁸ He did not get it quite right, however. He asserts that my proposal would result in permitting cases to go to the jury even though there is no evidence of necessary elements. *Id.* at 87 n.79. If there is no evidence of a necessary element, and that

means that the probability of that element is 0.0, then the probability of some story being true that contains that element is also 0.0. In that case, the defendant could not tell a less convincing story, and a directed verdict would be in order. If, in light of the evidence, the probability of the plaintiff's case exceeds 0.0, then the defendant generally should be required to proceed. I say generally only to permit the possibility of the unusual case where a special rule may be called for. He also asserts that under the reconceptualization a party could not rely on cross-examination to make a case. *Id.* at 87. I have no idea where he got that impression.

Ronald J. Allen,
Professor of Law,
Northwestern University School of Law.

L. Jonathan Cohen*

THE ROLE OF EVIDENTIAL WEIGHT IN CRIMINAL PROOF†

I

In *The Probable and the Provable,* and in a number of subsequent writings,[1] I have developed several lines of argument that support an interpretation of Anglo-American standards of forensic proof in non-Pascalian terms. And the thesis for which I have thus contended is essentially a normative one, concerned with answering the question "What is the legally correct way to judge proofs?," not a factual one, concerned with answering the question "What is the way in which proofs are actually judged?" So the thesis cannot be defended or refuted by an experimental investigation of the thinking that actually goes on in people's minds when they are called upon to decide issues of fact like those that come up in the courts. But neither can the thesis be defended or refuted by the ordinary methods of common law jurisprudence, since the concepts that have to be interpreted or clarified—viz. proof beyond reasonable doubt and proof on the preponderance of evidence—are concepts that lay jurors are charged to apply just as they would in their everyday lives outside the courts. Jurors could not be expected to do this if those concepts were saturated with the kind of technical legal meaning that requires reference to relevant cases or statutes for its proper determination. Rather, a juror's understanding of those concepts is not open to restriction or extension in the light of any legal authority, but only by reference to acceptable norms of practice in everyday life.

Accordingly, to ask for an interpretation or clarification of the standards of forensic proof is to call for the execution of a typically philosophical project. It is to call for an analytical reconstruction rather than for either a psychological description or a jurisprudential definition. We need to make explicit the principles that are tacitly presupposed in those everyday patterns of reasoning that are appropriate for deciding issues of fact under the particular constraints that the forensic situation imposes. Above all, we need to elucidate why those principles are what they are. So in the present paper I shall approach this task of elucidation yet again, but from a different angle. I shall offer here what is in substance an indirect argument, or argument by *reductio*. I shall argue that if you try to give an account of the standard of criminal proof in Pascalian terms—i.e., as requiring a very high value for a probability function that conforms to the axioms of the mathematical calculus of chance—you are inevitably driven to reserve the crucial place in your reasoning for the assignment of a high value to a *non*-Pascalian function for the assessment of evidential weight.

II

Let us begin therefore with the assumption (which will soon turn out to be untenable) that the requirement of proof beyond reasonable doubt is to be interpreted as meaning that the accused's Pascalian probability of guilt on the facts before the court is numerically very close to one. A probability of one is not obtainable for an inference from one set of empirical facts to a logically distinct set, it might then be said, and that is the point of asking for proof beyond reasonable doubt rather than for absolute certainty, though anyone who supposes that a proposition has been proved beyond reasonable doubt must suppose that the probability of the proposition on the available premises falls short of one by only a negligible interval.

Now the first correction that has to be made in this assumption is a relatively uncontroversial one. Initially, at least, it can be stated as being that what the standard of proof must require, if interpreted in Pascalian terms, is a high level of unconditional probability for the guilt of the accused, not a high level of conditional probability. The verdict is to be "Guilty as charged," not "Guilty on the facts actually put before the court." The jury must not be able to say to itself, for example, "If only we had been told more about the friendship between the victim and the accused, we might have judged that on the *totality* of the relevant facts the accused's guilt had a somewhat lower level of probability." So jurors must be entitled to treat a statement of all the facts actually before the court as a premise that authorizes them to detach a value for the unconditional probability of guilt. Unless it is reasonable to assert the high probability of the accused's guilt unconditionally, this guilt is not beyond reasonable doubt.

We must therefore look rather closely into the circumstances under which, if given E, we are entitled to detach an appropriate judgement of unconditional probability from a judgment of conditional probability that has the form $p(H|E) = n$, where H states that John Doe is guilty as charged and E states all the facts before the court.

Note first that, unless adequate elucidation is provided, it is misleading to symbolize the judgment of unconditional probability that is appropriate in this connection as $p(H) = n$. For, if in the circumstances of proof beyond reasonable doubt $p(H) = n$ is derivable from the conjunction of E and $p(H|E)$

$= n$, then in those circumstances p(H) must be equal in value to p(H|E), because both are equal to n, and E must therefore be irrelevant to H. But there cannot be any proof at all if the premises (E) are irrelevant to what is to be proved (H). Moreover, we need the expression "p(H)" to denote the prior probability, which is related to (H|E) by Bayes's Theorem, and it must therefore be misleading to use it also for the evidentially derivative judgment of unconditional probability. And yet, despite these undeniable points about the familiar technical terminology, in ordinary speech we seem to glide smoothly enough, without any obvious equivocation, from discussing the level of the conditional probability that, on the facts before the court, John Doe is guilty as charged, to discussing the level of the unconditional probability that John Doe is guilty as charged.

One way for a Pascalian analysis to deal with this situation is to suppose that "H" here may stand in for two different propositions, with a referring expression such as "John Doe" functioning somewhat differently in each case. In the case of the conditional probability (or of the prior) "John Doe" merely refers to a particular person as an instance of a certain class, while all the facts on which the probability of his guilt is being evaluated are stated in "E." The referential expression neither adds to nor subtracts from the facts on which the probability is conditional, and anyone else in such a situation would have the same probability of guilt. But in stating the putatively derivative unconditional probability we cannot avoid using the referential expression in a way that takes into account all the facts about John Doe that a reasonable person might think relevant, whether or not they are actually presented to the court. The statement is not now about John Doe as one instance of a certain class, namely the class characterized in "E," but rather about John Doe in all his particularity and individuality.

Another way for a Pascalian analysis to deal with this situation is to suppose that two different probability-functions are involved. That is, expressions referring to John Doe may be replaced uniformly by expressions referring to Richard Roe in a statement of the conditional probability "$p_1(H|E) = n$" (or in a statement of the prior probability "$p_1(H) = m$") without affecting the statement's truth-value. But they may not be so replaced in a statement of the putatively derivative unconditional probability "$p_2(H) = n$." Thus p_1 might be a measure of logical probability and p_2 a measure of justified belief, as in Carnap's treatment of the issue;[2] or p_1 might be a measure of relative frequency and p_2 some kind of epistemic probability as in the examples that Hempel discusses;[3] or p_1 might be a measure of avowed intensity of belief at an initial stage of Bayesian estimation and p_2 at a later one.

Yet another way for a Pascalian analysis to deal with the situation would be to eschew unconditional probabilities altogether, except as priors, and to speak of the desired standard of proof as a very high value for $p(H|T)$, where T states the totality of facts that a reasonable person might think relevant as a basis for proof in the case in question. In that way we need not suppose more than one probability-function to be involved nor need we suppose any equivocation in "H."

From a formal point of view it does not matter much which of these three modes of analysis we adopt. Substantially the same point emerges in each. On the first two modes of analysis our entitlement to move from the conditional to the derivative unconditional probability must depend on the extent to which the facts that are premised in E, and generate the conditional judgment $p(H|E) = n$, include all the facts that a reasonable person might think relevant to the case in question. On the third mode we can say that our entitlement to infer $p(H|T) = n$ from $p(H|E) = n$ depends on the extent to which E manages to approximate T. I shall for the most part assume that one or the other of the first two ways of dealing with the situation is adopted—i.e., that we speak of deriving an unconditional probability—because that seems closer to the actual language and purpose of the criminal courts.

But before I go on to consider what is involved in assessing the level of our entitlement to make this derivation (that is, in assessing the extent to which E manages to approximate T) I should like to anticipate a possible objection. Perhaps someone will say that if $p(H|E)$ has a sufficiently high value—a value sufficiently close to one—there is no room for E to fall short at all: a very high probability on the facts before the court is thus itself equivalent to proof beyond reasonable doubt and suffices to certify the legitimacy of the derivation in question.

This objection will not work. Imagine a lottery with as many tickets as you please and only one winning ticket. Then the proposition that John Doe's ticket is not the winning one has as high a conditional probability as you please on the evidence available about the total number of tickets participating. But you would certainly be rash to infer from this to a very high *un*conditional probability that John Doe's ticket is a loser. You would still have some reason to doubt the validity of that inference until you knew that the lottery was run fairly. For example, if it were also a fact that John Doe is a mobster whose gunmen supervise the draw, that fact would be part of the totality of facts that a reasonable person might think relevant to your conclusion. Consequently, the probability of your conclusion on the basis of this totality of facts would be a good deal lower than its probability on the evidence that was originally available. Of course there may sometimes be issues such that, by the time enough facts have been cited to uphold a very high probability for a certain outcome on those facts, there are no other facts that a reasonable person might still think relevant to the outcome. But, if so, this is a contingent feature of the issue concerned. It is certainly not the consequence of any general principle governing either the mathematical properties of the probability-function or the estimation of its value for particular fillers of its argument-places. Rather, it depends in each case on the completeness of the evidence relative to the outcome at issue.

III

By now, I hope, the thrust of my *reductio* argument is beginning to emerge. Even if you set out to judge the success of a criminal proof in Pascalian terms you cannot avoid using, implicitly or explicitly, an assessment of the completeness of the facts before the court. And then it turns out that, with an adequate mode of assessment for this, the Pascalian judgment is otiose. The Pascalian measure not only does not suffice to assess all we need to assess in meeting the standard of criminal proof, but it is not even necessary for assessing what it does assess. In the remainder of my paper I shall first say a few things about the appropriate way to assess completeness, and then (in IV and V) try to explain why it follows that in interpreting the standard of criminal proof we can, and should, dispense with any reference to Pascalian probability. And I need hardly add that nothing in the paper should be taken as denying the wide range of tasks that may be—or even have to be—accomplished by Pascalian reasoning in the courts. I am talking only about the evaluation of a proof as a whole, not about what is asserted in its premises or lemmas. The latter may often, in appropriate contexts, be usefully formulated in Pascalian terms, as when, for instance, a litigant leads expert testimony to establish the size of the probability that someone who has worked twenty years in an asbestos factory will contract asbestosis. And I am also not talking about any issue that goes beyond the epistemics of proof, such as issues about the proportioning of costs, or damages, or even punishment, in accordance with some established Pascalian probability.

A conditional probability's degree of evidential completeness is—very nearly—what Keynes called its "weight."[4] Many interesting questions can be raised, and answered, about this concept of weight. But I have recently written elsewhere at length about it.[5] So on the basis of what is said there I shall confine myself here to a few salient points.

First, by speaking of the weight of the probability of H on E, or of the weight of the evidence that E affords the probability of H, Keynes did not intend to treat weight as a property of a proposition like $p(H|E) = n$. In his view it is possible to know the weight of a probability without knowing its value, or to know its value without knowing its weight. So it would be more accurate to say that weight is a parameter of ordered pairs of propositions, like the ordered pair [H, E], than to say that it is a parameter of probabilities, and so I shall write the weight of the argument from E to H as $W(H|E)$. But, strictly speaking, we should say that weight is a parameter of ordered pairs of what Russell called propositional functions,[6] because $W(H|E)$ is unaffected if all expressions in H and E that refer to John Doe are replaced by expressions that refer to Richard Roe.

Second, though Keynes thought of conditional probabilities as measuring logical relations, the need for a conception of weight arises just as much if we think of conditional probabilities as measuring relative frequencies, causal propensities or intensities of belief. Thus the need for a probability with maximal weight, which in Carnap's logical-relation analysis of probability appears as a requirement of total evidence,[7] is expressed in Hempel's relative-frequency analysis as a requirement of maximal specificity for the reference class.[8] And within a Bayesian personalist analysis in terms of belief-intensities the requirement would be one for maximal conditionalization.

Third, though the standard of proof that has to be met is satisfied only by what it would be reasonable to regard as a maximization of weight, it does not follow that we need have no interest in any other levels of weight. In building up a proof beyond reasonable doubt we implicitly recognize differences of weight, as more and more relevant facts are taken into account. Also, in criticizing an alleged proof we may need to point out the existence of more than one ground for reasonable doubt. That is, we may need to point out that the weight of the probability in question is more than one level short of being maximal.

Fourth, the intuitive idea of weight requires that $W(H|E_1 \& E_2)$ be greater than $W(H|E_1)$ if $p(H|E_1 \& E_2)$ does not equal $p(H|E_1)$. But it would be rather paradoxical to claim that this is the only kind of situation in which a difference of weight arises. Suppose $p(H|E_1 \& E_2)$ does not equal $p(H|E_1)$ and $p(H|E_1 \& E_2 \& E_3)$ does not equal $p(H|E_1 \& E_2)$ but $p(H|E_1) = p(H_1|E_1 \& E_2 \& E_3)$. On the one hand E_2 and E_3 seem each to add an increment of weight to the argument for H that begins with the premise E_1. But, on the other hand, the conjunction $E_2 \& E_3$ seems to add no weight, if relevance is requisite for an addition of weight. It would therefore be preferable to say that $W(H|E_1 \& E_2)$ is greater than $W(H|E_1)$ if and only if E_2 entails a proposition R such that $p(H|E_1 \& R)$ does not equal $p(H|E_1)$. Even this turns out to require further qualification, as Keynes failed to notice.[9] And we also have to be careful not to suppose that increments of weight vary in size with the extent of relevance involved. It may be tempting to think that if the difference between $p(H|E_1)$ and $p(H|E_1 \& E_2)$ is greater than that between $p(H|E_1)$ and $p(H|E_1 \& E_3)$—so that E_2 may be said to have greater relevance to $p(H|E_1)$ than has E_3—then $W(H|E_1 \& E_2)$ is greater than $W(H|E_1 \& E_3)$. But that way we could end up with different weights for an argument to the same conclusion from logically equivalent premises just because we calculated the weights on the basis of different orderings for the addition of new premises. If we ordered the statement of those additions so as to maximize the violence of swing from favorable relevance to unfavorable relevance, we should get greater weight than if we ordered it so as to minimize this swing. It follows that, if weight is to remain invariant over logical equivalence, as we surely want it to be, increments of weight should not vary with the extent of relevance involved.

Fifth, we should note that Keynes's conception of weight should not be confused with various probabilistic or quasi-probabilistic measures to which the term "weight" has been applied in the literature.

Thus Reichenbach[10] held that, if we know the limit towards which the relative frequency of a certain kind of outcome tends within a sequence of events, then this value can be regarded as "the weight of an individual posit concerning an unknown element of the sequence." Or, in other words, "the weight may be identified with the probability of the single case." Clearly this is not the sense in which Keynes was using the term since he emphasized that an argument of high weight is not, as such, "more likely to be right" than one of low weight. It is easy to find arguments that have high probability but low weight, or low probability but high weight.

Good[11] has discussed weight in the sense in which the weight of evidence concerning H that is provided by E, given G, is equal to log $[p(E|HG)$ divided by $p(E|H\overline{G})]$. But the quantity of evidence relevant to a certain argument is independent of the probability of the evidence given the conclusion. A great quantity of evidence might have been collected in a murder trial, with most of it tending to incriminate the accused, but it might also include an unshakable alibi. In such a case the evidence available would have relatively low probability, given the innocence of the accused, but it would have a heavy Keynesian weight.

It is sometimes suggested that, if the probability that a person assigns to his belief may be quantified in terms of the odds, given specified evidence, at which (within a coherent betting-scheme) he would accept a bet on its truth, then the weight of that evidence may be taken to be reflected in the amount that he is prepared to bet: he may be expected to be willing to put a larger sum at risk when there is more evidence from which to estimate appropriate odds. But other considerations also may affect our attitude towards the size of a bet. Suppose that there is a great deal of evidence, and that this evidence suggests the appropriateness of very long odds. Would you really be willing to risk losing just as large a part of your fortune then as you would if the odds, on the same evidence, were much shorter? Of course, you may have some appropriate mechanism for taking the length of the odds into account here. But it has to be kept in mind that what is ultimately at issue is not so much the measurement of the risk as the nature of the reason for being willing to risk losing such-or-such a part of your fortune in such-or-such circumstances. Keynes raised a question about what makes some ensembles of evidence weightier than others, not just about how people may properly react to these differences of weight.

It is therefore sometimes suggested instead that the weight of an argument from E to H may be taken to vary inversely with the mathematical expectation of gain from a search for further relevant evidence for H. But this suggestion is open to at least two cogent objections. First, in order to avoid begging the question, the gain talked about must presumably be in some non-epistemic kind of utility. And this raises familiar problems about the evaluation of epistemic functions by reference to non-epistemic criteria.[12] Secondly, what are we to say when, for example, a vital eyewitness has died without ever disclosing what he saw? The expectation of any kind of gain from further research in that direction may then be zero, but the weight of the evidence about what actually happened is not increased because of the missing data. This is because the weight of the evidence obtained is being assessed by comparison with the supposed totality of relevant facts, not with the supposed totality of discoverable relevant facts. So, even if we had all the *available* evidence, our argument might still not have maximal weight. Sometimes the prosecution cannot prove guilt beyond reasonable doubt because some crucial issue happens in practice to be undeterminable.

Finally, it has recently been suggested that the amount of evidence in favor of a conclusion in forensic proof should be measured by the level of confidence it affords that the probability of this conclusion falls within such-or-such an interval of a specified figure.[13] For example, if from a large vat of blue and white marbles fifty drawings (with immediate replacement and reshuffling) have produced thirty white marbles and twenty blue ones, then the probability that the next marble chosen will be white is 95% certain to be .6 plus or minus .14, whereas the same ratio in 100,000 drawings would justify saying that this probability is 95% certain to be .6 plus or minus .0003.

More specifically, it is suggested that forensic proofs of fact should be supposed to aim at establishing their conclusions within a certain interval of the requisite probability, so that we may have some canonical level of confidence that they have established this if and only if the quantity of evidence is sufficient. Hence, in the paradoxical civil case against an alleged gatecrasher at a rodeo (where the only evidence is that 1000 people were on the seats and only 499 paid) it is argued that the plaintiff would lose because the probability interval allowed by the evidence extends too far below .501. Similarly, in a criminal case it is argued that the accused's probability of guilt would only fall within the requisite interval of some specified high probability if the evidence were sufficiently complete.

This method of estimating Pascalian probabilities is due to Neyman,[14] and its mathematical validity is guaranteed by James Bernoulli's famous theorem. But the price that has to be paid for the method's validity is that it is properly applicable only to an estimate of the probability of a particular kind of outcome within a certain population of outcomes and only when the evidence for that estimate consists of the appropriate ratio within a sample of outcomes that has been drawn from the same population. Quantitative evidence of this type is essential for the Neyman-Bernoulli method to get a grip on the situation, and "a greater quantity of evidence" in Neyman-Bernoulli terms means the ratio of the *same* kind of outcome within a larger sample drawn from the *same* population. By contrast, "a greater quantity of evidence" in the normal forensic situation means something quite different, viz. the provision of hitherto missing kinds of information so that evidence is now included on relevant issues that were previously undetermined. Perhaps a witness is now available who saw the defendant buy a handgun the day before the crime, or perhaps a garment has been found in the defendant's room with bloodstains that match the victim's. The Neyman-Bernoulli method of interval-estimation is quite inapplicable when we compare amounts of evidence in *this* way. Similarly, it has no bearing on a possible increase of evidence in the rodeo case whereby the alleged gatecrasher might be shown to have torn his clothes on the surrounding barbwire. It could perhaps be made to apply to the rodeo case if we compared that case with one in which 1000 people assembled and only 300 paid, or with one in which 100,000 people assembled in a much larger stadium and only 49,000 paid. But even if such evidence were to bring the mathematical probabilty within a supposedly appropriate interval of .51 it would still seem unjust for the plaintiff to win. Accordingly, the Neyman-Bernoulli method of interval-estimation cannot resolve the paradox of the gatecrasher and so justify a mathematicist conception of proof on the preponderance of the evidence, any more than it can measure the completeness of the evidence in a criminal case and so justify a mathematicist conception of proof beyond reasonable doubt. The only way in which it can legitimately enter into the process of forensic proof is within the testimony of an appropriate expert, about a relevant issue, and then the fact that he or she has given such testimony for a litigant constitutes not an evaluation of the litigant's proof but a premise for the proof which the trier of fact must evaluate in other terms.

IV

In fact, as I have shown elsewhere,[15] Keynesian weight turns out to be closely connected with the Baconian gradation of reliability, proximity to law or, as I now prefer to call it, "legisimilitude." On a proper reckoning the Keynesian weight of the argument from E to H, where $p_1(H_1|E) = n$, is, roughly, the ratio of the number of relevant facts covered by E to the total number of relevant facts, when these are counted in some appropriate way. So it should turn out equal to the Baconian legisimilitude of the conditional schema "If E, then $p_2(H) = n$."

The weight of $p_1(H|E)$, we may say, grades our entitlement to derive $p_2(H) = n$ from E—or, if you prefer it, our entitlement to derive $p_1(H|T) = n$ from E. So it looks at first sight as though, if we return for the moment to our original assumption about interpreting the standard of criminal proof in Pascalian terms, we cannot avoid supplementing our Pascalian criterion with a Baconian one. On the one hand, we assume that what is required is a very high Pascalian probability of guilt on the facts before the court; on the other hand, we find that such a requirement does not suffice and needs to be supplemented by a requirement of maximal weight.

The question that then arises is this: If we need in any case to bring in a Baconian mode of assessment in order to determine weight, do we still need to have a requirement spelled out in Pascalian terms? I have already argued that the Pascalian requirement is inadequate without the Baconian one. I shall now argue that the Baconian requiremant can be put into a form that makes the Pascalian one superfluous.

The point is that, instead of interpreting the standard of proof as requiring maximal legisimilitude for the conditional "If E, then $p_2(H) = n$," where E states all the facts before the court and H states that the accused is guilty as charged, we ought to simplify our interpretation by requiring maximal legisimilitude just for the conditional "If E, then H." This is because the proof of H from E is beyond reasonable doubt if and only if the conditional "If E, then H" has a level of legisimilitude that is maximal by normal human standards, while what has to be proved beyond reasonable doubt—by our Baconian reasoning—is just the truth of H. We do not need to introduce *two* qualifications, by first saying that the probative relationship between premises and conclusion need not be absolutely certain but only beyond reasonable doubt, and then adding that also the conclusion itself need not be the actual guilt of the accused but only a very high probability of his guilt.

Of course, if each stage of the proof were normally open to ready evaluation in Pascalian terms, there might be a case for retaining a mention of Pascalian probability in a statement of the ultimate standard to be attained. One would then be able to measure, as one went along, how far at any particular stage in the development of the proof it still fell short of the standard sought. But in fact no such Pascalian evaluation is normally available. As has often been remarked, very few arguments are offered in criminal courts that can be evaluated by objectively estimable Pascalian probabilities. Where those probabilities are relevant, they tend to form the subject matter of expert testimony which then constitutes a premise for the proof, not an evaluation of it. But the credibility of a witness, the cogency of a motive, the purpose of an action, or the significance of a letter are rarely issues that lend themselves to expert determination by reference to Bayesian calculation, social statistics or Carnapian language-systems.

Perhaps it will be objected that in certain methods of establishing personal identity, such as by fingerprint or voiceprint, forensic proof relies essentially on high Pascalian probability. The reason why the accused's finger on the murder weapon incriminates him, it will be said, is because of the enormously high probability that two different people have different fingerprints, where the size of the probability is generated by compounding the various independent probabilities for the fingerprints of two different people not to share this, that or the other particular characteristic.

But what actually counts in establishing that two sets of fingerprints derive from the same hand is that the prints should have been found to correspond in a sufficient number of characteristics—for example, ridge endings, bifurcations, short ridges, ridge dots, enclosures, etc.—within the same general class of pattern (arch, loop or whorl). Eight such points of resemblance are normally thought to be sufficient.[16] So, though some kinds of characteristics are thought to be more significant than others, no probabilities have to be stated or calculated. The assumption is always that the fingerprints of two different people must differ in some recognizable respect, not just that there is a very high probability that they so differ. Indeed, there may be an extremely high probability that two sets of prints derive from the same hand, given that they resemble one another in each of a certain list of parameters.

But, if there is one further assessable parameter in which they are recognizably dissimilar, the two sets of prints must still derive from different hands. So in the end, if the jury itself had to compare the prints (rather than listen to expert testimony on the matter), it would be a supposed sufficiency of weight that counted, not a supposed size of probability, if the question were whether the facts incriminated the accused beyond reasonable doubt.

Nor will it help at all to claim here that subjectivist methods of conceiving probabilities are always available. But a short digression will be needed in order to show that subjectivist methods will not do the trick.

V

We have to distinguish in this connection between two rather different questions. The first is whether the axioms of the Pascalian calculus can be given a subjectivist interpretation that fits the imaginable introspections of jurors and advocates. The second is whether such an interpretation can elucidate the reasonings appropriate to jurors and advocates. And I shall argue that at best it is only the former of these two questions that deserves an affirmative answer.

It is easy enough to defend the use of a subjectivist theory of probability to describe the partialness of partial beliefs about past events. Anti-subjectivists may object that stake size may affect the connection between the degree of partial belief and the choice of odds found acceptable. But this objection can be met, as Ramsey met it,[17] by assessing probability in terms of the difference between certain values of the bettor's preference-function, or as Mellor met it,[18] by supposing the believer compelled to bet by a "greedy but otherwise mysterious opponent" who "subsequently decides both the stake size and the direction of the bet," provided always that the stake is reasonably within what the gambler has been bank rolled to afford. Again, anti-subjectivists may object that, if a bet about a past event is made in the full knowledge of all available evidence, then the bet can never be settled, so that the choice of betting odds would not be adequately constrained by actual beliefs. But the determined subjectivist like Kaye[19] will argue that, even though no such bets are ever actually made, yet anyone with sufficient intellectual grasp of the issues can state the lowest odds that he *would* find acceptable *if* such a wager were to be considered seriously. Finally, anti-subjectivists may object that to declare one's willingness to bet on a given proposition at such-or-such odds is merely to describe one's own state of belief without justifying it. And the subjectivist will then reply, as Kaye does, that one degree of belief in the proposition may nevertheless be much more justified than all others.[20]

Thus it is certainly possible to interpret the Pascalian calculus in such a way that probability-functions come to operate as descriptions of the partialness of partial beliefs about past events. And what I mean by this—very roughly—is that a sentence accepting any bet at odds not less than of n to m minus n on H, given E, can be mapped onto each formula or part of a formula $p(H|E) = \dfrac{n}{m}$ in the calculus, even though H is a statement about some past event; and then all of the same relationships of equivalence, entailment, etc. that held between the various mathematical formulas will come to hold between these interpreting sentences, provided that all the bets described are mutually "coherent" in the requisite sense. But, though such sentences would be intelligible, it does not follow that one can easily imagine them as being uttered to report the introspections of jurors and advocates. It is tempting to suppose that assertions about how he would *bet*, if he were to bet with a "mysterious opponent" such as Mellor mentions on an issue that is in fact irresoluble, would appear to the average juror to constitute too sophisticated a way of describing his own state of mind, while at the same time perhaps appearing to the average lawyer as too frivolous a speculation.

Be that as it may, we have to remember the normative nature of the problem before us. The problem is not just to find an interpretation that fits the imaginable introspections of jurors and advocates. It is also to find an interpretation that elucidates what reasonings are appropriate to jurors and advocates in the context of currently acceptable procedures for forensic proof. A subjectivist interpretation is inherently incapable of achieving the latter objective.

The basic reason for this incapacity is that on a subjectivist interpretation each person's judgment of the probability of a particular proposition on given evidence does no more than describe that person's own state of mind. It implies a willingness to bet on the proposition at specified odds. Hence, on a subjectivist interpretation, if I think one of your judgements of probability may be incorrect, I am just doubting your description of your own state of mind, and any reasons that I advance in favor of my view must be reasons that bear either on your ability to introspect your own mind accurately or on the honesty with which you report your introspections. But arguments of such a nature are quite out of place in the context of forensic proof. If the defense wishes to call in question the alleged degree of validity with which the accused's guilt follows from the facts put before the court by the prosecution, it must demonstrate other facts or possibilities that bear on the circumstances of the crime or on the character, behavior, movements, etc. of the accused. Defending counsel will hardly be allowed to introduce testimony tending to prove the inability of prosecuting counsel to introspect accurately or his readiness to report his introspections falsely. Nor would it, or should it, profit the defense to introduce that testimony even if allowed to do so.

Subjectivists sometimes try to get around this difficulty by claiming, as does Kaye,[21] that it is quite possible for A to give reasons to B why B should not be satisfied with having such-or-such a number for his subjective probability. A has only to mention a relevant fact, they may say, that B has left out of his consideration and then ask B to conditionalize on that fact. But the trouble with such an account is that it offers no elucidation at all of how B is to judge the actual size of the probabilities involved. Subjectivism provides us with a way of putting guesses into mathematical uniform, but denies us a way to tell good guesses from bad ones. Subjectivists have to fall back on claiming, as Kaye does, "that subjective probabilities are well suited to the study of forensic proof while firmly maintaining a discreet agnosticism as to which camp of philosophers has found the correct answer to the question of what it really is that makes one degree of belief more justified than all other degrees of belief."[22] But what else can we usefully be discussing, when we analyze the standards of proof, than "what it really is that makes one degree of belief more justified than all other degrees of belief" when the jury comes to consider the issues on the facts before the courts? How else can we improve on the guidance that judges are already accustomed to give when they charge their juries? Those who try to defend a subjectivist interpretation here relapse inevitably into talking about the wrong issue. In trying to defend that interpretation they may tell us what can *cause* a juror to change his probability-assignments. But they cannot tell us what would objectively justify such a change. And this normative, epistemological problem has to be the *philosophical* question at issue, because in its official role forensic proof aims to convince by the validity of its arguments: any other mode of persuasion (for example, misrepresentation, hypnosis, emotional prejudice, etc.) is intrinsically open to challenge.

VI

A Baconian interpretation, on the other hand, provides a simple and commonsensical account of the basic nature of forensic proof. A trier of fact just has to apply to the instant case his general knowledge about regularities in nature and in human affairs and his appreciation of the circumstances under which those regularities admit of exceptions. Or, in other words, John Doe's guilt is provable from the facts before the court if and only if Richard Roe's guilt would be provable from analogous facts. Moreover, each stage in the proof of guilt can readily be evaluated in terms of the proof's progress toward completeness—completeness in the elimination of grounds for reasonable doubt. Each stage is marked, quite commonsensically, by an increase in the number of reasonably relevant issues already dealt with, as against the number of reasonably relevant issues still outstanding. Of course, a precise reconstruction of how such reasoning ideally operates—as attempted in *The Probable and the Provable*[23]—is bound to be quite an elaborate affair, and to bring to light many implications and presuppositions of legisimilitude that would otherwise remain unnoticed. But that is true in all areas of philosophical analysis. It holds just as much for philosophical reconstructions of the Pascalian reasoning that is practiced in casinos or at racetracks, for example.

In sum, then, my argument in this paper is as follows. Even if you set out to judge the success of a criminal proof in Pascalian terms you cannot avoid using also, implicitly or explicitly, an assessment of the completeness of the facts before the court. And then it turns out that with an adequate mode of assessment for such evidential weight (as Keynes called it), the Pascalian judgment is otiose. That is, to say that there is *reasonably* complete evidence that there is *reasonably* maximal probability of guilt is to overdo the hedging required. The appropriate standard of proof rests solely on the reasonable completeness of the evidence of guilt.

NOTES

† © 1986 by L. Jonathan Cohen.

* Fellow, Tutor & Praelector in Philosophy, The Queen's College, Oxford University.

[1] L. J. COHEN, THE PROBABLE AND THE PROVABLE (1977). *See, e.g.,* Cohen, *Freedom of Proof*, 16 ARCHIV FÜR RECHTS-UND SOZIALPHILOSOPHIE 1 (1983); Letter from L. Jonathan Cohen to Editor, 1980 CRIM. L. REV. 257; Cohen, *The Logic of Proof*, 1980 CRIM. L. REV. 91; Cohen, *On "Away with the Fuzz,"* 20 J. FORENSIC SCI. SOC'Y 222 (1980); Cohen, *The Problem of Prior Probabilities in Forensic Proof*, 24 RATIO 71 (1982); Cohen, *Subjective Probability and the Paradox of the Gatecrasher*, 1981 ARIZ. ST. L. J. 627; Cohen, *What is Necessary for Testimonial Corroboration?*, 33 BRIT. J. PHIL. SCI. 161 (1982).

[2] R. CARNAP, LOGICAL FOUNDATIONS OF PROBABILITY 211 (2d ed. 1962).

[3] C. HEMPEL, ASPECTS OF SCIENTIFIC ENQUIRY, AND OTHER ESSAYS IN THE PHILOSOPHY OF SCIENCE 55 (1965).

[4] *See generally* J. KEYNES, A TREATISE ON PROBABILITY ch. VI (1921). Keynes himself was not sure that the theory of weight had any practical significance.

[5] *See* Cohen, *Twelve Questions About Keynes's Concept of Weight*, 37 BRIT. J. PHIL. SCI. 263-78 (1986).

[6] B. Russell, Introduction to Mathematical Philosophy 155-66 (1920).
[7] R. Carnap, *supra* note 2.
[8] C. Hempel, *supra* note 3, at 399.
[9] *See* Cohen, *supra* note 5.
[10] H. Reichenbach, The Theory of Probability: An Inquiry Into the Logical and Mathematical Foundations of the Calculus of Probability 465 (1949).
[11] Good, *Corroboration, Explanation, Evolving Probability, Simplicity and a Sharpened Razor,* 19 Brit. J. Phil. Sci. 123-143 (1968).
[12] *See, e.g.,* P. Levi, Decisions and Revisions: Philosophical Essays on Knowledge and Value (1984).
[13] *See* N. Cohen, *Confidence in Probability: Burdens of Persuasion in a World of Imperfect Knowledge,* 60 N.Y.U. L. Rev. 385 (1985).
[14] Neyman, *Outline of a Theory of Statistical Estimation Based on the Classical Theory of Probability,* 236 Royal Soc'y London Phil. Transactions 348 (1937).
[15] *See* Cohen, *supra* note 5.
[16] A. Moenssens, Fingerprints and the Law 127 (1969).
[17] F. Ramsey, The Foundations of Mathematics 176-78 (1971).
[18] D. Mellor, The Matter of Chance 37 (1971).
[19] Kaye, *Paradoxes, Gedanken Experiments and the Burden of Proof: A Response to Dr. Cohen's Reply,* 1981 Ariz. St. L. Rev. 635, 642-43 (1981) [hereinafter Kaye, *Paradoxes*]. In a later article, *Do We Need a Calculus of Weight to Understand Proof Beyond Reasonable Doubt?, infra* at 129, Kaye apparently rejects a subjectivist analysis in this context: "Attorneys, courts and jurors do not—and should not—act as if one sincere probability is as good as another." *See id.* at 137. But he still supposed a mathematical probability function to be operative—indeed, a mathematical probability-function that will pay due attention to the completeness of the evidence as well as to its content. Unfortunately he thinks it unnecessary for him to specify the semantics of this function, i.e. to say exactly what is meant by 'probability' here. So he fails to respond to my claim that no function with appropriate semantics can satisfy the axioms of the mathematical calculus of chance.
[20] Kaye, *Paradoxes, supra* note 19, at 644.
[21] *Id.*
[22] *Id.* at 645.
[23] *See* L. J. Cohen, The Probable and the Provable, *supra* note 1.

L. Jonathan Cohen,
Fellow, Tutor and Praelector in Philosophy,
The Queen's College, Oxford University.

D. H. KAYE*

DO WE NEED A CALCULUS OF WEIGHT TO UNDERSTAND PROOF BEYOND A REASONABLE DOUBT?†

Nearly a decade ago, Dr. L. J. Cohen described a system of ordinal, inductive "probabilities" said to measure the strength of certain arguments and said to be very much in the spirit of Sir Francis Bacon.[1] In his current article, *The Role of Evidential Weight in Criminal Proof*,[2] Cohen explains how his inductive probabilities are closely connected to the work of John Maynard Keynes. Building on some tentative thoughts that Keynes expressed in 1921,[3] Cohen argues that probabilities *per se* are insufficient to describe the probative force of evidence and that a calculus of "weight" is required to supplement or supplant these probabilites. He asserts that one must attend to the completeness of the evidence used to estimate a probability, and that a measure of completeness cannot be incorporated into the probability itself or by means of another probability. Cohen develops his argument with reference to the burden of persuasion in criminal cases:

> [I]f you try to give an account of the standard of criminal proof in Pascalian terms—i.e., as requiring a very high value for a probability function that conforms to the axioms of the mathematical calculus of chance—you are inevitably driven to reserve the crucial place in your reasoning for the assignment of a high value to a *non*-Pascalian function for the assessment of evidential weight.[4]

I believe that Cohen's fundamental premise is correct. One must examine the completeness of a body of evidence and the circumstances under which the evidence was gathered if one is to assess its probative value. But I disagree with Cohen's conclusion that one is "inevitably driven" to construct a model of forensic proof that relies on a conception of the "weight" of a body of evidence that is logically distinct from the probability of an event conditioned on that body of evidence. Nor is it apparent to me that Cohen's method of expressing "weight" represents an advance over an analysis that relies on conventional, cardinal probabilities.

To establish that resort to a non-probabilistic measure of weight is unnecessary, we can criticize the details of Cohen's reasoning and comment on his examples. Alternatively, we can simply exhibit a model of forensic proof that (a) relies exclusively on conventional probability theory, and (b) accomplishes the tasks that Cohen thinks must be borne by an independent concept of evidential weight. I shall do a little of each, but mostly I shall try to pursue the second route. That is, I shall sketch an account of courtroom inference based strictly on probabilities and an interpretation of the burden of persuasion derived from decision theory.

If my elaboration on this conventional interpretation of the burden of persuasion is not flawed in the ways that Cohen believes any strictly probabilistic analysis must be, then his latest complaint against standard probability theory must fail.[5] Part I of this paper sketches the basic outlines of the conventional probabilistic interpretation of the burden of persuasion. Part II supplements this theory to allow it to recognize gaps in the state's case. Part III offers a few observations about a separate problem that Cohen mentions, namely, the failure of the decision-theoretic analysis to articulate interpersonal standards for arriving at subjective probablities.

I. THE CONVENTIONAL INTERPRETATION OF THE BURDEN OF PERSUASION

A. *Evidence, Contentions, and Stories*

To establish that the conventional decision-theoretic interpretation of the burden of persuasion is a viable theory, I must be more comprehensive and precise than most writers—myself included—have been. The detail is tedious, yet it is necessary to avoid confusion. For clarity, I shall use some symbols, but I have no formal proofs to give. Thus, while the analysis is not mathematically rigorous, it has the advantage of being accessible to any moderately determined reader.[6]

At the outset, I want to distinguish between "evidence," "factual contentions," and "stories."[7] An item of evidence is a stipulation, a fact established by judicial notice, or most frequently, the portion of the testimony of a witness or a documentary exhibit that is introduced to help prove a factual contention. With the exception of matters of which judicial notice is taken, evidence comes from two sources: the prosecution and the defense. The prosecution offers some number n of "items" or "pieces" of evidence (assertions of a witness, photographs, etc.), and the defense presents m items of its own. The prosecution's evidence can be represented as $\{E_1^P, E_2^P, \ldots, E_n^P\}$. The defense's evidence can be represented as $\{E_1^D, E_2^D, \ldots, E_m^D\}$, and the totality of the evidence is $E^P \cup E^D$.[8] The *j*th item of prosecution evidence may be denoted E_j^P, where j is between one and n. E_j^D has similar meaning.

If we think of the trial as analogous in some respects to an experiment[9] with various possible outcomes (events) whose probabilities can be computed (under various hypotheses) prior to running the experiment, then the presentation of the pieces of evidence comprising E are like the events (the experimental data) that the experimenter observes. E_j^P is the event that the prosecution presents E_j^P (the j*th* item of evidence), E^P is the event that the prosecution offers the items of evidence E_1^P through E_n^P, and so on. In other words, E^P, E^D, and the joint event E (which is the intersection of these two events) states what happened at trial, including features such as the identity of the party adducing the evidence and the demeanor of a witness, that may bear on the credibility of the propositions being asserted.

In an armed robbery case, for instance, E_1^P might be testimony of a store clerk that the defendant is the woman who robbed the store, E_2^P might be his testimony that she pointed a gun at him, E_3^P might be a police officer's testimony that he found a gun in the defendant's apartment, and so on. Of course, the defendant may also produce evidence. E_1^D might be the testimony of the defendant's mother that she and her accused daughter were at a movie theater watching *Gone with the Wind* during the time of the robbery, and the final item of evidence might be E_9^D, the testimony of defendant's minister to the effect that the defendant has a good reputation for being law-abiding and peaceable.

The evidentiary events differ from the outcomes of an experiment in many ways. For one, an impassive Mother Nature who determines the experimental results does not dictate the evidentiary outcomes. Rather, the parties introduce the items of evidence to convince the judge or jury of their factual contentions. These contentions are statements about what happened in the past, and we may think of them as events to which probabilities can be assigned, just as we described the presentation of evidentiary items as events transpiring at trial. The prosecution may contend, for example, that the defendant fled from the store in her car. Each such factual contention advanced by the prosecution can be denoted C_j^P, where, as with the items of evidence, j indexes each such contention. We may let r represent the total number of these contentions of the prosecution. Analogously, the defense's contentions of fact may be denoted C_j^D, and there are some number s of these.

The number of contentions, r + s, and the number of items of evidence, n + m, need not be equal. A single item of evidence may bear on several contentions. Alternatively, several items of evidence may bear on a single contention. In addition, some of the defense contentions may be simple negations of the plaintiff's contentions, while other contentions may be identical.

The totality of the factual contentions presented by each side make up the "stories" that the parties tell. S^D is defendant's story. It consists of the joint event C_1^D through C_n^D. If the defendant declines to make an opening statement and moves for a directed verdict at the close of the prosecution's case, S^D is not yet known.

S^P is the prosecution's story. It is the joint event C_1^P through C_r^P. For there to be a disputed issue of fact, S^P and S^D must differ as to at least one of their constituent contentions. Furthermore, because the stories are constructed in light of the pertinent law, S^P is such that all the elements of the offense are satisfied, while S^D is such that one or more elements are not satisfied.[10]

B. *The Posterior Probability of a Story and the Burden of Persuasion*

The conventional account of the burden of persuasion supposes that factfinders review all of the evidence and, aided by closing arguments and relying upon their conception of human behavior and the way the world works,[11] somehow arrive at a personal probability that plaintiff's story, S^P, is true, as opposed to the defendant's story, S^D. This personal probability can be represented as

$$\Pr(S^P|E) = p \qquad (1)$$

The vertical bar between S^P and E emphasizes that the probability of S^P is conditioned on the receipt of all of the evidence E. The factfinder is not allowed to rely on extra-judicial knowledge of the case. Because the probability p is formed after all the evidence is in, we may call it a posterior probability.[12]

The burden of persuasion is then the simple rule that the judge or jury should convict or render a judgment for plaintiff if and only if p > p,* where p* is some critical probability (some number close to one in criminal cases). Deciding cases according to this criterion minimizes expected losses, where the loss function is a linear function of the numbers of false convictions and false acquittals.[13] This model can be used to describe and evaluate evidentiary rules concerning the burden of persuasion and presumptions or inferences.[14]

II. THE POSTERIOR PROBABILITY AND "GAPS" IN THE EVIDENCE

Cohen's chief criticism of this account is that p is a posterior probability conditioned on a body of evidence E that may be incomplete or misleading. According to Cohen, the dispositive probability must be unconditional. In other words, it is necessary to move from the posterior probability $\Pr(S^P|E)$ to an unconditional probability $\Pr(S^P)$. While this phrasing is misguided,[15] the recognition that $\Pr(S^p|E)$ does not contain information about the completeness of E is an important insight. Either E must be interpreted differently or a different equation must be substituted for (1). This can be done within a framework of conditional probability without resorting to a distinct calculus of evidentiary weight.

A. *The Prevalence of Gaps*

The exposition of the burden of persuasion in Part I ignores the possibility of gaps in a litigant's evidence—gaps that make the party's story less believable. Any good trial lawyer knows that the jury will expect to hear certain items of evidence in certain cases, and that it may regard the failure to produce such evidence with devastating skepticism.[16] Let me offer an anecdote drawn from my wife's experience as a juror. The state charged a young man with driving while intoxicated (DWI). A police officer testified that he stopped the defendant after seeing him change lanes illegally at two a.m. He further testified that he smelled alcohol on the man's breath, and that a "field sobriety test" showed the defendant to be intoxicated. The defendant's story was that he was lost in an unfamiliar neighborhood, listening to his wife in the car and trying to determine which way to proceed. The jury believed that the defendant probably was drunk, but it acquitted because the state, which had to prove guilt beyond a reasonable doubt, did not measure the defendant's blood or breath alcohol level. The jury expected to hear "breathalyzer" evidence in a drunk driving case, and the absence of that evidence in the state's case raised a reasonable doubt in the minds of the jurors.[17]

These "negative inferences" extend to civil matters as well. Consider a paternity case in which the plaintiff concedes that two men could have been the father. Suppose the plaintiff compels the defendant to submit to immunogenetic testing, and inexplicably ignores the other man. Even if the genetic tests implicate the defendant, the plaintiff's story is weaker than it would be if both men had been tested and the nonaccused man excluded as a potential father.[18]

Any model of courtroom reasoning that cannot handle the negative inferences that arise from the obvious incompleteness of a party's evidence is deficient. As I have indicated, the crucial question is whether our probabilistic analysis of the burden of persuasion can be modified to reflect valid reasoning about gaps in a party's case.

B. *A Probabilistic Account of Evidence in the Context of a Story*

Evidence cannot be evaluated in a contextual vacuum. It is puzzling to read of "John Doe" holding a losing lottery ticket without being told how the "lottery" was conducted.[19] In the forensic arena, a party's evidence is evaluated in the context of the whole story that the party is presenting. The prosecution's story S^P will suggest that various items of evidence will exist if the state has been diligent in preparing its case. The jury in the DWI case, for instance, believed that if the state's story had been true and if the police had been diligent, then evidence of the concentration of alcohol in the defendant's breath would have been presented.

More generally, we may denote the collection of evidence that one might expect to hear in support of S^P (if S^P is what occurred) as $F^P = \{F_1^P, F_2^P, \ldots, F_t^P\}$. Also, let us use F^D to stand for the collection of evidence that one would expect to hear in support of S^D if S^D occurred. In our probabilistic model of the fact-finding process, we may say that the judge or jury should contrast E^P with F^P, and E^D with F^D. If F_j^P are missing from E^P, then these gaps G^P in the state's case are part and parcel of the event E conditioning S^P. A gap is as real as a "hole" in a semiconductor. We might say that E refers not to the presentation of evidence in a vacuum, but to evidence in context, that is, with G^P (and G^D—the gaps, if any in defendant's evidence). It will be convenient to let G stand for all these gaps (G = $G^P \cup G^D$) and to let G be the corresponding joint event that there were these evidentiary gaps G at the trial.

Then we can make the full meaning of the joint event E explicit by substituting for $\Pr(S^P|E)$ in (1) the expression $\Pr[S^P|(E \cap G)]$. In other words, the decision rule becomes: Find for the defendant if and only if

$$\Pr[S^P|(E \cap G)] = p < p^* \qquad (2)$$

Given the importance of perceived gaps in a party's evidence, the rule of evidence should afford litigants the opportunity to introduce evidence and to explain to the jury why there appear to be gaps in their cases, assuming that this does not implicate other important values.[20] If G^P is not explained satisfactorily, then at the close of the prosecution's case, $\Pr[S^P|(E \cap G)]$ will be less than $\Pr(S^P|E)$.[21] If G^P is a significant enough omission, then this difference may bring p below the critical value p*, precluding a conviction. Instead of turning to a separate calculation of "weight," we account for the completeness of the prosecution's case in the posterior probability itself.[22]

III. Estimating $\Pr[S^P|(E \cap G)]$

The decision rule summarized in (2) responds to Cohen's principal objection to a probabilistic interpretation of proof beyond a reasonable doubt. In the penultimate section of his paper, however, Cohen questions on other grounds the use of personal probabilities in a normative model of forensic proof. Previously, he insisted that such probabilities were wholly inadmissible, since they presupposed a hypothetical lottery or wagering procedure that was meaningless inasmuch as the bets could not be settled and the personal probabilities elicited in this fashion would be affected by the stakes of the bets.[23] In his current article, he at last abandons (or retreats from) these arguments. The view that Cohen once dismissed as "grossly fallacious"[24] is now "easy enough to defend,"[25] although he continues to complain that the very idea of determining the personal probabilities of real jurors is "too sophisticated" and "too frivolous."[26]

In this regard, whether eliciting personal probabilities with the aid of a hypothetical wagering scenario is a plausible description of a juror's "own state of mind" or of "imaginable introspections of jurors and advocates"[27] is beside the point. No one imagines that jurors see themselves as stating their indifference points for wagers of indeterminate stakes. The wagering formulation of subjective probability is merely one way to motivate the axioms for subjective probability.[28] A suitably contrived betting procedure, like the ubiquitous meter rods and clocks of special relativity, is but an idealized measuring device, constructed to induce a certain type of consistency in hypothetical psychological judgments.[29]

The inquiry, as Cohen finally recognizes, is prescriptive rather than descriptive. He seeks a model that specifies how jurors *should* reason and act. The decision-theoretic analysis focuses on what actions should be taken under risk. It presumes that action must be taken (a verdict must be returned), and it prescribes a rule for doing so that minimizes the expected losses from erroneous verdicts. To this extent, it has normative content. For instance, a juror who sincerely determines that, as far as he or she can tell, $\Pr[S^P|(E \cap G)]$ is only one-half, would be unwarranted in voting to convict.

At the same time, the decision-theoretic analysis offered here says much less about belief formation. As Cohen now seems to admit, probabilities can serve to quantify partial beliefs, but they are only the raw material in an idealized scheme of decisionmaking. Consequently, the treatment of the burden of persuasion encapsulated in (2) hardly provides a full account of how jurors should evaluate evidence. For one thing, I have not discussed how p itself should be estimated. I have merely said that if judges and juries are to minimize an appropriate expected loss function, then they must evaluate the evidence to find the probability p and contrast it with p* to reach a verdict. And I have argued that the p in (2) can and should reflect the completeness of each side's evidence.

I have not offered a "Bayesian" interpretation of the burden of persuasion, if by "Bayesian" one means probabilities deduced through Bayes's rule.[30] The p in (2) might come from an entirely different reasoning process. For instance, it might reflect an intuitive judgment of the totality of the evidence E assessed in the context of S^P and S^D, as made with the aid of the various heuristics identified by cognitive psychologists.[31] Wherever it comes from, as long as the jury's value for p correctly expresses the probability that S^P is true given the evidence and its context, a decisionmaker who follows the rule given by (2) does the best that we can expect in the risky world of litigation.

But it is clear that Cohen, even if he were to agree that this account obviates the need for an independent measure of evidential weight, would not be satisfied. How do we know whether the judge or jury has arrived at the correct value for p? The analysis thus far gives no insight into the way in which judges or jurors *should* estimate this crucial quantity. It expresses this probability as a conditional probability—one determined in light of the entirety of the evidence E and its context—but it says nothing about how this conditional proability should be formed. A Bayesian explication of how a factfinder should successively revise an initial prior probability to arrive at p would give the model more structure, but it would merely push the objection one level down.[32]

Yet, Cohen slides from the legitimate perception that the existing theory is not all-inclusive to the much more problematic contention that "a subjectivist interpretation is inherently incapable" of elucidating "what reasonings are appropriate to jurors and advocates."[33] According to Cohen, "[s]ubjectivism provides us with a way of putting guesses into mathematical uniform, but denies us a way to tell good guesses from bad ones."[34]

These are puzzling if not disappointing claims. The transition from the statement that "theory T does not provide us with X" to "theory T denies us X" is obviously unsound. Knowing how to solve partial differential equations does not tell a physicist how to arrive at the equations that model the motion of fluid in a container, but the mathematical theory of partial differential equations certainly does not deny him the physical theory needed to discern the correct set of equations and their boundary conditions. Similarly, knowing that idealized jurors can express partial beliefs as conditional probabilities does not bar us from discussing or understanding the epistemology that governs these probabilities.[35]

Although it is plainly wrong to think that a theory is unsuitable merely because it is not all-inclusive, the question whether the jury's partial beliefs can, in principle, be quantified so that they lead to a suitable probability p in (2) is by no means trivial. Many probability theorists within the personalist or subjectivist school treat personal probabilities as Cohen does. They hold that the degrees of belief quantifed as probability judgments need only obey certain axioms concerning preferences.[36] This formal property of coherence may be a necessary feature of a satifactory normative theory of forensic proof, but I agree with Cohen (and many other philosophers) that we ought to demand more than this. Attorneys, courts and jurors do not and should not act as if one sincere probability is as good as another. They worry and argue about what warrants a given degree of belief, which is why a major part of the typical closing statements of counsel consist of exhortations to common sense and generalizations about human behavior and credibility.

It bears reiterating, however, that the use of subjective probability in an idealized theory of forensic proof does not preclude the attempt to formulate a philosophically adequate account of the interpersonal and logical standards that promote accurate estimation of the probability on which, I have argued, a verdict should turn.[37] Indeed, Cohen's doctrinaire claim that subjective probability theory is inherently incapable of elucidating such standards dismisses all too cavalierly the efforts of the many probability theorists and students of inductive logic who have attempted to systematize personal judgments of probability in ways that could elucidate such standards.[38] While it is possible that all attempts to explicate the logical or empirical bases for justifiable subjective probabilities will fail, I think it is too early to foresake this line of inquiry. Thus, the message that I have sought to convey in my exchange with Cohen is not that his own theory is false, but that his campaign against theories that employ cardinal probabilities to analyze factfinding in the legal process has yet to score a decisive victory. Such theories already have sufficent richness to produce insights on the law of evidence, and they are amenable to further refinement and supplementation.

NOTES

† © 1986 by D. H. Kaye. All Rights Reserved.

* Professor of Law and Director, Center for the Study of Law, Science and Technology, Arizona State University. Michael White and Laurence Winer provided useful comments on a preliminary draft of this paper. My treatment of this topic was stimulated by conversations with Ronald Allen, Ira Ellman, and Peter Tillers, and refined by observations from the faculty who attended workshops at the University of Pennsylvania and Arizona State University.

[1] L. COHEN, THE PROBABLE AND THE PROVABLE (1977).
[2] Cohen, *The Role of Evidential Weight in Criminal Proof*, supra at 113.
[3] J. KEYNES, A TREATISE ON PROBABILITY 71-78 (1921). Keynes prefaced his discussion of "weight" with the remark that "[t]he question to be raised in this chapter is somewhat novel; after much consideration I remain uncertain as to how much importance to attach to it." *Id.* at 71. Keynes further observed:
> [I]n deciding on a course of action, it seems plausible to suppose that we ought to take account of the weight as well as the probability of different expectations. But it is difficult to think of any clear example of this, and I do not feel sure that the theory of "evidential weight" has much practical significance.

Id. at 78. Cohen, it appears, does not share these doubts.

[4] Cohen, *supra* note 2.
[5] More precisely, his claim that he demonstrated that reference to non-probabilistic weight is necessary to an explanation of the burden of persuasion in criminal cases will have been shown to be false. This is not to say that a theory that does use non-probabilistic weight is necessarily implausible or inferior to a totally probabilistic theory. I do not deal with the question of comparative merit in any depth, however, because I do not feel qualified to judge whether Cohen correctly elucidates Keynes's idea of evidentiary weight or whether Cohen's inductive probabilities solve the problem he perceives. *Cf.* Allen, *A Reconceptualization of Civil Trials*, *supra* at 21, (questioning the capacity of Cohen's calculus to resolve other "paradoxes").

[6] I use some symbols that are standard in set theory. These include the curly brackets "{" and "}" to indicate that the items listed between the brackets form a set, "∪" to denote the union of two sets, and "∩" to indicate the intersection of two sets. The elements of some of the sets that I describe are events in a sample space. The symbols referring to sets that are within the sample space are italicized. The probabilities to which I refer are values of a non-negative, additive function, whose maximum value is one, defined on this sample space. I belabor these points because of Cohen's concern that the probability function that I use to measure rational, partial beliefs may lack "appropriate semantics." Cohen, *supra* at 124.

Although the definition of probability as a function whose domain is a sample space can be found in virtually all mathematics texts, logicians often speak of probability as being distributed over sentences in a formal language. *E.g.*, R. JEFFREY, THE LOGIC OF DECISION (2d ed. 1983). If my analysis were to be recast in these terms, the same conclusions would obtain. *Cf.* P. SUPPES, PROBABILISTIC METAPHYSICS 187 (1984) ("[T]here is a classical tradition of ambiguity in probability theory of whether to speak of the probability of events or of propositions, and I do not think it important which one we choose.").

[7] Professor Nesson's brief but provocative treatment of the conjunction of evidence as opposed to the conjunction of elements of a cause of action prompted me to think in these terms. *See* Nesson, *The Evidence or the Event? On Judicial Proof and the Acceptability of Verdicts*, 98 HARV. L. REV. 1357, 1387-90 (1985). Cohen does not attend to such distinctions. He speaks of "facts" sometimes to mean the evi-

dence before the court, and sometimes to mean factual contentions that are accepted as true.

[8] Where the individual pieces of evidence are distinct, this union is $E = \{E_1^P \ldots, E_n^P, \ldots, E_1^D, \ldots E_m^D\}$.

[9] See I. GOOD, GOOD THINKING: THE FOUNDATIONS OF PROBABILITY AND ITS APPLICATIONS 11-12 (1983).

[10] Some factual contentions in S^P may not be essential to satisfying the elements of the offense. For instance, the prosecution may argue that the defendant acted with a particular motive, but the jury may convict even though it concludes that the defendant acted for a different reason. Perhaps S^P should be thought of as the minimal body of contentions along the lines suggested by the prosecution that, if believed, would warrant a verdict of guilty.

The stories play the role of hypotheses in our analogy to a statistical experiment. Since we will want to speak of conditional probabilities for the competing stories, the mathematical formalism requires that they be treated as constituting a set analogous to the parameter space for such an experiment.

[11] See, e.g., Mansfield, *Jury Notice*, 74 GEO. L.J. 395 (1985).

[12] Also, because we are considering the normative question of how jurors ought to assess evidence, we do not deal with the personal probabilities that might be entertained by real jurors affected by passion, prejudice or misinformation, but with the probabilities that ideal jurors would come to have on the basis of the evidence E. See Lempert, *Modeling Relevance*, 75 MICH. L. REV. 1021 (1977). This prescriptive model can serve as a standard for evaluating the likely behavior of real jurors and for constructing rules of evidence that (to the extent desired) will promote accurate assessment of the probabilities. The problem of interpreting the personal probabilities is considered further in Part III.

[13] For a discussion of the difficulty of defining this loss function precisely, see Kaye, *And Then There Were Twelve: Statistical Reasoning, the Supreme Court, and the Size of the Jury*, 68 CALIF. L. REV. 1004 (1980). It is also important to recognize that the decision-theoretic interpretation does not mean that actual losses necessarily are minimized. Rather, the rule is constructed to minimize the *expectation* of the loss function. The law of large numbers establishes the link between expected values and long run frequencies. Not all commentators are careful to distinguish between expected values and actual outcomes. E.g., Nesson, *supra* note 7, at 1377 n.67.

[14] E.g., Kaye, *The Limits of the Preponderence of the Evidence Standard: Justifiably Naked Statistical Evidence and Multiple Causation*, 1982 AM. B. FOUND. RES. J. 487; Kaye, *Probability Theory Meets Res Ipsa Loquitur*, 77 MICH. L. REV. 1456 (1979); Kornstein, *A Bayesian Model of Harmless Error*, 5 J. LEGAL STUD. 121 (1976); Lempert, *supra* note 12; Note, *A Probabilistic Analysis of the Doctrine of Mutuality of Collateral Estoppel*, 76 MICH. L. REV. 612 (1978).

[15] Cohen urges that "we speak of deriving an unconditional probability . . . because that seems closer to the actual language and purpose of the criminal courts." Cohen, *supra* at 116. It is not clear to me what he has in mind. The "actual lan-

guage" of the courts usually includes an instruction to the jury to decide the case on the strength of the evidence admitted. This does not suggest an unconditional probability. As for the "purpose of the criminal courts," that is not one, but many things. It is not obvious that basing decisions on unconditional probabilities would better promote these functions than returning verdicts on the basis of conditional probabilities. This phrasing is not meant to suggest that I advocate instructing jurors to articulate any kind of a probability as an aid to decisionmaking.

Perhaps my confusion arises from an idiosyncratic use of the phrase "unconditional probability." There are indications in Cohen's paper that he does not use the unconditional probability of an event A to represent the probability of A prior to the receipt of a body of evidence E relevant to A, but rather that he means to indicate the probability of A conditioned on all conceivable relevant evidence about A (the "totality of the facts"). If so, and if we let T stand for the presentation of this totality, then the "unconditional" probability of which he speaks is $Pr(A|T)$. Of course, we could omit the explicit reference to T, and merely write $Pr(A)$, but this abbreviated notation would not make the probability of A, assessed in light of T, into an unconditional probability. *Cf.* J. KEYNES, *supra* note 3, at 40 (remarking on the errors that arise when notation fails to make the conditional character of a probability explicit); *id.* at 119 (urging the adoption of notation that makes the premises of an argument explicit to avoid "endless confusion" about probability).

[16] My former colleague, Wendell Kay, an eminent trial lawyer, used to tell his students of a news reporter's interview with a juror in 1979, reported in an Associated Press story, that indicates that the ordinary scenario, as the jury commences its deliberations, goes something like this:

Juror #1: That guy looked guilty to me. Did you ever see such shifty eyes, and did you notice the way he slouched at the table?
Juror #2: Well, I don't think that has anything to do with it, but you know darn well that the police must have had something to go on or they never would have arrested the guy in the first place.
Juror #3: Well, the Judge told us not to pay any attention to the fact that he got indicted, but I sure say that I never saw a cloud of smoke without some fire burning somewhere.
Juror #4: When do we go to lunch?
Juror #5: The guy sounded pretty good to me, but if this fellow Wilson was actually with him like he said, why didn't he call Wilson as a witness?

Kay, Final Argument (unpublished manuscript). The remark of Juror 5 is the type that this comment tries to analyze.

[17] There are two distinct reasons why the jury might have reached this result. One is that it actually found the other evidence less persuasive. The "negative inference" would be that the police officer decided not to collect scientific evidence because he wanted to issue a citation but his "field sobriety test" was really not so conclusive, and hence he feared that the more scientific tests might have exonerated the defendant. The other possibility is that the jury decided to "punish" the police for being "lax," that is, for not following procedures that the jury felt the state should follow in dealing with citizens suspected of drunk driving.

In previous articles, I have emphasized the latter systemic concern as a justification for some putative rules of law that preclude liability on the strength of evidence that seems to satisfy the pertinent burden of persuasion. Kaye, *The Laws of Probability and the Law of the Land*, 47 U. CHI. L. REV. 34 (1979); Kaye, *Paradoxes, Gedanken Experiments and the Burden of Proof: A Response to Dr. Cohen's Reply*, 1981 ARIZ. ST. L.J. 635. Despite my window into the jury room, I cannot say whether the jury in this case declined to convict because it drew a "negative inference," or because it adopted a systemic perspective of the sort that might explain certain common law rules, or both. The jurors did not articulate why they found the gap in the state's case so troubling.

[18] *Cf.* Ellman & Kaye, *Probability and Proof: Can HLA and Blood Testing Prove Paternity?* 54 N.Y.U. L. REV. 1131, 1158-61 (1979) (advocating a rule of law that would bar plaintiffs from recovering when they choose not to introduce more revealing evidence which is reasonably available to them). For examples of judicial recognition of the impact of the negative inference resulting from a gap in the case, see Galloway v. United States, 319 U.S. 372, 385-88 (1942); Hirst v. Gertzen, 676 F.2d 1252, 1259-60 (9th Cir. 1982).

[19] I am referring to the following hypothetical case presented by Professor Cohen: Imagine a lottery with as many tickets as you please and only one winning ticket. Then the proposition that John Doe's ticket is not the winning one has as high a conditional probability as you please on the evidence available about the total number of tickets participating. But you would certainly be rash to infer from this that to a very high *un*conditional probability John Doe's ticket is a loser. You would still have some reason to doubt the validity of the inference until you know that the lottery was run fairly. For example, if it were also a fact that John Doe is a mobster whose gunmen supervise the draw, that fact would be part of the totality of facts that a reasonable person would think relevant to your conclusion. Consequently the probability of your conclusion on the basis of the totality would be a good deal lower than its probability on the evidence that was originally available.

Cohen, *supra* note 2, at 638.

For clarity, let us define some terms. At the outset, there are three facts to consider: that John Doe holds a lottery ticket, that there are N tickets, and that exactly one is the winning ticket. Let L be the event that Doe's ticket is a loser. The probability of L then can be denoted $\Pr(L)$. To estimate this probability, we need to adopt some model of how Doe acquired his ticket. The three background facts are not part of any sample space needed to define the probability that Doe holds a losing ticket. They merely bear on the probability model that gives the unconditional probability $\Pr(L)$.

Evidently, Dr. Cohen believes that in the absence of any more information about the distribution of tickets, one should presume that each ticket is equally likely to be the winner. On this assumption of random sampling from a finite population, the probability distribution is uniform, and $\Pr(L) = (N - 1)/N$. But the validity of the random sampling model is an empirical, not a logical matter, and we must consider other plausible models before embracing $(N - 1)/N$ as the probability in question.

Perhaps John Doe is a lottery official who knows the winning ticket in advance and has appropriated it for himself. Perhaps he is the mobster who secretly controls the lottery, as Dr. Cohen posits. In the context of such models, the probability $Pr(L)$ would seem to be zero, or close to it.

If we must estimate $Pr(L)$ without learning more about the model that actually pertains, then we must rely on our background knowledge about the prevalence of fair lotteries and about John Doe's character. If all lotteries are fair, we should take $Pr(L)$ to be $(N - 1)/N$. If all are rigged for John Doe, then $Pr(L)$ approaches 0. Our best guess probably would lie somewhere between these extremes. In sum, a thoughtful statistician or probabilist would not maintain that $Pr(L)$ must be $(N - 1)/N$, for he or she would realize that this value presupposes the random sampling model, which may not hold for the hypothetical lottery. The facts that Dr. Cohen initially enumerates must be viewed in some context.

This is not to deny the main point of Dr. Cohen's hypothetical crooked lottery case. An estimate of $Pr(L)$ derived from a probability model that has very few facts to give it structure may be unstable in the sense that learning more facts could cause us to change our model, and hence, our estimate.

[20] *See* Saltzburg, *A Special Apsect of Relevance: Countering Negative Inferences Associated with the Absence of Evidence*, 66 CALIF. L. REV. 1011 (1978). Sometimes evidence will be unavailable due to the passage of time or for other innocent reasons. Sometimes evidence will be unavailable because it is privileged or the witness who has it is immune from appearing or testifying. Neither the physically unavailable nor the legally unavailable evidence should diminish the probability of a story.

Inasmuch as Keynesian weight relates to the "amount" of evidence, J. KEYNES, *supra* note 3, at 74, it might seem as if a "weight-based" theory leads to a contrary and unacceptable result. Yet, Keynes clearly restricts the computation of weight to "relevant evidence." *Id.* at 71-72. Since he defines relevant evidence as that which alters the prior probability, *id.* at 55, it is tempting to argue that physically or legally unavailable evidence is irrelevant and does not count in the computation of weight.

Still, it is not clear that this defense of Keynes's theory works. His definition of relevance seems to distinguish between a collection of evidence that is irrelevant as a whole but is nevertheless individually relevant. The former adds to weight, but the latter does not. As Glenn Shafer suggested in his remarks at the Boston symposium, perhaps the weight can be computed as the sum of the absolute values of the log-likelihood ratios for each evidentiary event, although this measure may have some technical difficulties. Here the existence of gap G reduces the prior probability (has a negative log-likelihood ratio), but the subsequent evidence of unavailability (having a positive log-likelihood ratio of equal magnitude) boosts us back to where we started. This should add to the weight.

If so, reliance on "weight" does not seem to explicate correct reasoning. Take the case of an alleged purse snatching in New York City in which S^P includes the event that the victim's husband was beside her when a juvenile grabbed the purse and ran straight into the arms of a nearby policeman. The state does not call the husband to testify, and the jury may wonder why this gap exists. So the state establishes that this witness is unavailable because he is the United Nations ambassador from Iran, who has claimed diplomatic immunity. Now consider a similar case in which the woman was walking without her husband. I would think that $\Pr(S^P|E \cap G)$ is the same in both cases and that a juror's partial belief in S^P also should be identical. Even though the Keynesian weight of S^P given the evidentiary events in the first case looks like it might be greater than the second, the requirement of proof beyond a reasonable doubt seems satisfied (or not satisfied) to the same degree.

Either I misunderstand Keynesian weight—which is quite possible—or, like Keynes himself, I have great difficulty imagining an instance in which this weight has much practical significance. In this regard, it may be worth noting that Cohen also gives no clear example of a case in which attending to Keynesian weight would warrant a different action than reliance on a well-founded probability.

[21] Cf. Lindley & Eggleston, *The Problem of Missing Evidence*, 99 LAW Q. REV. 86 (1983) (Bayesian analysis).

[22] Michael White has suggested to me that one can account for the completeness of the evidence with what I would construe as a second-order probability. The idea is to obtain an appropriate measure of the invariance of $\Pr(S^P|E)$ under conditionalization on F. We want to know whether $\Pr(S^P|a)$ is some number a. If the probability would not change much when S^P is conditioned not merely on E but on E together with the presentation of every subset of the additional items of evidence in F, then we should take a number very close to one to express the probability of $\Pr(S^P|E)$ being a. Cf. B. SKYRMS, CAUSAL NECESSITY: A PRAGMATIC INVESTIGATION OF THE NECESSITY OF LAWS 12-13 (1980) (defining the "resiliency" of a conditional probability). This line of inquiry leads us into the literature on second-order probabilities. Although there is much work to be done in this field, it now appears that the arguments of the advocates of subjective probability about the coherence of ordinary beliefs can be generalized to demonstrate the coherence of higher-order beliefs. Skyrms, *Higher Order Degrees of Belief*, in PROSPECTS FOR PRAGMATISM 109 (D. Mellor ed. 1980).

I mention this approach to the problem because I think it could lead to an interpretation of the burden of persuasion that supplements $\Pr(S^P|E)$ (as first presented in (1)) with a fully probabilistic measure of the weight of E. Second-order probabilites also might be capable of capturing the notion of subjective confidence that Professor Neil Cohen wrongly ascribes to a frequentist confidence interval. Cohen, *Confidence in Probability: Burden of Persuasion in a World of Imperfect Knowledge*, 60 N.Y.U. L. REV. 385 (1985). For a more careful explanation of the meaning of a confidence interval, see Kaye, Mixing Apples and Oranges: Confidence Coefficients and Significance Levels Versus Posterior Probability and the Burden of Persuasion (1986) (unpublished manuscript).

I shall not pursue the analysis of second-order probabilities here, in part because I lack the technical expertise, and in part because the first-order model that I have sketched should give the same results. *See* Skyrms, *supra*, at 167-68 ("first-order models are good medicine for acrophobia").

[23] L. COHEN, *supra* note 1, at 90; Cohen, *Subjective Probability and the Paradox of the Gatecrasher*, 1981 ARIZ. ST. L.J. 627, 629-32. For a previous rejoinder on this

topic, see Kaye, *Paradoxes, Gedanken Experiments and the Burden of Proof: A Response to Dr. Cohen's Reply*, 1981 ARIZ. ST. L.J. 635.

[24] L. COHEN, *supra* note 1, at 90.

[25] Cohen, *supra* at 124.

[26] *Id.* at 125.

[27] *Id.*

[28] *See* Skyrms, *Higher Order Degrees of Belief*, *supra* note 22, at 115-16.

[29] Quantifying partial beliefs by reference to what a person would say if he had to decide under a veil of ignorance about which side of a bet of unknown stakes he will be forced to take is an idealization. The procedure is even more hypothetical if the bet cannot be settled, like a bet on the existence of an afterlife. But even this extreme case (more extreme, I think, than the notion of betting on the accuracy of a verdict in a criminal case) does not preclude asking an individual to contemplate the fair odds for such a bet on the strained assumption that a definitive answer to the question of an afterlife will emerge soon. *Cf.* I. HACKING, LOGIC OF STATISTICAL INFERENCE 215-16 (1965) (acknowledging but questioning the possibility that "you might even imagine yourself pretending to bet on theories and hypotheses, as if you were gambling with an omniscient being who always pays up if you are right").

[30] The criterion of minimizing expected loss is called a "Bayes risk criterion." J. MELSA & D. COHN, DECISION AND ESTIMATION THEORY 44 (1978). In the context of statistical hypothesis testing, a procedure that meets this criterion is called a "Bayes test procedure." M. DEGROOT, PROBABABILITY AND STATISTICS 382 (1975). The phrase "Bayesian" is used in so many ways in the statistical and philosophical literature that it is hard for the uninitiated to know what it means. Kaye, *The Laws of Probability and the Law of the Land*, *supra* note 17, at 45 n.37, 51 n.57. Thus, one could say that my treatment of the burden of persuasion adopts "Bayesian decision theory" but not necessarily "Bayesian learning theory." E. EELLS, RATIONAL DECISION AND CAUSALITY 12 (1982).

[31] *E.g., Judgment Under Uncertainty: Heuristics and Biases* (D. Kahneman, P. Slovic & A. Tversky ed. 1982); R. NISBETT & L. ROSS, HUMAN INFERENCE: STRATEGIES AND SHORTCOMINGS OF SOCIAL JUDGMENT 117, 241-45 (1980). For that matter, in the abstract at least, it could even be a relative frequency figure, although there is no obvious reference class for the events in question at a trial.

[32] *See* I. GOOD, PROBABILITY AND THE WEIGHTING OF EVIDENCE v (1950).

[33] Cohen, *supra* at 125.

[34] *Id.* at 126.

[35] It also does not suggest that the best epistemological theory includes Cohen's original measure of the strength of an inductive argument.

[36] *E.g.*, de Finetti, *Foresight: Its Logical Laws, Its Subjective Sources*, in STUDIES IN SUBJECTIVE PROBABILITY 97, 111-18 (H. Kyburg & H. Smokler ed. 1964); Ramsey, *Truth and Probability*, in STUDIES IN SUBJECTIVE PROBABILITY, 63, 69-82 (H. Kyburg & H. Smokler ed. 1964); L. SAVAGE, FOUNDATIONS OF STATISTICS 27-40 (1954).

[37] *Cf.* D. MELLOR, THE MATTER OF CHANCE xii (1971) (remarking with respect to

his propensity theory of physical probability that "nothing in the characterization and measurement of partial belief excludes further empirical constraints of rationality upon it"). According to Mellor, "[p]ersonalists have admittedly tended to accompany their theories with a view of probability statements as merely subjective. But that is an incidental defect of personalism." *Id*. at 2. *See also* Skyrms, *supra* note 22, at 20 (using the phrase "epistemic probability" rather than subjective probability to denote partial belief, and observing that "constraints of rationality no doubt require a good deal more [than coherence]; to say what more is the chief business of epistemology").

[38] F. BENENSON, PROBABILITY, OBJECTIVITY AND EVIDENCE 44-45 (1984) (distinguishing between "wholly subjective theory," "logical relation theory," and "credal probability" theory). I do not feel qualified to judge which of these approaches is the most promising, and I continue to be agnostic about which camp of philosophers or logicians has discovered the "real" nature of probability.

David H. Kaye,
Professor of Law,
Arizona State University College of Law.

LEA BRILMAYER*

SECOND-ORDER EVIDENCE AND BAYESIAN LOGIC†

The current literature on the uses of probabilistic inference in law evokes a sense of deja vu. It is reminiscent of an earlier debate in which proponents of formal logic argued for its greater usage in law. This movement gained sufficient momentum to found its own journal, Modern Uses of Logic in Law,[1] and to find representation in elite law reviews.[2] If only lawyers would adopt mathematical symbolism, so the arguments went, they could improve precision, eliminate ambiguities and avoid logical fallacies.[3] One author called for lawyers to hire mathematicians to go over their contracts.[4] We have even seen articles asking whether we might not replace judges with computers.[5]

Now we find ourselves in another battle, although this time the techniques being promoted are probabilistic. As this symposium illustrates, the current battle is over the suitability of Bayesian probabilistic reasoning for legal processes. Surprisingly, the participants on both sides of this battle agree on the existence of certain common problem cases. One of these cases, the famous "blue bus case," also known as the "paradox of the gatecrasher," severely tests the applicability of Bayesian reasoning to legal reasoning processes. Such examples challenge the notion that legal decisions can or should be based upon naked statistical evidence.

This paper attacks the solution that Bayesian legal scholars have proposed to this paradox. It suggests that their solution cannot work because it is based upon second-order evidence. Such second-order evidence cannot adequately be accommodated within the usual Bayesian framework, but plays an important role in legal decisionmaking. Moreover, other types of second-order reasoning are equally important to legal decisionmaking, but also not subject to Bayesian analysis. The blue bus problem, in other words, is just the tip of the iceberg.

The analogy to legal usage of symbolic logic is instructive for two reasons. First, whatever the value of symbolic logic as a field of intellectual endeavor, it has not proven directly useful to lawyers. Although we sharpen our minds by studying logic in college, our legal goals are not necessarily advanced by phrasing things symbolically. The same is true, interestingly, of the use of symbolic logic in mathematics. Theorems are not discovered through the manipulation of formal symbolism. This suggests a rather indirect and sophisticated interplay between formal and informal reasoning, with the latter perhaps having priority. One might doubt, analogously, whether probabilistic reasoning will be directly useful.

The second point of the comparison is perhaps more troubling. Even aside from its relative uselessness in discovering the theorems to be proved, formal logic has been shown to be profoundly flawed. Over the last hundred years, there has been something of a crisis in the foundations of mathematical logic and set theory.[6] Logicians have found ways to cope with these foundational issues, but the solutions have never been incorporated into Bayesian reasoning. Instead, Bayesian probability theory is founded upon a form of logic that has been recognized as inadequate since the turn of the century. My claim is that some of the problems in the legal applications of Bayesian decisionmaking arise out of these foundational difficulties. In short, it is not the distinctiveness of legal reasoning that accounts for the unsuitability of its incorporation of probabilistic logic, but rather those features it holds in common with other modes of rationality.

These general problems will be addressed only after a series of particular legal problems is laid out. The first section discusses the blue bus and gatecrasher paradoxes. The second discusses the problem of weighing evidence. The third addresses the issue of irreversible reasoning. The final section discusses the general problems of second-order evidence, and how it relates to foundational issues of Bayesian probability logic.

I. Blue Buses and Gatecrashers

The contours of the blue bus problem and the gatecrasher paradox are probably familiar. In the blue bus problem, the plaintiff has uncontroverted eyewitness testimony demonstrating that he was run over by a blue bus. However, he does not have a license plate number or any other evidence tending to show which blue bus it was. The defendant in the hypothetical owns four-fifths of the blue buses in town. From a strictly probabilistic standpoint, then, there is a greater than one-half likelihood that it was one of the defendant's buses that injured the plaintiff. Also from a strictly probabilistic standpoint, it would seem that the civil proof standard of a preponderance of the evidence should be one-half. Hence, it seems recovery should be allowed.

The gatecrasher paradox is comparable. A rodeo operator has sold 499 admission tickets to the rodeo; however, there are one thousand people sitting in the rodeo stands. The explanation, it turns out, is that there was a hole in the fence, through which 501 people sneaked in. From a strictly probabilistic point of view, the likelihood that any one of the individuals attending the rodeo is a trespasser is greater than one-half, namely, 501/1,000. This proof seems to satisfy a preponderance of the evidence standard, and the rodeo operator should be able to recover in a suit against any one of the spectators.

The blue bus case is probably the less persuasive of the two illustrations because of the uncertain relevance of the prior assessment of probability. Prior to hearing the evidence about bus ownership, the jurors may have entertained probabilities that the defendant's bus hit the plaintiff. So when the ownership evidence is introduced it must be combined with the earlier probabilities, and the resulting probability might still be less than one-half. Thus, argumentative Bayesians might escape the paradox by claiming that, since the true probability is less than one-half, it is proper to deny recovery. However, such a gambit is not available on the gatecrasher facts. Regardless of whether jurors entertain prior estimates about the likelihood that a civil defendant brought to trial is liable, the posterior estimate must still be .501 if the jurors are unable to differentiate among the one thousand potential trespassers. Since such differentiation is impossible in the example, the probability of each defendant's liability must be the same—.501. This is true regardless of any extra assumptions one might make about whether the owner or the defendants should have better access to additional evidence. Because of its added clarity, we shall focus primarily on the gatecrasher example in the discussion below.

Of course, the law would not allow recovery in either of these cases. In fact, as Professor Nesson points out,[7] the judge would probably not even allow the case to go to the jury. The explanation seems to lie in the law's unwillingness to base a verdict upon naked statistical evidence. The problem is that the evidence in question does not deal with each defendant's guilt individually. In the given hypotheticals there is no evidence to suggest that the ABC company "really" was the owner of the bus that caused the injury; there is no evidence to suggest that Sally Smith "really" was one of the trespassers. There is only a background statistic about the number of buses owned, or the number of tickets sold.

From a probabilistic standpoint this reasoning is dubious. There is no reason to be suspicious of evidence merely because it is based upon naked background statistics. Numerical evidence is like any other. Furthermore, to deny recovery would increase unnecessarily the number of errors in the long run. Holding each rodeo spectator liable for trespass will result in 501 correct decisions and 499 incorrect decisions. Disallowing liablility will result in only 499 correct decisions but 501 incorrect ones.

Notice the type of legal intuition that, in both cases, is rejected by the probabilistic analysis. The legal intuition in question is that naked statistical evidence is not enough. The inferential or decision-making principle that naked statistical evidence is not enough is a second-order principle. First-order statements are propositions about the real world. The eyewitness testimony that there were one thousand persons at the rodeo is a first-order proposition, as it concerns the empirical state of affairs. But a proposition that the testimony about how many persons attended the rodeo was interesting, or convincing, or relevant, is a second-order proposition. It is a proposition about a proposition, not a proposition about the world.

The distinction between first and second-order propositions (or third-order, fourth-order, and so forth) is familiar to all students of symbolic logic.[8] It corresponds to the differentiation, in set theory, between sets and sets of sets. For example, the set containing the Boston University Law School building is not the same thing as the Boston University Law School building itself. And neither is equivalent to the set containing the set containing the Boston University Law School building, and so forth. So it is with statements made in court as well. The witness's statement about the color of the bus is not the same as a characterization of the witness's statement about the color of the bus. The evidence that four-fifths of the blue buses belong to the ABC company is different from the proposition that such a piece of evidence is "naked statistical evidence." Legal unwillingness to rest a verdict upon naked statistical evidence is second-order because it is based upon a characterization of the available evidence.

If the Bayesian legal scholars wish to assert that the law more or less conforms to Bayesian precepts, the appropriate strategy is to devise a Bayesian explanation of the legal system's distaste for naked statistical reasoning. Professor Kaye is at the forefront of this effort.[9] His solution consists of two related points. First, he says, the law's aversion to naked statistical proof reflects the belief that plaintiffs ought to be forced to supply the best possible evidence. Discounting naked statistical evidence supplies such an incentive:

> A rule of law denying the plaintiff recovery in such a case would not establish that probabilistic reasoning is inapplicable or that jurors would not be well advised to find facts according to their best estimates of the relevant probabilities. It would merely reflect the policy that where individualized evidence is likely to be available—evidence which would typically permit better estimates of the probabilities than can be had from background statistics alone—plaintiffs should be forced to produce it. In the long run, fewer mistaken verdicts should result under this rule of law.[10]

This rationale suggests that it may be appropriate to sacrifice a few cases to achieve better long run results. It raises a number of problems, however.

First, why should the law attempt to motivate persons to provide more evidence? This policy seems to reflect a rather non-Bayesian hostility to naked statistical evidence. It concedes, in other words, that naked statistical evidence is in some sense inferior. In the gatecrasher problem, arguably, we might wish to have more evidence because we are concerned about the slim margin by which the case against a particular defendant has been made out, namely, by a likelihood of only 501/1,000. But this is not a consequence of the fact that the case is made out by statistical evidence. Other purely statistical cases might have a much higher likelihood and conversely there might be cases based upon nonstatistical eyewitness testimony that produced a probability of only 501/1,000. There seems to be no good Bayesian reason for penalizing plaintiffs merely because their case consists of naked statistical evidence.

A second, and related, point is that once this reasoning is employed there is no stopping it. Precisely because there is no good Bayesian reasoning for differentiating between naked statistics and other evidence, this reasoning perhaps could be extended to deny recovery where the plaintiff has produced a nonstatistical case based upon what is seen as less than ideal evidence.[11] Why should we not similarly penalize individuals who have produced only enough (nonstatistical) evidence to support a likelihood of .501? The legal system, after all, might enjoy greater popular approval if the results it reached were clearly the right ones, and a verdict based upon a .501 probability is not so clearly right. Why not force plaintiffs to provide better evidence?

Furthermore, the hypothesis that plaintiffs will be motivated by such a rule is questionable. It depends upon the assumption that individualized evidence is readily available, but that the plaintiff chose not to produce it. I would be interested to see the empirical evidence suggesting that plaintiffs engage in such maneuvers. More frequently, one imagines, evidence is not presented because it is simply not available. And why would this only be true in cases involving naked statistical evidence? The generalization that cases based upon statistical evidence are the ones where plaintiffs could readily have produced more evidence is unsupported and conclusory.

In some circumstances, admittedly, burdens of proof are allocated by assumptions about the availability of evidence. For instance, the defendant might be given the burden of disproving negligence where he or she is likely to have the best access to the evidence about whether negligence occured. This is one basis for the legal doctrine of res ipsa loquitur. This example does not support Professor Kaye's gatecrasher reasoning, however. The rule of res ipsa loquitur does not assume that parties will choose not to come forward with evidence that supports their position. Instead, it assumes that the defendant will not come forward with evidence that helps the opposing party prove its case (a fairly plausible assumption, albeit somewhat dated in light of modern discovery techniques). As such, the doctrine's purpose is not to motivate a party to come forward with favorable evidence that he or she otherwise would not produce.

Finally, it is erroneous to assume that fewer mistaken verdicts should result under the rule of law that denies recovery to the rodeo operator. Assume that Professor Kaye is correct on the earlier points, and that in some instances the plaintiff has access to some additional nonstatistical evidence. The rule against relying upon naked statistical evidence—call it the "insistence" rule—is supposed to cause the plaintiff to produce this evidence. Assume also that Professor Kaye is correct about the incentive effects so that the plaintiff, because of this rule, provides the additional evidence. Although the verdict may be more psychologically satifying, it is no more correct when based upon the larger amount of evidence than when based upon the smaller amount of naked statistical evidence. Whichever body of evidence the verdict is based upon, either a verdict for the plaintiff is correct, or a verdict for the defendant is correct. To the extent, in other words, that an insistence on more than mere statistical evidence produces the same pattern of results as a willingness to base verdicts on naked statistical evidence, the insistence cannot be said to foster a more accurate decision-making process. If the results are the same, the accuracy is the same.

Thus, if the insistence rule fosters more accurate decisionmaking, the greater accuracy must stem from those results that are different under the "insistence" rule than under the alternative rule—call it the "naked statistics" rule. The cases in which the results will vary depending on which rule is used are those in which the plaintiff cannot or does not offer the additional nonstatistical evidence which the "insistence" rule requires. In such cases, the plaintiff wins under the "naked statistics" rule, but not under the "insistence" rule. It is on the results in these cases that the validity of the claim to greater accuracy depends. We can illustrate this diagramatically:

Is the Insistence Rule More Accurate?

1. Cases Where Recovery Allowed Under Either Rule
   a) plaintiff would adduce nonstatistical
      evidence under either rule   no improvement

b) plaintiff adduces nonstatistical evidence because of insistence rule, but would not otherwise	in accuracy
2. Cases Where Recovery Allowed Under Naked Statistics Rule But Not Insistence Rule (Those Where Plaintiff Cannot Adduce More Evidence).	arguably increased accuracy
3. Cases Where Recovery Denied Under Either Rule	no improvement in accuracy

Phrased in terms of this diagram, it is category two that must account for the supposed greater accuracy of the insistence rule. These cases, however, are the ones that pose the most problems for that rule. There are two reasons for this. First, these are the cases in which we ought to be most sympathetic to the plaintiff who loses under the insistence rule. After all, in these cases the plaintiff provided the best possible evidence; that evidence established a likelihood of more than one-half that the plaintiff should prevail, and yet the judge nevertheless directed a verdict for the defendant. The rationale for sacrificing such plaintiffs cannot be that it led to better results in other cases; as just noted, the result is the same under both legal rules. Second, these are cases in which the insistence rule seems to produce less accurate results. The probability was greater than one-half, and yet the plaintiff lost. If the result was greater accuracy in the other case, there might still be an overall improvement in the decision-making process. But to repeat: the result, in the other cases, is the same under both schemes. To the extent, therefore, that the two rules vary, the insistence rule seems less accurate.

Now, there is one possible way to avoid this line of reasoning. As just noted, the increase in accuracy must stem from those cases in which the results are different, namely the cases in which the insistence rule would not bring about greater production of evidence by the plaintiff. Arguably, in such cases it is more accurate to direct a verdict for the defendant. This is a different argument from the one about the value of motivating plaintiffs in the other cases to present greater evidence, which justifies the rule in terms of its beneficial effect in the case where recovery is denied. Instead, it attempts to claim greater accuracy in category two itself. This argument constitutes Professor Kaye's second response to the gatecrasher paradox.

The argument proceeds as follows. Professor Kaye suggests that the plaintiff's failure to adduce evidence that is more probative (which I think, in this context, he takes to mean "not merely statistical") suggests that any other evidence which the plaintiff may have will not support his claim. There is reason to believe, therefore, that the probability of 501/1,000 is not accurate. It is probably overstated, so the argument would go, considering that there is probably other evidence which has not been produced and which supports a verdict for the defendant. Accuracy is increased in these cases because the true probability is less than one-half, and recovery is properly denied. This argument is puzzling in several respects, some of which resemble the problem with Professor Kaye's incentive argument. For one thing, it is not clear why it is only in statistically-based cases that we assume that the failure to produce more evidence demonstrates that there is other evidence that could be produced, and that it would support the defendant if it were produced.

Perhaps the most telling objection, however, is the consequence of applying the argument to all of the defendants. In the hypothetical we know there are 501 trespassers and 499 legal entrants. Should we hold Sally Smith liable when the only evidence that we have against her is this naked statistic? Professor Kaye suggests that we should not because we ought to revise the probability to account for the fact that there is only statistical evidence against her. But if we do that, we would also have to revise the probabilities against all of the other spectators; there is no reason to distinguish Smith from any of the other spectators. The result of such across-the-board revisions, however, is that the probabilities will fail to add up to one, thus violating the Bayesian axioms.

This argument is better visualized if the numbers are simplified. Assume there are two spectators, and only one ticket was sold. The case against Smith has a probability of one-half, but it is based solely upon naked statistical evidence. Hence it is overstated. If the probabilities of the alternatives must add up to one, then reducing Smith's probability to, say, .47 must have the consequence of increasing the probability of the liability of Jones, the other spectator, to .53. Thus a verdict against Jones is appropriate. But that cannot be, since there is neither more nor less evidence against Jones than against Smith. In other words, although the evidence against Smith is merely statistical, the evidence exonerating her is nakedly statistical also. Yet both probabilities cannot be less than one-half.

It is interesting to highlight those features of the example that give rise to this difficulty. First, the "evidence" to the effect that the initial estimate is based upon mere statistics is equally pertinent to the Smith estimate and the Jones estimate. Second, it supposedly supports the same inference in either case, namely that the initial estimate must be revised downwards. It is for this reason that once we use the "evidence" to revise the Smith estimate downwards, we must also use it to revise the Jones estimate downwards. It is equally applicable to both estimates, and it conveys the same message.

But when we treat the two estimates symmetrically, the result is that they do not add up to one.

Bayesian logic requires that whatever decreases the likelihood of one alternative must increase the probability of the other (or, in a world of more than two alternatives, it must increase the probability of at least one of the others). It does not contemplate evidence that would have identical effects on all of the probabilities, except for evidence that leaves all of them the same. Conversely, any evidence that by its logical form can be seen to have identical consequences for the likelihood of one alternative as it does for all the others must be irrelevant; that is, it can treat them all the same only by leaving them all untouched. Any evidence about the totality of evidence that has been previously offered is normally irrelevant because it applies equally to all alternatives based upon that body of evidence.

The observation that mere statistical evidence is involved in formulating the initial distribution of probabilities is "evidence" of this type. It speaks equally to all probabilities based upon statistical evidence. It is merely a characterization about the body of evidence that supports the initial distribution of estimates. But the body of evidence that supports one likelihood out of that distribution—say, the likelihood of Smith being the trespasser—also supports the remaining estimate(s) in that distribution—the likelihood of Jones being the trespasser. Thus, it will always be problematic to revise a likelihood estimate on the basis of some characterization of the evidence that first gave rise to it. Because the same body of evidence gave rise to the probabilities assigned to the alternative hypotheses, it would seem that they would have to be revised in the same way.

It was noted earlier that a characterization of a body of evidence as "naked statistics" is a second-order proposition, a proposition about other pieces of evidence. This feature is implicated in the aforementioned "revision" difficulty. The symmetry of treatment of the probabilities of the alternatives arises from two facts: first, the new "evidence" is about evidence, and second, that the same evidence supports the entire probability distribution. Because in the Bayesian system probability estimates must be complementary rather than symmetric (i.e., they must add up to one), evidence with only symmetric consequences must be disregarded as irrelevant.

That one type of second-order evidence is unpalatable to the Bayesian system does not mean that second order evidence is unpalatable to the Bayesian system generally. We will examine two more instances of second-order evidence that Bayesian theory cannot accommodate, and then speculate upon the reasons that second-order evidence is problematic.

II. Weighing the Evidence

In one sense, evaluating the weight of evidence is similar to characterizing available evidence as "naked statistics." Both are second-order propositions. Both are about evidence, and not about real world events. The difference between them is that evaluating the weight of evidence is quantitative, while characterizing evidence as "naked statistics" is qualitative. Despite this difference, incorporating the weight of evidence into Bayesian calculation poses the same problems as incorporating the type of evidence.

For instance, assume that I flip a coin that is not known to be fair; that is, it is not known whether the probability of it landing heads is one-half. In order to gather evidence, I perform ten trials. Based upon these ten trials and whatever other background evidence I may possess about coins in general, I assign a probability of six-tenths to the event that the outcome of another toss will be heads.

In some sense, I have based this estimate of .6 on "not very much evidence." After all, I could have performed one hundred trials and used all of that data in arriving at an estimate. The question is whether a Bayesian could use the fact that the .6 probability was based upon not very much evidence to revise the .6 estimate. This problem is identical to the problem with basing revisions upon the characterization of available evidence as "naked statistics." If one were to lower the .6 "heads" estimate because it was based upon very little evidence, one would also have to lower the .4 "tails" estimate; it also was based upon very little evidence. After all, both estimates were based upon the same evidence.

Historically, Bayesians have shown little interest in using propositions about the weight of evidence to revise probability distributions. The weight of the evidence is irrelevant, they claim.[12] In particular, they argue that the weight of evidence does not affect betting behavior.[13] In other words, if an initial probability is the result of very little evidence, and subsequent accumulation of evidence were to result in the same probability, then betting behavior would be the same under the probabilities regardless of the fact that they were based upon different amounts of evidence. To be more concrete, it is possible that if I perform another ninety trials with my coin, the probability of heads will still be .6 and the probability of tails will still be .4. The Bayesian argument is that since my betting behavior will be the same after the hundred trials as after the ten trials, there is no need and no capability for the probability calculus to accomodate the different weights.[14]

One reason why it is difficult to evaluate this claim is the inherent vagueness of the concept of the weight of the evidence. In particular, we tend to confuse the concept with a related notion, namely, the extent to which the evidence points towards a prediction of heads on the next toss. This latter notion might be referred to as weight, as for instance where I say, "There is a lot of evidence suggesting that the next flip will come up heads." Such evaluations are already accommodated in the Bayesian scheme, for to say that there is a lot of evidence suggesting that the next toss will be heads amounts to saying that the probability of heads is high. What is not accounted for is the second-order notion that *a substantial amount* of evidence suggests that the probability of heads is .6. This second-order fact, according to Bayesians, is irrelevant.

If we are talking about total *amounts* of evidence, however, as opposed to the extent to which the evidence points in one direction or another, then there seems to be no clear way to assign a numerical weight to a particular amount of evidence.[15] This is, perhaps, one reason for the reluctance to take weights of evidence into account. Although there may not be a method for assigning weights, it is at least possible to partially order sets of evidence through set inclusion.[16] One body of evidence would be weightier than another if it includes every item of evidence that the latter does, but also includes at least one item of evidence that the latter does not. This is only a partial and not a total ordering, as it will not be the case that, for every two sets of evidence, either they are equal or one is weightier than the other. In other words, it is not the case that any two sets of evidence are comparable. This definition may not be totally adequate, for in some situations the first set does not contain every element the second set does, though we would wish to say that it was weightier.[17] Nevertheless, the intuitively plausible notion of set inclusion is adequate to show that there are some circumstances in which we ought to take weight into account.

Assume, for example, that my husband and I are watching the election returns. As of eleven o'clock, the Democrat is leading the Republican by some particular margin. All the returns are not yet in, however, and based upon my experience with elections I realize there is a certain probability of a last-minute upset. On this information, I assign a probability of .8 to victory by the Democrat, and .2 to victory by the Republican. I then go to bed, but my husband stays up later to watch more returns. In the morning, before I have the chance to read the paper or turn on the radio, he asks me to give him odds about who won the election.

Based upon my best information, I would have to give him four to one odds in favor of the Democrat. Of course, I would be unwilling to do that, however. I know that he knows more about the election results than I do; indeed, he may have stayed up until the returns were all in and the result definitively determined. He knows which way to bet because of his superior evidence; he knows everything that I do about the election, and then some. Yet I cannot use the second-order fact that he knows more than I do to revise my estimate. How would I know whether to revise it up or down? If I were to revise my probability of a democratic victory downward because I have "too little evidence" then I would have to revise my republican probability downward also, for the same reason. Then they would add up to less than one.

The Bayesian scheme is designed to prevent an opponent from making "Dutch book" against you.[18] If your probability distribution is non-Bayesian, a clever opponent can take advantage of that fact to guarantee that he or she will make money in the long run regardless of the outcome in the particular case. But having a Bayesian probability distribution does nothing to protect you against an opponent with superior information. Such an opponent can guarantee a long run advantage by using his or her weightier evidence. Indeed, much betting behavior is best explained in terms of an individual's beliefs that he or she has superior information.[19]

The Bayesian framework, then, does not even explain all aspects of its own paradigm, betting behavior. An individual who blindly followed Bayesian precepts would be fleeced. Now, perhaps the Bayesian would say that I have violated the ground rules when setting up the example. My example assumes, after all, that individuals have a right to refuse to offer odds. Admittedly, I have not shown that if an individual does bet he or she should not bet in accordance with a Bayesian distribution. But my argument does at a minimum suggest that there are decisions that are not susceptible to a strict Bayesian analysis, namely, decisions whether or not to bet.

Decisions of this sort, moreover, are necessary in the legal system. The legal system is not artificially constrained in its available options. One alternative to finding for a rodeo operator-type plaintiff is to dismiss a case, on the grounds of insufficient evidence, after the plaintiff presents the case-in-chief. Such a dismissal does not necessarily indicate that, given the evidence that has already been admitted, the probability that the plaintiff is correct is less than fifty percent. The probability may be based on such a small quantum of evidence that the defendant will not even be put to the burden of refuting it. Similarly, in a regulatory proceeding the quantity of available evidence is an important factor. A small probability that a particular drug is carcinogenic will be more persuasive if based upon years of testing than if it is based only upon a few experiments. In law, the possibility of basing a decision upon the insufficiency of the evidence is a viable alternative, as is a refusal to bet in any realistic betting example.

I anticipate the response that since the dismissal of the court case, at least, will be on the merits, it is tantamount to a finding of the defendant's factual innocence. Thus, it might be said, it is not weight but probability that should matter. But the prejudice to future actions will result not from any factual finding about the probability of guilt, but from the fact that a plaintiff only gets one bite at the apple. Having had one opportunity to present the case the plaintiff is barred even if he or she later returns with better evidence.

Consider the following analogy. A math professor gives a take-home examination that the students can pick up at any point during a two-week exam period, and can then take twenty-four hours to complete. Problem one is "prove Cantor's Theorem." A student picks up and completes the exam within the time limit, but submits a faulty proof. Once the exam has been handed in the student will not get a chance to submit a second proof. But this disposition with prejudice does not mean that the professor believes that Cantor's Theorem is false. It simply means that the student has failed to carry his or her burden of proving that it is true.

Oddly, Professor Kaye himself recognizes the legal system's interest in ensuring that decisions be based upon the appropriate (and, presumably, amount) of evidence. This concern is evident in his discussion of the desirability of insisting upon more than mere statistical evidence. If we insist that the plaintiff provide nonstatistical evidence, Professor Kaye believes, then rodeo operators who might have chosen to present only statistical evidence, if given the option, would instead also present some nonstatistical evidence. As we saw earlier, if such plaintiffs could have recovered under the rule allowing verdicts based on naked statistical evidence, and actually do recover under the insistence rule, then the results under the insistence rule cannot be said to be more accurate. Under either rule, the same pattern of results emerges.

But in some legal or intuitive sense, the pattern of results under the insistence rule may be more satisfying. Even if the insistence rule does not increase accuracy, it does yield results that are based upon a larger quantity and higher quality of evidence. It is this greater intuitive satisfaction, perhaps, that accounts for our insistence on more evidence. Such insistence has its costs: the plaintiff must go to greater expense to prove his or her case, and the legal system may penalize plaintiffs who cannot provide such evidence by granting verdicts against them even though the probability that they deserve to win is greater than one-half. The costs are justified by collateral values about weight and quality of evidence, and not by improved accuracy in the limited Bayesian sense. Such values as increased weight and quality do not fit easily within the Bayesian framework.

III. CORRECTING ERRORS

We will consider one final example of second-order evidence which cannot be accommodated by the Bayesian framework: evidence which contradicts prior evidence, showing it conclusively to have been mistaken or misleading. Bayesian reasoning, in other words, cannot correct certain errors.

Imagine a murder prosecution in which the defendant relies heavily upon an alibi defense. He claims that he was in Albany at the time that the murder was committed in New York City, and that he was there for at least several days before and after the murder occurred. A friend backs the story up. Neither one is completely convincing, but there is some force to the testimony. Now the prosecution puts on its surprise witness, a waiter from a restaurant in Scarsdale. The waiter testifies that he saw the defendant in Scarsdale the day before the murder. The waiter is completely credible; he is able to describe the defendant in great detail because he had extensive opportunity to observe the defendant in the restaurant. Furthermore, he has no apparent motive to lie or exaggerate.

The waiter's testimony convinces the jury completely; indeed, they are probably completely rational under these circumstances in being so convinced. This testimony totally destroys the alibi defense that the defendant had sought to establish. The jury adjusts their probabilities accordingly, assigning to the proposition "The defendant was in Albany the entire week of the murder," a probability of zero. Next, however, the defense puts on a rebuttal witness. It turns out that the defendant has a twin brother who lives in Scarsdale. The jurors can see for themselves how similar the two brothers look and act. The credibility of the defendant and his friend is restored, and the jurors wish to reassign a positive probability to the defendant's having been in Albany all week.

Bayesian reasoning, however, does not allow them to do that. Because of the form of the mathematical formula, it will never be possible to raise a probability from zero to a positive number. Arguably, the probabilities of the other alternatives that were positive (e.g., the defendant was in Scarsdale, the defendant was in New York City, the defendant was in Acapulco) can be reduced once the evidence about the twin brother is presented. This is because they are positive numbers, and Bayes's Theorem might revise them downward. But this only adds to the puzzle; for if we have a list of all the available alternatives, and reduce the probabilities of all the ones with non-zero probabilities without increasing the probability of the one that had a zero probability, then the total will no longer add up to one.

From one perspective, the inability to augment a zero probability makes sense because zero probability indicates belief in impossibility, and no amount of evidence can show an impossible thing to be possible. Thus, no amount of evidence will show a tautology, or a theorem of arithmetic, or any logical truth, to be false (or its opposite to be true). From another point of view, however, we surely want to be able to raise zero probabilities to positive numbers. We might want to do so when it turns out that the assignment of zero, in the first place, was erroneous. Even with regard to an apparent logical truth, we might find an error such as a mistake in the proof of a mathematical theorem, or a linguistic confusion accounting for an apparent tautology. With empirical reasoning, the need to revise zero probabilities is even clearer, since we may acquire contradictory evidence or even uncover an arithmetic error in calculation of the probabilities. This is second-order reasoning. It is not reasoning about the phenomenon that the probability statement described; it is reasoning about the probability statement itself.

I anticipate one objection to this argument,[20] namely, that the jury never should have allowed the probability that "The defendant was in Albany the entire week of the murder" to drop to zero. The jury should instead have allowed for the possibility that there was a twin brother, or that the waiter was hallucinating, or that there was some sort of prosecutorial plot. Then the probability would only have dropped to some small but positive number, and it would have been possible to retrieve a higher positive probability once the "twin brother" testimony was presented.

There are two responses to this objection. First, the concession that the jurors made a mistake does not exonerate the Bayesian system. The very fact that Bayesian reasoning does not allow a decisionmaker to correct mistakes within the system is highly problematic. To correct the mistake, the jurors need to put themselves back to the status quo before the waiter gave his testimony, and retrieve their prior probabilities. They need some sort of "override rationality" such that they can step outside the Bayesian framework and turn back the clock. Bayesian theory is irreversible because of the arithmetic consequences of allowing a probability to vanish to zero. Even if the jurors made a mistake in setting the probability to zero, Bayesian reasoning is problematic in that it does not allow the jurors to shift into reverse.

Second, it is not at all clear that the jurors did make a mistake. It seems fairly rational to take the evidence as it appears, and then go back and revise in the unlikely event of a twin brother or prosecutorial plot. Our reasoning processes would be paralyzed if at every juncture we had to consider outrageously implausible events. Any process of estimation or simplification would be hazardous, because mistakes could never be corrected.

Furthermore, even Bayesian reasoning must proceed by reducing the probability of some events to zero. When a piece of data is transmitted, at a minumum that datum must be taken to have been correctly transmitted if it is to be taken into account. For example, the juror might assign a probability of one to "the waiter says that he saw the defendant in Scarsdale" even if he recognizes that there may be a twin brother or a prosecutorial plot. Yet, even this may turn out to have been mistaken if, for instance, the other eleven jurors later convince this juror in the jury room that he misheard the waiter's testimony.

Since reasoning must proceed through the reception of data, and since receiving data necessarily entails assigning zero probabilities, Bayesian reasoning is irreversible. The only way that it could be reversible would be if no probabilities were ever allowed to be reduced to zero. This would not only be cumbersome, it would be impossible. What data are so epistemologically certain that no later quantity of evidence might prove them false? The epistemological foundations of Bayesian reasoning seem to assume that there are some data that we can safely conclude will never be refuted, so that we will never later have to retrace our steps and revise a zero probability upward. This seems shaky; it would be much better to have a system that allowed us to make assumptions, and then go back and correct them later if they turn out to be wrong.

To put the problem another way, the Bayesian system does not allow an individual to revise any belief either that something is definitely true or that it is definitely false. Assume I believe that P is false. This means that I assign a probability of zero to P. Under Bayesian analysis I can never raise that probability to a positive number, i.e., I can never come to believe that P is possible. Next assume that I believe that P is true. Then the probability of P is one, and the probability of not-P is zero. Since the probability of not-P cannot thereafter be raised to a positive number, the probability of P cannot be lowered to less than one. But of course this is absurd; it is not uncommon to alter a state of complete belief or complete disbelief. And it is implausible for Bayesians to try to avoid this problem by saying that we never should have believed (or disbelieved) in the first place.

IV. Second Order Evidence

The Bayesian system is not set up to accommodate second-order propositions. The Bayesian framework is based upon Boolean algebra and the first-order propositional calculus, which includes only statements of the first order. In fact, it cannot include statements of the second order, intermingling them indiscriminately with first-order statements, without running the risk of generating paradoxes. This was well established by developments in the foundations of mathematics around the turn of the century.[21] Logicians and set theorists recognized that such constructs as the set of all sets, and the set of all propositions, could not be posited without creating logical contradictions within their theories.[22]

Consider for example the famous paradox in which a slip of paper contains the following two sentences:

(1) The sentence below this one is true.
(2) The sentence above this one is false.

The paradox arises when one asks whether sentence (1) is true or false. If it is true, then what it says it true. This means that sentence two is true. But sentence two says that sentence one is false—a contradiction, and disproof of the assumption that (1) is true. Consider, on the other hand, the possibility that sentence one is false. If it is false, then what it says is false. Thus, sentence two is false; but this means that sentence one is true. Either way, a contradiction is established.

A typical device for avoiding these paradoxes is to carefully separate propositions (or sets) of one order from those of a different order.[23] Propositions, for example, might refer to other propositions, but only ones of a lower logical order. For one sentence to refer to another, it must be of a higher logical order. This is not possible where (1) refers to (2), and (2) refers to (1). For this reason, the two sentences above would be deemed nonsensical or ungrammatical. The paradox cannot be stated within a language that observes the constraints about reference only to sentences of a lower logical order. This is the approach which most logicians and set theorists have taken, allowing reference only from the meta-language to the object-language and not the reverse.

The paradox described above can be easily phrased in terms of probabilities, if one ignores the admonition to keep first and second-order statements separate. The slip of paper reads:

(1) The statement below has a probability of one.
(2) The statement above has a probability of zero.

These two sentences are, of course, the equivalents of the earlier sentences about truth and falsity. Truth and falsity are simply translated into the language of probability, namely, as probabilities of one and zero. It is so easy to re-create the paradox within the Bayesian system because Bayesian logic contains within it a complete propositional calculus.[24] Bayesian logic, after all, is merely an extension of a propositional calculus of truth and falsity to a calculus of probabilities ranging on a continuous spectrum from one to zero. Therefore, any problems contained within the propositional calculus will also be contained within the probability calculus.

Therefore, it is a wise strategy for the probability calculus to exclude second-order statements; if these were included, all of the logical problems that plagued early versions of logic and set theory would be re-created.[25] This does, however, impose certain difficulties for probabilistic reasoning. If the probabilists steadfastly refuse to take such second-order statements into account at all, then they have arbitrarily excluded from the human reasoning process certain sorts of evaluative statements. As the examples above demonstrate, in some circumstances it is necessary to take into account that a probability estimate is based on little evidence, or on false evidence.

Alternatively, the Bayesians might concede the limited scope of Bayesian reasoning, restricting its application to first-order statements and recognizing that higher order methods of reasoning will sometimes override strict Bayesian principles. This is probably the best approach. I have not argued so much that Bayesian reasoning is erroneous as that it is incomplete. There are certain types of arguments and pieces of information that it cannot accommodate. Yet a properly constructed meta-system could deal with such information. While it is unclear to me how this meta-system might be formalized,[26] it is not certain that a formal second-order system would be necessary. It may be that our intuitive reasoning processes handle these problems well enough without formalization.

This possibly highlights, however, the dangers of trying to import Bayesian reasoning into the legal system. The limits of the utility of the Bayesian approach are not well understood. As a result, it is assumed by the proponents of that system that everything must be somehow put into a form where the Bayesian system can digest it; if information cannot be put into that form, then it must be discarded. For example, Professor Kaye attempted to vindicate Bayesian reasoning by showing how it would take into account the fact that a case was based upon naked statistical evidence. Other proponents of Bayesian reasoning might assert that such facts are simply irrelevant. Neither conclusion would be sound. Such arguments based upon the nature of evidence are perfectly appropriate, but simply are not suitable for incorporation through Bayesian reasoning.

The law is not, of course, perfect. There may be much for the legal system to gain through attention to formal mathematical models, even Bayesian ones. But there is a danger in being mesmerized by precision and formalization. That danger is that all counter-intuitions which suggest ways in which the model may be incomplete or erroneous will be swept aside as "unscientific" or imprecise. When the tendency to abandon the intuitions which run counter to the "scientific" way of doing things first asserts itself, the proper response is caution. The intuitive way may in some situations be the more logical.

NOTES

† © 1986 by Lea Brilmayer.

* Professor of Law, Yale Law School. J.D. Boalt Hall School of Law, 1976.

[1] *Modern Uses of Logic in Law* was later renamed the *Jurimetrics Journal*. Its first issue was in 1959. Typical articles dealing with the use of symbolic logic include Cobb & Thomason, *Law, Logic, and Rationality*, 11 JURIMETRICS J. 1 (1970); Delagrave, *Can the Formulas of Modern Logic Help Solve Legal Problems?*, 1964 MOD. USES LOGIC L. 98; and Mason, *The Logic Structure of a Proposition of Law*, 11 JURIMETRICS J. 99 (1970).

[2] *See, e.g.*, Allen, *Symbolic Logic: A Razor-Edged Tool for Drafting and Interpreting Legal Documents*, 66 YALE L.J. 833 (1957); Allen & Orechkoof, *Toward A More Systematic Drafting and Interpreting of the Internal Revenue Code: Expenses, Losses, and Bad Debts*, 25 U. CHI. L. REV. 1 (1957); *see generally* I. TAMMELO, OUTLINES OF MODERN LEGAL LOGIC (1969).

[3] Allen & Orechkoof, *supra* note 2, at 61-62.

[4] Pfeiffer, *Symbolic Logic*, SCIENTIFIC AMERICAN 22-24, Dec. 1950, *quoted in* Allen, *supra* note 2, at 879.

[5] D'Amato, *Can/Should Computers Replace Judges?* 11 GA. L. REV. 1277 (1977).

[6] For a description of this phenomenon, see Part III below.

[7] Nesson, *The Evidence or the Event? On Judicial Proof and the Acceptability of Verdicts*, 98 HARV. L. REV. 1357, 1379 (1985). *But see* Brook, *The Use of Statistical Evidence of Identification in Civil Litigation: Well Worn Hypotheticals, Real Cases and Controversy*, 29 ST. LOUIS U.L.J. 293 (1985).

[8] For a detailed and technical account of second-order logic, see G. HUNTER, METALOGIC: AN INTRODUCTION TO THE METATHEORY OF STANDARD FIRST ORDER LOGIC (1971). Hunter defines metalogic as "the theory of sentences-used-to-express-truths-of-logic." *Id.* at 3.

[9] Kaye, *The Laws of Probability and the Law of the Land*, 47 U. CHI. L. REV. 34 (1979); Kaye, *The Paradox of the Gate Crasher and Other Stories*, 1979 ARIZ. ST. L.J. 101; Kaye, *Naked Statistical Evidence*, 89 YALE L.J. 601 (1980).

[10] Kaye, *The Laws of Probability and the Law of the Land*, 47 U. CHI. L. REV. 34, 40 (1979).

[11] *Cf.* Brook, *The Use of Statistical Evidence of Identification in Civil Litigation: Well Worn Hypotheticals, Real Cases and Controversy*, 29 ST. LOUIS U.L.J. 293, 325 (1985) (claiming that plaintiff may not have to present exhaustive proof in order to prevail).

[12] *See, e.g.*, I. LEVI, GAMBLING WITH TRUTH: AN ESSAY ON INDUCTION AND THE ARMS OF SCIENCE 139-40 (1967).

[13] *Id.* at 141.

[14] Weight of evidence is relevant in the Bayesian system in the following way. If a probability of .6 is based on substantial evidence, it will be more robust in the face of new evidence and will change more slowly as further evidence accumulates than the same probability based on little evidence. This fact, however, has no necessary influence on the question of how one ought to act based on the probability distribution one already has. For example, if the probability of guilt is .999, the rapidity with which that probability would change given new evidence would not matter to a Bayesian gambler.

[15] For one interesting (but I believe ultimately unsuccessful) effort to use weights of evidence with subjective probabilities, see Cohen, *Confidence in Probability: Burdens of Persuasion in a World of Imperfect Knowledge,* 60 N.Y.U. L. REV. 385 (1985).

[16] A partial ordering is a relationship R such that it is not necessarily true for every pair of sets A and B that either A(R)B, B(R)A, or A = B. For a technical definition, see P. HALMOS, NAIVE SET THEORY 54 (1970).

[17] For example, set A might contain a very minor piece of evidence that set B did not, but set B contains a great deal of probative evidence that A did not.

[18] A. BURKS, CHANCE, CAUSE, REASON: AN INQUIRY INTO THE NATURE OF SCIENTIFIC EVIDENCE 213-16 (1977).

[19] Betting, in this odd sense, can be conceptualized as a positive-sum game. Both parties expect to actually make money, because each thinks that he or she has better information than the other.

[20] It is an inadequate response to say that the jury should delay the calculations to the end of the trial. First, this objection is of no use in non-legal contexts, where probabilities are revised continually. Second, even a legal decision ought to be subject to revision on collateral attack. Third, the evidence cannot be taken into account sequentially even if its consideration is delayed until after all of the evidence is presented—yet Bayes's Theorem seems to contemplate sequential incorporation of evidence.

[21] For a historical account of these developments, see M. KLINE, MATHEMATICS: THE LOSS OF CERTAINTY ch. 9 (1980).

[22] *See generally* S. HAACK, PHILOSOPHY OF LOGICS ch. 8 (1978). *Compare* L. SAVAGE, THE FOUNDATIONS OF STATISTICS 10 (2d ed.) (relying on a "universal set") *with* P. HALMOS, NAIVE SET THEORY 6-7 (1970) (proving that there is no universal set).

[23] *See* S. HAACK, *supra* note 22, at 138-45.

[24] For typical efforts that build the probability calculus upon functional, propositional, or Boolean logic, see A. BURKS, CHANCE, CAUSE, AND REASON 42 (1977); I. HACKING, LOGIC OF STATISTICAL INFERENCE 32-134 (1965); R. JEFFREY, THE LOGIC OF DECISION 54-59 (1965). Burks concedes that, strictly speaking, the first and higher-order statements should not be intermingled, but asserts that this is unnecessarily complex. A. BURKS, *supra*, at 42.

[25] Not all violations of the separation of first and second-order statements create paradoxes. Even certain self-referential statements, as Haack points out, are perfectly harmless. S. HAACK, *supra* note 22, at 139. The problem, she notes, is to delineate principles about which should be disallowed.

Some of the second-order evidence that I have described is not harmless. Consider, for instance, the observation that "All of the evidence in this case is naked statistical evidence." If this statement is itself taken to be a piece of evidence, then is it true or false? If it is true, then what it says is true, and it must itself be nakedly statistical—which it does not seem to be. But if it is false, then it does not seem that it should be taken into account. A similar argument can be raised against "There are only five pieces of evidence in this case."

But there is a broader point to be made here. The very existence of *some* paradoxes sent logicians scurrying for solutions. The paradoxes threatened the entire foundations of their discipline. The impact of these paradoxes is described in M. KLINE, *supra* note 21. My point is merely that subjective probabilists ought likewise to be examining skeptically the foundations of their own field.

[26] Some probabilists have struggled with these issues. *See, e.g.*, J. KEYNES, A TREATISE ON PROBABILITY 76 (1921); H. REICHENBACH, THE THEORY OF PROBABILITY (2d ed. 1949); Skyrms, *Higher Order Degrees of Belief,* in PROSPECTS FOR PRAGMATISM 109 (D. Mellor ed. 1980).

Lea Brilmayer,
Professor of Law, Yale Law School.

Anne W. Martin*

A COMMENT IN DEFENSE OF REVEREND BAYES †

In her paper for this volume,[1] Professor Brilmayer argues that Bayesian inferential methods are inadequate for treating second-order evidence[2] in legal reasoning processes. She builds her argument from three major points:

1. A Bayesian approach can be applied to "naked statistical evidence," despite the law's unwillingness to base a verdict on such evidence.
2. Bayesian approaches do not take the weight of the evidence into account in revising opinions.
3. The Bayesian framework cannot accommodate evidence that contradicts prior evidence.

All three arguments reflect a certain innocence regarding Bayesian theories and applications, both in law and elsewhere:

1. In discussing "naked statistics," she ignores Bayes's Theorem and its application; in fact, the conclusions she draws from her examples imply a misapplication (or no application) of Bayes's Theorem.
2. She seems to confuse "weight of evidence" with "sample size," or the number of data points on which a statistical estimate is based; in asserting that a Bayesian framework cannot accommodate weight of evidence in the opinion revision process, Brilmayer both ignores the Bayesian statistical methods, which treat sample size explicitly, and denies the alternative view of weight of evidence as the probative value of the evidence.
3. In discussing contradictory evidence, Brilmayer ignores Professor Schum's work[3] modeling various species of evidence; she also begs the question of how Bayesian modeling can reasonably be expected to be applied to juridical inference. (She seems to imply that Bayesian methods are proposed for real-time, sequential application to evidence as it is presented in the course of a court trial.)

I will discuss these points in order.

"Naked Statistics"

Professor Brilmayer seems to argue that Bayesian inference cannot treat naked statistical evidence in the same way the legal system does. This is hardly clear. In fact, I am not even certain I know what "naked statistical evidence" is. I infer from Brilmayer's usage and examples, however, that a naked statistic is simply a measure of the relative frequency of some phenomenon that might be of interest at trial. (The examples she uses are the relative frequencies of gatecrashers versus ticket buyers, and the proportion of the blue buses in town that are owned by a given bus company.) In applying probability theory, a phenomenon's relative frequency can function as its probability. This is so even when the application uses Bayes's Theorem, which is a prescription for revising opinions about the relative likelihood of some hypothesis based on observable evidence. Perhaps the simplest form of Bayes's Theorem is the odds-likelihood ratio version:

$$\frac{P(H_1|D)}{P(H_2|D)} = \frac{P(D|H_1)}{P(D|H_2)} \times \frac{P(H_1)}{P(H_2)}$$

D represents the evidence to be used in opinion revision, and H_1 and H_2 are mutually exclusive hypotheses. $P(H_1|D)$ is read, "the probability of H_1 given D." It is also called the posterior probability of H1 when D is known. The ratio $P(H_1|D)/P(H_2|D)$ is called the posterior odds. On the other side of the equation, $P(D|H_1)$, is called the likelihood of D under H_1, and the ratio $P(D|H_1)/P(D|H_2)$ is called the likelihood ratio. Finally, $P(H_1)$ is the prior probability of H_1 (before D is known), and $P(H_2)$ is the prior probability of H_2, and the ratio $P(H_1)/P(H_2)$ is called the prior odds.

To apply Bayes's Theorem we need two probabilities—prior probabilities and likelihoods. We can estimate these probabilities using relative frequency data. In the gatecrasher paradox,[4] the relative frequency information about gatecrashers versus ticket buyers (501/499) is an estimate of prior odds—also called a "base rate."[5] This ratio should represent our opinion about the relative likelihood that any person present at the rodeo is a gatecrasher *before* we obtain any observable evidence (e.g., eyewitness testimony) about whether or not the person purchased a ticket. There is nothing particularly "Bayesian" about these prior odds. They are derived not from Bayesian analysis, but from a population base rate. Since only this base rate is provided, Bayes's Theorem is not applicable to the gatecrasher problem for the same reason that a case based on this problem would probably be thrown out of court—there is no observable evidence.

This raises an interesting question about what a jury's opinion *should* be before observable evidence is presented. In criminal cases, the jury is instructed to presume the defendant's innocence. Accordingly, it is probably safe to assume that the prior odds for guilt to innocence are less than fifty/fifty. In civil cases, however, the jury is not instructed on pre-evidence presumptions. It seems only fair, though, that the factfinder start without bias toward either party. Thus, the prior odds ought to be fifty/fifty. In the gatecrasher problem, though, the prior odds are 501/499. This conflict raises a critical question: if some observable evidence were produced, which set of prior odds should be used to apply Bayes's Theorem? Although I will not answer this question here, there is precedent for excluding such evidence, even though it would help to establish the truth of some fact-in-issue, when its admission would not serve the ends of justice.

The blue bus problem[6] differs from the gatecrasher paradox in that there is observable evidence—eyewitness testimony that the bus that hit plaintiff was blue. We would like to use Bayes's Theorem to revise our opinion about the likelihood that this bus belonged to the defendant. Unfortunately, the only information provided is that the defendant owns four-fifths of the blue buses in town, and this is insufficient to apply Bayes's Theorem. First, we have no information about what the state of our opinion should be before we learn that the bus was blue. (This is not a serious problem if my speculation that prior odds in a civil case ought to be fifty/fifty is taken seriously.) Second, and more significant, there is insufficient information to estimate a likelihood ratio. The likelihood ratio in this case is composed of the probability that the bus would be blue given that it belonged to the defendant, and the probability that the bus would be blue given that it did not belong to the defendant. There may be an initial impulse to estimate these likelihoods at 80% and 20%, respectively, so that applying Bayes's Theorem would yield posterior odds of four to one (an eighty percent probability that the bus belonged to defendant), assuming prior odds of one to one. A second glance should convince us, though, that there simply is not enough information to determine the likelihoods. For example, suppose there are 100 blue buses and 300 red buses in town, and that defendant owns 80 and 280 of them, respectively. Then the likelihood ratio would be (80/360)/(20/40), or 4/9. If prior odds are one to one, then the posterior odds would be 4/9, so that the probability that defendant owned the bus that hit plaintiff would be .44—less than a preponderance of evidence against the defendant. On the other hand, there might be 100 blue buses and 100 red buses in town, with defendant owning 80 blue buses and 20 red buses, for a total of 100 buses. Now, the likelihood ratio would be (80/100)/(20/100), yielding posterior odds of 80/20, and a posterior probability of .80 that the bus in question belonged to defendant—a clear preponderance against the defendant.

In two of Brilmayer's examples, the only information available is naked statistical evidence. Here, she asserts, Bayesian analysis would yield a certain finding even when a court would refuse to try the case. We have just seen, however, that these cases do not provide sufficient information to apply Bayes's Theorem. Actually, Bayes's Theorem could apply to the blue bus problem if sufficient statistical (relative frequency) information were provided, but I suspect the evidence would be considered insufficient to bring suit against the defendant. This is equally likely to be true, however, of any potential suit based solely on circumstantial evidence provided by eyewitness testimony, even without a statistical basis for evaluating the evidence.

The Weight of Evidence (AKA Sample Size)

Professor Brilmayer insists that Bayesian approaches do not treat the "weight of evidence." In so doing, she rejects a definition of "weight of evidence" that probabilists have used for more than thirty years[7]—that the weight of evidence is the extent to which the evidence favors one hypothesis over another or, as jurists would say, the probative value of the evidence. But this is precisely the focus of Bayesian approaches. Brilmayer avoids this conclusion only by defining it away.

The meaning of "weight of evidence," however, is only tangential to her main point that Bayesian approaches ignore sample size in the process of opinion revision. In a sense, this is true. Bayes's Theorem may yield a stronger result based on ten good pieces of evidence than on a hundred weak ones—a desirable feature in law, where some pieces of evidence are more probative than others. In Brilmayer's coin-toss example, however, all the pieces of evidence (the flips) convey the same amount of information. Bayesian statistics easily addresses this problem under the Beta-Bernoulli paradigm,[8] in which the objective is revision of a probability distribution over p, the binomial probability of a given outcome of a binary process, based on repeated sampling from the binary process (like flipping a coin). Sample size is important in this paradigm because it is used in determining the variance of the probability distribution over p. In deciding whether to bet on the outcome of a coin toss, one should use this variance measure to determine the probability that p, the probability of a heads on the next flip, is sufficiently high or low to justify the bet.

To the extent that sample size is important, the opinion revision process must be based on sampling (e.g., performing experiments), and the sample results should be analyzed statistically. Bayesian statistical techniques are more than adequate for analyzing and drawing inferences based on sampling data, including sample size. Much legal reasoning, however, is not statistical but is based on evidence about unique events that are not subject to sampling. Such reasoning is still amenable to Bayesian inferential analysis, but not to statistical inference (Bayesian or other); the number of pieces of evidence available is likely to be subordinate to their probative value in determining the overall weight of the evidence (as defined by probabilists).

Correcting Errors

Professor Brilmayer asserts that "Bayesian reasoning cannot correct errors," because when Bayes's Theorem is applied to opinion revision about the likelihood of a hypothesis with a probability of zero or one, the probability of the hypothesis is unchanged. Furthermore, when opinion is revised based on evidence that, under a given hypothesis, has a likelihood of zero, the probability of the hypothesis becomes zero. As Professor Brilmayer explains, however, this is not unreasonable in theory: if a hypothesis is impossible, no evidence can prove it possible, and if a hypothesis is certain, no evidence should be able to prove it less than certain. Brilmayer is concerned, though, that in practice a probability of zero is likely to be incorrectly assigned. If Bayesian inference is to be an appropriate model for legal reasoning, she argues, it must provide some formal means for correcting such errors.

One possible response is that factfinders should hedge rather than allow probabilities to drop inappropiately to zero. Applied to Brilmayer's example of a defendant's alibi and a waiter's testimony, this means the jury should allow for the possibility that the waiter's statement could be a "false positive"—for example, testimony that the defendant was in the restaurant in question at the crucial time when in fact the defendant was not there. If the jury recognizes the possibility of a false positive, then the waiter's testimony will not be taken as conclusive, and no probability zero problem will arise. Brilmayer anticipates this response and dismisses it because the probabilities of *some* events must be reduced to zero (or taken to be one). This raises the question of whether our analysis is normative or descriptive. Brilmayer seems to assume that it must be descriptive: since people make mistakes and must correct them, Bayesian inference should model the making and correction of mistakes. All the Bayesians I know, however, have long since given up the idea that Bayes's Theorem is descriptive of how people actually revise their opinions. Instead, we take it as a normative model—opinion revision *ought* to be consistent with the axioms of probability theory, and evidence *ought* to be evaluated taking all possibilities into account.

Professor Brilmayer's objection is rooted in her assumption about how Bayes's Theorem is applied. In her examples, she seems to apply the theorem sequentially to each new piece of evidence as it is presented in the course of a trial. This is not necessary to Bayesian reasoning. As noted earlier, Bayes's Theorem can be applied to opinion revision based on either a collection of data or on an individual piece of evidence. Thus, it could be considered just as "Bayesian" to delay revising opinion until all the evidence has been presented as it is to revise opinion sequentially. This approach has two advantages. First, it is consistent with the jury instruction to "keep an open mind." Second, this "open mind" approach can be used to avoid the difficulties Brilmayer finds in the sequential Bayesian approach. Brilmayer's error correction problem is caused by the multiplicative nature of Bayes's Theorem coupled with the fact that any number multiplied by zero is zero. We avoid the problem under the "open mind" approach by delaying the multiplying and first assigning only provisional likelihoods for individual pieces of evidence. Subsequently, when all the evidence is presented, we edit those likelihoods based on pertinent new evidence. Thus error correction is achieved.

To the extent that Professor Brilmayer bears the burden of proving the inadequacy of Bayesian methods for treating second-order evidence in the legal context, the Reverend Bayes is entitled to a directed verdict. Much of the evidence she offers is irrelevant or incredible, and while her examples raise interesting questions, I have shown that Bayesian methods supply satisfactory answers.

Notes

† © 1986 by Anne W. Martin.

[1] Brilmayer, *Second-order Evidence and Bayesian Logic*, supra at 147.

[2] Professor Brilmayer terms this "evidence about evidence." *Id.* at 155.

[3] Schum, *On the Behavioral Richness of Cascaded Inference Models: Examples in Jurisprudence*, in 2 Cognitive Theory 149, 149-153 (N. Castellan, Jr., D. Pisoni, & G. Potts eds. 1977).

[4] In the gatecrasher paradox, the plaintiff holds a rodeo that 1000 people attend even though only 499 tickets are sold. The likelihood of any one of these people being a gatecrasher is thus 501/499. The question is whether this naked statistical evidence

alone can support a verdict for the plaintiff promoter against any individual audience member defendant. *See* Brilmayer, *supra* note 1, at 148–49.

[5] Tversky & Kahneman, *Evidential Impact of Base Rates*, in JUDGMENT UNDER UNCERTAINTY: HEURISTICS AND BIASES (1982).

[6] In Brilmayer's version of the blue bus hypothetical, the plaintiff is hit by a bus and injured. The only thing he can remember about the bus is that it was blue. The only other evidence presented is that defendant owns 80% of the blue buses that operate on the road where plaintiff was hit. Is this sufficient to sustain a verdict for the plaintiff? *See* Brilmayer, *supra* note 1, at 148.

[7] I. GOOD, PROBABILITY AND THE WEIGHING OF EVIDENCE (1950).

[8] *See generally* R. WINKLER, AN INTRODUCTION TO BAYESIAN INFERENCE AND DECISION (1972).

Anne W. Martin,
Senior Analyst, Decision Science Consortium.

D. H. KAYE*

A FIRST LOOK AT "SECOND-ORDER EVIDENCE"†

In *Second-Order Evidence and Bayesian Logic*,[1] Professor Lea Brilmayer touches on many intriguing issues in the fields of the philosophy of probability, mathematics, and logic, as well as issues in contemporary theory about legal proof. In the pages allotted to me, I cannot hope to address all these matters in any detail. At best, I can sketch a quick and conclusory summary of some of the difficulties that I see with her paper.[2]

At the outset, however, I should like to identify a fundamental issue on which we agree. One cannot hope to analyze all features of the rules of evidence and proof in terms of what lawyers think of as formal, mathematical reasoning. A fortiori, one cannot expect an uncompromisingly formal analysis in the spirit of the *Principia Mathematica* to yield much jurisprudential wisdom. Informal mathematics is merely a tool that casts light on certain aspects of some legal rules.[3] It makes sense only within a framework originating from a set of propositions lying beyond the mathematics itself.[4] This conclusion applies not merely to the Bayesian reasoning that Brilmayer attacks, but also to other theories of statistical inference and inductive logic.[5] To recognize this limitation of mathematical modeling is not to condemn Bayesian inference or any other mathematical reasoning. It is to appreciate and understand it.

So Brilmayer and I arrive at more or less the same conclusion: Bayesian analysis and formal logic are but a small part of the legal picture, and to be "mesmerized by precision"[6] would be to dwell in a fool's paradise.[7] However, many of the specific arguments that lead her to these conclusions strike me as puzzling and problematic.

An initial problem concerns Brilmayer's argument about "irreversibility." She points out that if an individual whose coherent partial belief in a sentence S were, at one moment in time, to take on a probability of zero (or one), then later relevant evidence E about S (or not-S) could not raise (or lower) that probability.[8] This, she says, puts the subjectivist in the awkward position of being unable to correct a mistaken extreme belief.

There are two obvious parries. The first is that, as a practical matter, we can recognize our errors and redistribute the probability mass accordingly.[9] Brilmayer correctly notices that Bayes's rule does not specify how to do this. One must step outside of the mathematics to use the formalism.

The other reply is that one should not be so rash or prejudiced as to think that the probability of any argument or proposition is zero or one.[10] Brilmayer rejects this constraint as an implausible attempt to avoid the problem, but the full-blown Bayesian model is not put forward as a plausible description of human behavior. Those who appeal to this model usually do so to establish the notion that hypothetical intellects with plenty of time would have partial beliefs that conform to the laws of probability—including Bayes's Theorem.[11] Brilmayer alludes to the Dutch book argument,[12] but she ignores the variant known as a semi-Dutch book argument. This variation is important here because it establishes that strictly coherent beliefs preclude probabilities of zero and one.

Brilmayer chides Bayesian statisticians for not showing sufficient interest in "the weight of evidence to revise probability distributions."[13] Her analysis confuses data with sample size and disregards actual Bayesian methods, which are capable of handling the problem cases she constructs.[14]

Brilmayer suggests that because Bayesian decision theory does not accommodate decisions not to bet, it cannot account for the fact that "[i]n law, the possibility of basing a decision upon the insufficiency of the evidence is a viable alternative, as is a refusal to bet in any realistic betting example."[15] The premise is false—decision theory has no difficulty evaluating a decision not to gamble along with decisions to accept various gambles. From the perspective of this theory, a directed verdict granted at the close of a plaintiff's case is a statement that the plaintiff's evidence, viewed in context, is insufficient to justify a reasonable degree of belief in the plaintiff's story. In other words, the directed verdict is a statement that it is very clear that the plaintiff has not proven his story to be sufficiently probable.[16]

Cases of naked statistical evidence arise when the only evidence linking a defendant to an act (or, more generally, identifying a cause) is overtly statistical. According to Brilmayer, the courts invariably are unwilling to countenance verdicts based on naked statistical evidence.[17] Bayesian decision theory, she argues, is unable to account for this rule of law. The true explanation is to be found in the fact that the statistical evidence does not suggest that the defendant "really" committed the act.[18]

To begin with, Brilmayer's account of the law is questionable. The oft-heard claim that there is a categorical rule against basing a verdict on naked statistical evidence is a myth.[19] What needs explanation or justification is neither the existing law (which is obscure and underdeveloped) nor some putative rule against all naked statistical evidence (which may not be at all suitable as a rule of law). Rather, it is our intuitions about hypothetical cases that flourish only in the fertile soil of law reviews.

In the papers that Brilmayer criticizes, I have attempted to explain these intuitions by suggesting that they may reflect a concern that the proponent of the evidence should have done more to make out his case.[20] I do not maintain that this concern is appropriate in every case, imaginable or real, of naked statistical evidence. In fact, my analysis builds on the distinction between justified and unjustified naked statistical evidence.[21] I maintain only that both a static spoliation argument and a dynamic incentive argument suggest that it may be appropriate to deny victory to the proponent of unjustified naked statistical evidence.

An analysis that applied these spoliation and incentive arguments correctly would show that the important value of accuracy in factfinding goes a long way toward justifying a rule that normally denies recovery to the proponents of unjustified naked statistical evidence. The mathematical theory is rich enough to accommodate the intuition, articulated long ago by Wigmore,[22] that there are many instances in which the failure to produce certain evidence diminishes the plausibility of a party's story. When this reasoning applies to naked statistical evidence—when the plaintiff relies on unjustified naked statistical evidence—a directed verdict may well be appropriate. Whether every conceivable case of unjustified naked statistical evidence should be met with a directed verdict depends on how far one wants to push the dynamic incentive argument.

This leaves two major objections to the resolution of the puzzle of naked statistical evidence. "Perhaps the most telling"[23] is that if we use Bayes's rule to conclude that the posterior probability in a case of unjustified naked statistical evidence is less than that in a corresponding case of justified naked statistical evidence, then "symmetry"[24] implies that "the probabilities will fail to add up to one."[25] This objection is clever, but the cleverness comes at a grave cost—a crippling lack of clarity. Brilmayer never quite tells us which probabilities are supposed to add up to one. The confusion this can cause is apparent if we ignore the negative inference and treat the probability that a defendant x who was randomly selected from the crowd is a gatecrasher as $P(Gx) = .501$. Since this is the probability without the allegedly improper negative inference, nothing should be wrong with it. Yet, the symmetry of the spectators implies that this probability pertains not only to the defendant, but to all 1000 spectators. The sum of these probabilities greatly exceeds one.

As far as I can tell, such conundrums arise from confusing certain probabilities taken from distinct cases with other probabilities within a case. While the sum of certain probabilities within a case must be one, summing across cases need not give the same result.[26]

The last major objection to the Bayesian analysis of naked statistical evidence is that the reference to missing evidence in these cases is a second-order proposition that cannot be mixed with first order propositions without creating a paradox or contradiction somewhere in the system.[27] The only example of a paradox that Brilmayer provides has nothing to do with the analysis of naked statistical evidence.[28] She starts with a trivial variation on the hoary Liar Paradox. Brilmayer advocates a Tarskian resolution that posits a hierarchy of truth-predicates for a natural language, each applying to the levels below the level to which it belongs. Although orthodox, this resolution of the problem is controversial and has come under heavy attack in recent years. Unlike the paradoxes of set theory, the Liar Paradox is not generally regarded as solved. Whatever the solution may be, if the Liar Paradox is a problem for probability theory,[29] it is equally a problem for any use of natural language in legal discourse.

Thus, the Liar Paradox, the Russell Paradox, and other semantic and mathematical paradoxes do not sound the death knell for probability theory in general or for Bayesian inference in particular. Let us suppose, however, that it is not only crucial for the sake of conceptual clarity to distinguish sentences that express evidence from sentences that comment on evidence, but also that we must extirpate all manifestations, however benign, of self-reference, direct or indirect at every level, lest insidious contradictions or paradoxes take root in the nooks and crannies of our language. This demand does not threaten the Bayesian treatment of naked statistical evidence. It merely requires a more precise formulation than that which Brilmayer criticizes. Elsewhere I have tried to treat the issue in this fashion by conditioning it on what can be called gap-sentences.[30] In addition to describing the events at trial, these sentences include a negation operator. Thus, contrast typical sentences like "Ms. Witness A says that she observed defendant Smith at the rodeo," and "The books of the rodeo company showed that 499 persons paid the admission fee," to the gap-sentence "Neither witness A nor B, nor any other witness, testified that he saw Smith climb the fence."

Of course, we attach significance to gap-sentences and include them in our list of evidentiary sentences because of our reasoning about the evidence that we expect or demand a party to bring forth. This background reasoning could be said to occur within a meta-language of sentences about evidentiary sentences. If so, we have come full circle. I began these remarks by agreeing that the probability calculus itself does not supply the raw material needed to reach the best decisions or to derive wise rules of law. No mathematical theory is self-applying. To apply it, we must step outside the theory. If it is appropriate and desirable to say that this pops us into a meta-language, then we need to explore this meta-language. One can only hope that Brilmayer will embark on this voyage of discovery.

NOTES

† © 1986 by D.H. Kaye. All Rights Reserved.

* Professor of Law and Director, Center for the Study of Law, Science and Technology, Arizona State University.

[1] Brilmayer, *Second-Order Evidence and Bayesian Logic*, supra at 147.

[2] In quickly reaching conclusions without full argumentation, I run the risk of sounding cryptic or dogmatic. To deal more fully and fairly with Brilmayer's paper, I plan to develop my underlying arguments in a subsequent paper.

[3] Professor Richard Lempert analyzes the question of naked statistical evidence and other such issues from this perspective in Lempert, *The New Evidence Scholarship: Analyzing the Process of Proof*, supra at 61. Still, Brilmayer asserts that "it is assumed by proponents of that [Bayesian] system that everything must somehow be put in a form where the Bayesian system can digest it." Brilmayer, supra at 164. This characterization may apply to some unnamed Bayesian, but it does not fairly describe the work of most legal scholars who try to use probability theory to explicate selected features of the rules of proof. Indeed, in my previous published exchange with Brilmayer, I emphasized the inability of any mathematical theory to capture all the pertinent legal rules. Kaye, *The Laws of Probability and the Law of the Land*, 47 U. CHI. L. REV. 34, 41 (1979).

[4] Perhaps this is what Brilmayer has in mind when she concludes with a statement that I otherwise would be inclined to dismiss as mystical—namely, that "[t]he intuitive way may in some situations be the more logical." Brilmayer, supra at 691.

[5] *Cf.* Shafer & Tversky, *Languages and Designs for Probability Judgements*, 9 COGNITIVE SCI. 309, 311 (1985) (treating belief functions and Bayesian analyses as "probability languages" that do not contain "formal criteria or . . . general empirical procedures for evaluating designs.").

[6] Brilmayer, supra at 164.

[7] This admonition is not new. See Kaye, *Mathematical Models and Legal Realities: Reflection on the Poisson Model of Jury Behavior*, 13 CONN. L. REV. 1 (1980); Kaye, *The Paradox of the Gatecasher and Other Stories*, 1979 ARIZ. ST. L.J. 101, 109; Tribe, *Trial by Mathematics: Precision and Ritual in the Legal Process*, 84 HARV. L. REV. 1329 (1971).

[8] Brilmayer believes Bayesian arithmetic creates this rigidity. Brilmayer, supra at 162. The only arithmetic reason she offers is that "[b]ecause of the form of the mathematical formula, it will never be possible to raise a probability from zero to a positive number." *Id.* at 160. However, Bayes's rule cannot be applied to a prior probability of zero or one. If $Pr(S) = 0$, then $Pr(E|S) = Pr(ES)/Pr(S)$ is undefined,

and if Pr(S) = 1, then Pr(E|not-S) is undefined. Presumably, the formula that Brilmayer finds wanting is the very definition of conditional probability.

[9] Fienberg, *Gatecrashers, Blue Buses, and the Bayesian Representation of Legal Evidence*, 66 B.U.L. REV. 695 (1986). Bayesian statisticians are cognizant of the need to cope with mistakes. *See* H. JEFFREYS, THEORY OF PROBABILITY (3d ed. 1961).

[10] *See* D. LINDLEY, MAKING DECISIONS 104 (2d ed. 1985).

[11] One can then consider the limitations of real jurors and ask what rules of evidence and procedure seem likely to promote factually accurate verdicts (to the extent that we are interested in such verdicts) on the part of real jurors. *See, e.g.,* Lempert, *supra* note 3.

[12] Brilmayer, *supra* at 158. Brilmayer writes that a Bayesian probability distribution protects against a Dutch book. *Id.* The word Bayesian is superfluous at best. Any distribution of partial beliefs that conforms to the probability calculus—that is, any valid probability distribution—has this property. In addition, Brilmayer is mistaken about the way a Dutch book works.

[13] Brilmayer, *supra* at 156.

[14] *See* Martin, *A Comment on* Second Order Evidence and Bayesian Logic, 66 B.U.L. REV. (1986).

[15] Brilmayer, *supra* at 158.

[16] This is a decision-theoretical interpretation of a directed verdict against the party who has the burden of persuasion. Some commentators believe that a verdict in favor of this party is a statement, not about what probably occurred (at an appropriate level of subjective probability), but of a belief in what actually occurred. *See* Nesson, *The Evidence or the Event? On Judicial Proof and the Acceptability of Verdicts*, 98 HARV. L. REV. 1357 (1985). Since the belief can only be a partial belief, these two statements denote the same thing. Nevertheless, the connotation is different, so if the trial broadcasts a message to the public, this message may vary with the phrasing that is announced.

[17] Brilmayer, *supra* at 149. Referring to the blue bus and the gatecrasher hypotheticals, Brilmayer announces that "[o]f course, the law would not allow recovery in either of these cases." *Id.* In the next breath, she says that "the judge would *probably* not even allow the case to go to the jury." *Id.* (emphasis added). I suppose that she means that the propriety of a directed verdict is so clear that most courts would grant it, and that courts that denied such a motion nevertheless would grant judgment n.o.v. Whatever the procedural vehicle, Brilmayer is clear about the "law's unwillingness to base a verdict upon naked statistical evidence." *Id.*

[18] *Id.* This explanation does not get us very far. At best, it is highly elliptical; it could be circular. At worst, it is mystical.

[19] *See* Allen, *A Reconceptualization of Civil Trials*, *supra* at 21, Brook, *The Use of Statistical Evidence of Identification in Civil Litigation: Well-Worn Hypotheticals, Real Cases, and Controversy*, 29 ST. LOUIS U.L.J. 293, 299–305 (1985); Kaye, *The Limits of the Preponderance of the Evidence Standard: Justifiably Naked Statistical Evidence and Multiple Causation*, 1982 AM. B. FOUND. RES. J. 487, 488.

[20] Lempert, *supra* note 3, adroitly pursues the problem of justifying our intuitions, and Brook, *supra* note 19, is sensitive to the ease with which factual modifications in the hypotheticals can cause dramatic changes in our intuitions.

[21] Kaye, *supra* note 19.

[22] 2 WIGMORE ON EVIDENCE § 285, at 162 (3d ed. 1940).

[23] Brilmayer, *supra* at 154.

[24] *Id.* at 155.

[25] *Id.* at 154.

[26] My meaning would be clearer if I defined "probabilities within a case" and "summing across cases" with appropriate notation. But until Brilmayer specifies an explicit probability model, with a clear definition of the sample space and the conditional probabilities that do not sum to one, I am not inclined to do so. *Cf.* Fienberg, *supra* note 9. Brilmayer should tell us more about her "most telling objection . . . ." *Id.*

[27] *See id.* at Part IV.

[28] *Id.* at 163. Her example is not an illustration of specific contradiction or paradox implied by the putative indirect self-reference in my analysis of naked statistical evidence. A mildly frivolous response to Brilmayer's assertion that such problems arise would be to apply the logic of the negative inference, which Brilmayer disputes in cases of naked statistical evidence, to her argument about symbolic logic: (1) if there were a paradox or contradiction implied by the spoliation analysis, Brilmayer would have discovered it, and (2) if Brilmayer had discovered it, she would have presented it, but (3) Brilmayer did not present it. Therefore, (4) no paradox or contradiction exists. Of course, this spoliation argument is not conclusive proof. Spoliation arguments never are.

[29] Brilmayer thinks it is "easily possible to recreate the paradox within the Bayesian system." Brilmayer, *supra* at 164. In fact, it is impossible to do so without resorting to the extreme probabilities of zero and one. In these limiting cases, however, the language of probability is superfluous. Furthermore, a strict coherence criterion excludes these values from Bayesian deliberations about testimony.

[30] Kaye, *Do We Need a Calculus of Weight to Understand Proof Beyond a Reasonable Doubt?*, *supra* at 129.

David H. Kaye,
Professor of Law,
Arizona State University College of Law.

GLENN SHAFER*

THE CONSTRUCTION OF PROBABILITY ARGUMENTS[†]

I. INTRODUCTION

Though this book has been assembled by legal scholars interested in probability, I will address my contribution to a broader audience—students of the foundations of probability and statistics. This group includes statisticians, philosophers, and psychologists, as well as law professors. My primary purpose will be to suggest ways that the foundations of probability can benefit from the legal perspective on problems of evidence and argument.

Incorporating the legal perspective on probability does not necessarily mean seeking instruction from judges and law professors. It does mean taking seriously the situations of the various actors in the legal setting: the lawyer who must make a case, the opposing lawyer who must criticize it, and the judge or juror who must evaluate the conflicting arguments. Within the statistical tradition, at least, most thought about the foundations of probability has been inspired by the situations of rather different actors: the gambler in a pure game of chance or the scientist disentangling systematic from random variation. Putting the metaphor of the courtroom on a par with the metaphors of the gambler and the scientist can help us achieve a fuller understanding of probability judgment.

The courtroom metaphor forces us to pay more attention to the constructive nature of probability judgment. The lawyer in the courtroom must make a case. She must construct an argument; she cannot merely present facts. She must establish the relevance of the facts she presents. The judge and jurors must do more than understand the conclusion of the argument. They must decide how well the argument supports the conclusion.

Contrast this with the rhetoric of Bayesian statisticians. They have little to say about relevance, argument, or evaluation. Fundamentally, everything is relevant; we are supposed to condition on all the evidence. A probability analysis is supposed to be all-inclusive, and hence not subject to another level of evaluation. The final probabilities are the last word.

My purpose is not to caricature the practical statistician—Bayesian or non-Bayesian. Thoughtful statisticians are perfectly capable of judicious analyses. But they do so only by standing back from basic theory and rhetoric. I am suggesting that the courtroom metaphor can help integrate a more judicious attitude into this basic theory and rhetoric.

To put this thought in perspective, recall that the legal tradition had a hand in creating numerical probability. Numerical probability was first invented by legal scholars.[1] These scholars did not develop the calculus of probability we know today because they did not connect their idea to games of chance. But Leibniz and Bernoulli, the seventeenth-century scholars who did connect probability to the mathematics of games of chance, also appreciated the legal tradition. In his famous *Ars Conjectandi*,[2] Bernoulli based probability squarely on the concept of argument. Many of his basic examples concerned legal problems and, as Garber and Zabell have shown,[3] some came from the Ciceronian legal tradition.

Unfortunately, Bernoulli's eighteenth-century successors did not preserve the tie between probability and the legal tradition. They were intent on developing the mathematical vistas that Bernoulli's innovative ideas had opened up, and this led to a renewed emphasis on the gambling metaphor. Legal ideas never subsequently regained a central role in probability theory; I argue that they should.

The next section of this article outlines the constructive philosophy of probability developed in more detail by Shafer and Tversky,[4] and relates this philosophy to the courtroom metaphor. The following sections consider some specific issues that are illuminated by the courtroom metaphor. Section III discusses how the significance of evidence depends on the ground rules for its acquisition. Finally, Section IV discusses the constructive nature of likelihood, using as an example the problem of assessing probabilities of paternity.

There are two disclaimers that need to be included in this introduction. First, I can claim no originality when I urge that students of probability pay more attention to the legal perspective; the thought has been formulated many times. I should call particular attention to the work of David Schum, who is commenting on this article. Professor Schum has done more than anyone else in recent years to bring insights from the law into other contexts in which probability is used.[5] My message here is that these kinds of insights are part of the foundations of probability. The issues involved in justifying Schum's cascaded inference schemes are at least as fundamental as the Dutch-book arguments that philosophers of probability are so fond of debating.

Second, the interaction between law and the world of probability and statistics has many strands, and this article is not concerned with all of them. The increasing use of conventional statistical evidence in legal settings constitutes one important strand.[6] I regard this development positively, but it is not directly related to the theme of this article. Another strand is the use of legal examples by scholars who have advanced alternatives to the usual calculus of probability. Two scholars come to mind in this connection—L. Jonathan Cohen and Per Olof Ekelöf. Cohen has argued that the usual calculus is less satisfactory in the legal setting than a simple numerical scoring of hypotheses.[7] Ekelöf has argued that the legal setting requires calculations similar to those made in the theory of belief functions.[8] I have made similar arguments,[9] but this is not my theme here. Here I am arguing not for any particular theory of probability, but for an incorporation into all these theories of the constructive attitude represented by the courtroom metaphor.

II. Constructive Probability

In *The Foundations of Statistics*,[10] L.J. Savage distinguished three categories of probability: the objectivistic, personalistic, and necessary interpretations. According to the objectivistic interpretation—also called frequentist—a probability is an objective fact about a repeatable event; it is the long-run frequency with which the event happens. According to the personalistic interpretation—also called subjective or Bayesian—a probability is a particular person's opinion; it can be deduced from the person's behavior when she chooses among bets or other acts with uncertain outcomes. According to the necessary interpretation—also called logical—a probability measures the extent to which one proposition, out of logical necessity and apart from human opinion, confirms the truth of another.

The personalistic interpretation, which Savage advocated, has gained wide support in recent decades, and it is now the most vigorous and self-confident of the three interpretations. Most probabilists who offer advice to the legal profession are Bayesians. Their message is that mathematical probability in general, and Bayes's theorem in particular, can help judges and jurors evaluate evidence.[11]

Unfortunately, none of the three interpretations emphasizes the constructive character of probability judgment. The objectivistic and necessary interpretations both treat probability as objective, making it independent of human action. The personalistic interpretation treats probability as a form of opinion, but it neither emphasizes nor requires that probability opinions be deliberately constructed. Indeed, much Bayesian writing gives the impression that these opinions are ready-made in our minds, waiting to be "elicited."[12]

A. *The Constructive Interpretation*

If we want an interpretation of probability that emphasizes its constructive character, we must explain what people are doing when they construct probabilities. Shafer and Tversky suggest that they are matching problems to canonical examples.[13] When we make Bayesian probability judgments, we are matching an actual problem to a scale of canonical examples from physics or games of chance. In these canonical examples, the possible answers to questions of interest have well-defined and known probabilities.

Of course, we do not match all the evidence we have about a problem to a complex canonical example in one fell swoop. Instead we match parts of the problem or parts of the evidence to more modest canonical examples, and then try to fit these partial matches together. In so doing we construct an argument, an argument that draws an analogy between our actual evidence and the knowledge of objective probabilities in a complex physical experiment or game of chance.

There are many choices in the design of a probability argument: we must decide how to break down our evidence and how to put it back together in a probability model; we must choose what to think of as fixed when making numerical probability judgments, and how much detail to include; and we must determine on what, if anything, to condition.

One advantage of the constructive interpretation is that it pulls us down from the fantasy that a numerical probability analysis can take all evidence into account and hence provide the final word on a question, to the reality that any probability analysis must be treated as just another argument. As an argument, it must be evaluated, and the result of the evaluation may be negative for a variety of reasons: our evidence may fail to fit the scale of canonical examples to which we are trying to fit it; our evidence may be inadequate to justify some of the numerical probabilities in our argument (traditionally, we worry about whether our evidence is adequate to justify prior probabilities for statistical hypotheses, but there is nothing special about these probabilities; we need evidence for every probability judgment); or our evidence may be inadequate to justify some of the judgments needed to put individual numerical judgments together.

B. *Subjective Aspects*

The constructive interpretation of probability preserves some of the features of the personalistic interpretation. It preserves the connection with betting behavior because the canonical examples to which we are matching the evidence are examples in which probabilities are reasonable betting rates. Since, however, we are only drawing an analogy to these examples, the interpretation is less dogmatic. There is no pretense that we would really offer to bet all comers.

Subjectivity also enters the constructive interpretation in a more fundamental way. We often must make subjective judgments about which canonical example on the scale best matches the evidence in a problem, and about whether this best match is good enough to constitute a sound argument.

One important way in which the constructive interpretation differs from the personalistic interpretation is in its ability to acknowledge that the parts of a probability argument may differ in quality. The personalistic interpretation is based on the demand that a person should have betting rates, or preferences from which such rates can be derived, for all questions. This demand is made categorically; a person cannot be excused from the demand because of the poor quality of her evidence. The constructive interpretation, on the other hand, directs attention to the evidence. It asks about the adequacy of the evidence for each of the probability judgments we make.

Since one part of a probability argument may be better than the rest, we are led to ask whether conviction should ever be carried by partial arguments. I believe it should. An argument that can be seen as only part of a Bayesian argument, but is based on high quality evidence, may be more cogent than a completed argument that draws on weaker evidence. Section IV of this article argues for the cogency of partial likelihood arguments. Other partial arguments that are often cogent include frequentist arguments (interpreted subjectively) and belief-function arguments.[14]

C. *Objective Aspects*

The constructive interpretation also preserves some aspects of the frequentist or objectivistic interpretation. The probabilities in the canonical examples to which we compare our evidence are objective frequencies as well as subjective betting rates, and the analogy a probability argument draws to one of these canonical examples is therefore stronger when the probabilities we use in the argument are clearly relevant frequencies.

Since clearly relevant frequencies are seldom available, dogmatic statements of the frequentist interpretation make it seem that frequency ideas have very limited applicability. The constructive interpretation permits us to recognize a broader field of applicability for frequency ideas because it allows consideration of cases in which there is only an analogy to the random sampling that the frequentist interpretation requires. The constructive interpretation allows us to make arguments that turn on the implausibility of certain results were a situation the product of random sampling, without pretending that it actually is.

One way to explain the analogy to random sampling is to say that the probability model we have constructed represents a thought experiment. In this thought experiment, we imagine that facts were determined by a random drawing from a certain reference class. The results of such a thought experiment may constitute a persuasive argument, even after we acknowlege that the selection of the reference class was somewhat arbitrary. Section IV will apply this idea of a thought experiment to the problem of assessing the evidence provided by blood tests in paternity cases.

D. *The Courtroom Metaphor*

The constructive interpretation of probability was not developed with the courtroom metaphor in mind, but the metaphor can reinforce and add to the interpretation. One contribution the courtroom metaphor can make is to shift our attention from the idea of fixed evidence to the idea of argument. Most of what is written about the foundations of probability, including the preceding paragraphs, seems to take for granted that when a probability judgment is at issue, the evidence on which it is based is fixed and well-defined. But the courtroom metaphor encourages us to think of evidence as something that develops as an argument is made. Evidence is introduced into court. When a witness testifies, she gives evidence that was not there before, and when a lawyer cross-examines a witness, they together create evidence.

The creative nature of argument is relevant to many issues in the foundations of probability. Consider, for example, the suggestion sometimes made that we should base certain probabilities on our background knowledge. This suggestion assumes the background knowledge is well-defined. Certain things are stored in our heads—or on our library shelves—and others are not. What is stored, however, may not be so well-defined, and what we will find when we look certainly is not well-defined in advance. Rather than speak of the probability defined by our background knowledge, we should speak of the probability or the arguments that we happen to discover when we search our background knowledge.

A related contribution of the courtroom metaphor is to draw attention to how evidence is acquired. We will study this in the next section. Another important contribution the courtroom metaphor can make stems from its separation of the roles of creator and evaluator of argument. This separation forces us to think about both processes. We will return to this point in Section V.

III. The Acquisition of Evidence

Modern textbooks on mathematical probability teach that the conditional probability of an event A given an event B, denoted by Pr[A|B], is given by the following formula:

$$Pr[A|B] = Pr[A\&B]/Pr[B]$$

Figure 1.

They say that this formula is a mathematical definition. Mathematicians have simply decreed that the Pr[A&B]/Pr[A|B] should be called the conditional probability of A given B.

When we are taught the personalistic or subjective Bayesian interpretation of probability, we are further told that when you learn that B is true, you should "condition on B"—i.e., you should change your probability for A from Pr[A] to Pr[A|B]. This is a fundamental personalistic doctrine. One should always take new evidence into account through conditioning. But why? Surely there is something missing here. An arbitrary mathematical definition cannot determine how you should change your beliefs.

A. *Freund's Puzzle*

The suspicion that something is missing is confirmed by a number of puzzles and paradoxes. My favorite is Freund's puzzle of the two aces.[15]

I show you a deck containing only four cards: the ace and deuce of spades, and the ace and deuce of hearts. I shuffle them, deal myself two of the cards, and look at them, taking care not to let you see them. You realize that there are six equally likely possibilities:

> Ace of spades and ace of hearts,
> Ace of spades and deuce of spades,
> Ace of spades and deuce of hearts,
> Ace of hearts and deuce of spades,
> Ace of hearts and deuce of hearts,
> Deuce of spades and deuce of hearts.

If A denotes the event that I have two aces, B_1 denotes the event that I have at least one ace, and B_2 denotes the event that I have the ace of spades, then your initial probabilities are $Pr[B_1] = 5/6$, $Pr[B_2] = 1/2$, and $Pr[A] = Pr[A\&B_1] = Pr[A\&B_2] = 1/6$.

Now I smile and say, "I have an ace." You are supposed to react to this new information by conditioning on the event B_1. Thus you change your probability for my having two aces from 1/6 to

$$\Pr[A|B_1] = \Pr[A \& B_1]/\Pr[B_1] = (1/6)/(5/6) = 1/5$$

The information that I have at least one ace has increased your probability that I have two.

Now I smile again and announce, "As a matter of fact, I have the ace of spades." You are supposed to condition again, this time on the event B_2, obtaining the new probability

$$\Pr[A|B_1 \& B_2] = \Pr[A|B_2] = \Pr[A \& B_2]/\Pr[B_2] = (1/6)/(1/2) = 1/3.$$

The more specific information that I have the ace of spades has increased even further your probability that I have two aces. Is this second change reasonable? Should my decision to identify a suit make any difference?

Most people will agree, on reflection, that what is reasonable depends on what ground rules are established in advance regarding what I was supposed to tell you. If it had been agreed that I would first tell you whether I had an ace and then *whether* I had the ace of spades, then the change from 1/5 to 1/3 is reasonable (your probability that I had both aces would go down from 1/5 to zero if I told you that I did not have the ace of spades, so it is reasonable that it should go up from 1/5 to 1/3 when I tell you that I do have the ace of spades). But if it had been agreed that I would first tell you whether I had an ace and then tell you the suit of an ace that I had, if I did have one, then the change from 1/5 to 1/3 is not reasonable. Once I told you I had an ace, I had to tell you either spades or hearts, so it makes no sense for you to raise your probability from 1/5 to 1/3.

B. *A Personalistic Treatment of the Puzzle*

The personalistic interpretation helps us to understand the puzzle. The key is to observe that you should condition not only on the event that I have the ace of spades, but also on the event that I have told you so. You should calculate $\Pr[A|B_1 \& B_2 \& B_3] = \Pr[A|B_3]$, where B_3 is the event that I tell you I have the ace of spades. This event is part of your evidence, and you should condition on all your evidence.

Suppose, for example, that it is settled that I am going to tell you whether I have the ace of spades. Then B_3 is equivalent to B_2. The event that I tell you I have the ace of spades is equivalent to the event that I have it. So the change from 1/5 to 1/3 is correct.

On the other hand, suppose it is settled that if I have at least one ace, then I am going to tell you a suit. If I have only one, I will tell you its suit. If I have both, I will secretly flip a fair coin to decide which suit to tell you—spades if heads, hearts if tails. In this case, the event that I tell you I have the ace of spades is equivalent to the event that either it is my only ace or else the coin came up heads. When we work out the details, we find that your probability of 1/5 for my having both aces should remain unchanged.[16]

But suppose there are no ground rules; I just deal the cards and volunteer the information. How can you use Bayesian conditioning in this situation? The personalistic answer is that even though you do not know what ground rules I am following, you should have probabilities for the different possible sets of ground rules I might be following. Your probability model should model what I am going to tell you, not just what cards I have.

C. *A Contribution from the Constructive Interpretation*

This personalistic treatment of the puzzle gives us important insights, but it stops short of explaining why it is so valuable to you to know the ground rules.

If we stand back from the personalistic interpretation, it is clear that probability calculations are worth a lot more if ground rules are established in advance. If no such rules were established, then you would not know what was going on, and the assertion that you should have probabilities for the rules I might have been following would not help much. Even the assertion that I was following some unknown set of rules might not have much content. I did what I did and said what I said, and, in the absence of any prior agreement, the question of what I might have said lies more in the realm of imagination than of fact.

The personalistic treatment of the puzzle seems to deny this. Since it insists that we be able to supply a subjective probability for any event, regardless of the quality of our evidence, the personalistic interpretation is always unable to acknowledge that the strength of a probability argument depends on the quality of evidence. In this case, it is unable to acknowledge that the strength of the argument depends on the quality of the evidence for ground rules for the acquisition of evidence on which to condition.

Moreover, the personalistic treatment of the puzzle fails to deal with the question with which we began: Why is it imperative that your beliefs should change in accordance with the formula $Pr[A|B] = Pr[A\&B]/Pr[B]$ if this formula is merely an arbitrary mathematical definition?

The constructive interpretation of probability can do better. If probabilities are interpreted as betting rates, and if there are ground rules that single out B as one of the possibilities for what you might have learned at some time, then the formula $Pr[A|B] = Pr[A\&B]/Pr[B]$ is more than just a definition—it can be derived as a theorem.[17] Moreover, there are such ground rules in games of chance. Card games, for example, have ground rules for when the values of various cards are to be revealed to other players. Hence the existence of ground rules governing our acquisition of evidence in a practical problem is one aspect of the quality of any analogy that we can draw to a game of chance, and hence one aspect of the quality of any Bayesian probability argument that we can construct for the problem.

D. *A Contribution from the Courtroom Metaphor*

Teachers of probability and statistics have not dealt with these issues well.[18] This is due in part to the dominance of traditional metaphors: physical experiments and games of chance. Ground rules for the acquisition of evidence are so integral to both these metaphors that their presence is not even noticed. In a game of chance, the very rules of the game constitute such ground rules. In an experiment, such ground rules are implicit in the set-up of the experiment, in the specification of what is to be measured or observed. These metaphors do not, therefore, prepare students to deal with problems for which ground rules for the acquisition of evidence are problematic.

The courtroom metaphor might help us do better. Courtroom ground rules are not complete, and they are not there automatically; the contending parties must fight for them. The courtroom metaphor therefore forces us to pay attention to the presence or absence of ground rules.

There are many aspects of the courtroom metaphor that a teacher might use to reinforce this point. For example, jurors must often struggle with the significance of the absence of evidence. If a plaintiff fails to present evidence that should have been available had her claims been true, and if she fails to account plausibly for such absence, then a juror may conclude that the claims are probably false.

We can also point out the significance of give and take in the courtroom presentation of evidence. In order to evaluate a witness's testimony, a juror does not limit herself to what the witness has said. She also considers what questions the witness was answering and how well the cross-examining lawyer has tested the witness's credibility. She considers what the witness did not say and how the witness said what she did say. The importance of this give and take is recognized in law, for example, by the prohibition against hearsay testimony. It should also be recognized by the philosophy of probability.

E. *The Completeness of Evidence*

Professor L. Jonathan Cohen, in his contribution to this book, contends that a person is justified in changing the probability for A from Pr[A] to Pr[A|B] (where B is the evidence) only if the evidence B is complete. This contention will seem obtuse to most Bayesians because there are many very different scenarios in which conditioning is equally justified, and the evidence will be much less complete in some of these than in others. Professor Cohen's contention does have a kernel of truth, however, because the legitimacy of conditioning depends on the existence of ground rules for the acquisition of evidence. It is completeness relative to these ground rules that is needed. We cannot justify conditioning on B if B, instead of representing one possible body of evidence permitted by the ground rules, represents just part of such a body of evidence.

IV. The Quality of Likelihoods

Ancillary to the personalistic doctrine that new evidence should always be taken into account by conditioning is the doctrine that this conditioning should be carried out using Bayes's Theorem. It is often asserted that Bayes's Theorem provides a recipe for dealing with new evidence: determine the "likelihoods" associated with the evidence, and multiply prior probabilities by these likelihoods. This means that the likelihoods fully capture the import of the new evidence.

The idea that evidence should be assessed in terms of its likelihood has influenced even those who are hesitant to rely on full Bayesian thought experiments. We are often urged to think about the import of given evidence in terms of likelihood even if we are unable to assess prior probabilities.

The constructive interpretation of probability questions these dogmas. Bayes's Theorem is one tool for constructing probability arguments, but it is not always the best tool. There is no reason to think that probability judgments corresponding to likelihoods will always be the most meaningful or effective probability judgments we can make. In some cases, the analogy that a given likelihood judgment represents may be too poor to make an effective argument. We must seek other arguments in these cases. The arguments we find may involve only partial likelihoods, likelihood judgments based only on part of the evidence initially considered, or they may not involve likelihood ideas at all.

This section will illustrate the limitations of likelihood using as an example the problem of assessing the evidence for paternity provided by blood tests. It turns out that in this problem the most effective arguments usually do have a likelihood form, but that in some cases it is best to rely on partial likelihoods.

A. Bayes's Theorem

It will suffice for our example to consider Bayes's Theorem in its simplest form, where prior odds for a hypothesis are multiplied by a likelihood ratio for that hypothesis. A likelihood ratio in favor of A on evidence B may be denoted by L[B|A]. By definition,

$$L[B|A] = Pr[B|A]/Pr[B|\text{not-}A]$$

Figure 2

Bayes's Theorem says that,

$$O[A|B] = L[B|A]O[A]$$

Figure 3

where O[A] denotes prior odds for A, a ratio Pr[A]/Pr[not-A], and O[A|B] denotes posterior odds, Pr[A|B]/Pr[not-A|B].

B. The Probability of Exclusion

In paternity disputes, blood tests are often used to check whether a particular man could be the child's father. They are also used, when the possibility of the particular man's being the father is not excluded, to calculate relevant probabilities. These calculations are the subject of a large literature, dating back to Essen-Möller.[19]

The blood tests determine an individual's types for a set of antigens—the individual's phenotype. The law of genetics rules out certain combinations of phenotypes for mother, child, and father. It is possible, therefore, that by testing a mother, child, and alleged father we can exclude the possibility that he is the real father. If this does not happen, then it would seem that the evidence against him has been strengthened. How can this be measured?

One simple approach is the following.[20] Suppose the phenotypes of the mother and child have been determined but that the alleged father has not yet been tested. We calculate a probability p that he will be excluded, given that he is not the real father. If p is nearly one and yet, after testing, the man is not excluded, then p can be thought of as a measure of the doubt cast on his not being the father.

This is equivalent to a likelihood argument because the ratio $p/(1-p)$ is approximately equal to a likelihood ratio. Indeed, if B denotes the event that the alleged father is not excluded, and A denotes the event that he is the father, then Pr[B|not-A], calculated with the phenotypes of the mother and child fixed, is equal to $1-p$. And since $Pr[B|A] = 1$, the likelihood ratio

$L[B|A] = Pr[B|A]/Pr[B|\text{not-}A]$ is equal to $1/(1-p)$. If p is near 1, this is approximately equal to $p/(1-p)$.

How do we calculate p, or equivalently, $Pr[B|\text{not-}A]$? Here we must rely on the fact that large numbers of individuals from different populations have been tested, and tables of the frequencies of different phenotypes have been constructed. The populations are usually racial groups, because the frequencies of different phenotypes do differ for these groups. If the alleged father belongs unequivocally to a racial group for which frequencies of phenotypes have been recorded, then we add up the frequencies for that racial group of all the phenotypes that are inconsistent with his being the father, given the known phenotypes of the mother and child. This total frequency is the probability that the alleged father will be excluded, given that he is not the father, and assuming that his phenotype is drawn at random from the phenotypes of his racial group.

When we talk about the alleged father's phenotype being drawn at random from the phenotype of his racial group, we invoke a thought experiment. The alleged father is not a random man; he is a particular man in a dispute marked by many other particulars. But before we have typed his antigens, assuming he is not the father, our only knowledge concerning his phenotype is provided by frequencies for his racial group. Thus, we can compare our state of knowledge to what it would be if we knew his phenotype were drawn at random from a population having these frequencies. This comparison is the basis for the thought experiment. We imagine how surprised we would be and how much doubt would be cast on the randomness of the drawing if the phenotype drawn were from a subset singled out beforehand as having a very small total frequency. We argue that failure to exclude the alleged father casts similar doubt on the hypothesis that he is not the real father.

In this thought experiment, the phenotype of the mother and child are taken as fixed, and we imagine the phenotype of the alleged father being drawn at random. An obvious alternative is to imagine instead that the phenotypes of all three people—a mother and child and an unrelated man—are drawn at random. Or we might fix the phenotype of the alleged father and imagine those of the mother and child being drawn at random. The difficulty with these latter thought experiments is that they cannot be performed without a further assumption. In order to specify the probability with which a mother-child pair of phenotypes will be drawn, we must specify the racial group of the real father, and in order to draw our analogy to a game of chance, we must think of the real father as being drawn at random from this group. The strength of the thought experiment in which the phenotypes of the mother and child are held fixed lies in its simplicity. It assumes only that the phenotypes of the mother and child have no causal connection with the phenotype of the alleged father if he is not the real father, and it requires only that we think of this alleged father's phenotype as random.

There are situations, of course, in which we would question even a thought experiment in which the phenotypes of the mother and child are fixed and the phenotype of the father is drawn at random. If the alleged father is related to the mother or to the real father, then there is a causal connection between his phenotype and the phenotypes of the mother and child, and the thought experiment loses its cogency.

C. *A Better Argument*

It turns out that there is a cogent thought experiment that accounts for more than the mere fact that the father is not excluded. This thought experiment accounts for the particular non-excludable phenotype the alleged father turns out to have. Since the probability of this particular phenotype will be small whether the man is the real father or not, accounting for it will involve not just the contemplation of one small probability, but also the comparison of two small probabilities. This requires considering their ratio, which is a likelihood ratio.

It will be convenient, in the exposition of this more inclusive probability argument, to use a notation that makes explicit the fact that we are holding the phenotypes of the mother and child fixed. Let M, C, and S denote the phenotypes of the mother, child, and alleged father, respectively, and let A again denote the event that the alleged father is the real father. The argument I suggest compares two probabilities, Pr[S|M&C&A] and Pr[S|M&C&(not-A)].

To calculate Pr[S|M&C&(not-A)], we use the same thought experiment as before. We imagine determining the father's phenotypes by drawing a phenotype at random from the population of phenotypes for his racial group. This requires taking Pr[S|M&C&(not-A)] as the frequency of S in that group. Since we are assuming that the alleged father is unrelated to the mother and is not the child's real father, M and C are irrelevant.

The calculation of Pr[S|M&C&A] is more complicated. Here we assume that the alleged father is the real father, so the child's phenotype is relevant. The connection between the phenotypes of parent and child can only be understood, however, in terms of their genotypes. An appropriate thought experiment must consider the distribution of genotypes in the racial groups of the mother and alleged father. Fortunately, these distributions can be estimated from the distributions of phenotypes.[21] We consider the population of triples of genotypes that result from drawing genotypes for the mother and father at random (still assuming they are unrelated), and then crossing these two genotypes to obtain the genotype for a child. We then consider the subset of this population consisting of those triples of genotypes that give the observed phenotypes for mother and child. We imagine determining the father's genotype by drawing at random from this subset; Pr[S| M&C&A] is the frequency in this subset of genotypes for the father that give the phenotype S.

In this thought experiment, we are accounting for more than the mere fact that the alleged father is not excluded. We are using more genetic theory and more detailed statistical data. Assuming we have the data, the fact that we are taking it into account makes this a better thought experiment and a stronger probability argument.

The ratio Pr[S|M&C&A]/Pr[S|M&C&(not-A)] is widely used; it is often called the "paternity index" in the literature. If it is very large, we say that the alleged father's phenotype is much more likely if he is the real father than if not, and we construe this as a strong argument for his being the real father.

Writing Pr[S|M&C&A]/Pr[S|M&C&(not-A)] for the paternity index makes it explicit that we are holding the phenotypes of the mother and child fixed in the thought experiment. If we are content to leave this implicit, then we can write Pr[S|A]/Pr[S|not-A] for the index. This makes it clear that it is a likelihood ratio; it is the likelihood ratio L[S|A], calculated by holding the phenotypes of the mother and child fixed.

D. *A Yet More Inclusive Likelihood*

It might be argued that the thought experiment of the preceding section does not yet account for all the evidence blood tests provide. When we compare Pr[S|M&C&A] with PR[S|M&C&(not-A)], we are accounting for the phenotypes M and C to some extent—we are considering them fixed in the thought experiment--but we are not taking into account that M and C may suggest genotypes for the real father that are more common in some racial groups than others. This point becomes clear when we recognize that the total evidence provided by the three phenotypes S, M, and C would be fully accounted for if we could calculate joint probabilities for these phenotypes under the two hypotheses, A and not-A. Thus, instead of considering only the ratio

$$Pr[S|M\&C\&A]/Pr[S|M\&C\&(not\text{-}A)]$$

Figure 4

we should also consider the ratio

$$\Pr[S\&M\&C|A]/\Pr[S\&M\&C|\text{not-}A] =$$
$$[\Pr[M\&C|A]/\Pr[M\&C|\text{not-}A]][\Pr[S|M\&C\&A]/\Pr[S|M\&C\&(\text{not-}A)]]$$

Figure 5

In order to calculate this ratio, we need thought experiments that will yield Pr[M&C|A] and Pr[M&C|not-A]. The probability Pr[M&C|A] is obtainable from the thought experiment already used to calculate Pr[S|M&C&A]. We imagine genotypes for the mother and alleged father being drawn at random from their racial groups and then crossed to produce the genotype of the child, and calculate the probability that the phenotypes M and C will result from this process. But the probability Pr[M&C|not-A] poses problems. A thought experiment for this probability must specify the racial group of the real father, or at least specify a probability distribution for his racial group. If we assume that the real father belongs to the same racial group as the alleged father, then Pr[M&C|not-A] = Pr[M&C|A], and Figure 5 reduces to Figure 4. But if we cannot confidently assume this, then we probably will not be able to construct a convincing thought experiment for the selection of the real father's racial group.

In view of the difficulties encountered in carrying out the thought experiment represented by Figure 5, we should not necessarily prefer it to the one represented by Figure 4. The former does try to take more evidence into account, but it does not always succeed. A lawyer trying to construct the strongest possible argument against the alleged father might be better advised to rest content with a large ratio for Figure 4 than to weaken her argument by resting it on a questionable thought experiment about the racial group of the real father.

This, of course, leaves a possible role for Figure 5 in rebuttal. The defense, wishing to discredit the argument based on Figure 4, may be able to suggest a plausible racial group for the real father for which Pr[M&C|not-A] is much greater than Pr[M&C|A], resulting in a low value for Figure 5. If the racial group is only plausible, then this thought experiment might not count as a direct argument against the allegation of paternity, but it would effectively rebut the plaintiff's argument based on Figure 4.

E. *A Bayesian Thought Experiment*

A full Bayesian thought experiment would go one step further than the thought experiment represented by Figure 5. It would multiply that likelihood by prior odds for paternity based on other evidence such as the mother's allegation, the man's denial, and evidence bearing on the trustworthiness of their respective testimonies.

Some suggest that only such a full Bayesian thought experiment is truly useful because other thought experiments do not yield a number properly labelled "the probability of paternity." From the constructive viewpoint, this is unconvincing. The Bayesian thought experiment does not yield *the* probability of paternity in any absolute sense. It is merely one argument; other thought experiments may be more persuasive arguments, even though on the surface they address slightly different questions. In fact, full Bayesian arguments in paternity cases are usually weaker than arguments based solely on likelihoods from blood tests because the analogy to the picture of chance using the evidence from the blood tests is stronger than the analogy using other evidence.

F. *Where Do We Stop?*

We have studied a series of successively more complex thought experiments. Each takes more evidence into account, but by doing so risks weakening the analogy to the picture of chance. Where in this progression should we stop? When is the advantage of taking more evidence into account outweighed by the disadvantage of treating it less convincingly? There is no general answer; we must consider the circumstances of the individual case.

Some readers, although willing to stop short of the full Bayesian argument, will insist that the only reasonable stopping point is the argument that calculates the full likelihood—the likelihood given by Figure 5. These readers should recognize that there really is no such thing as a full likelihood because the evidence to be taken into account by the likelihood is not well-defined. We first studied a likelihood that accounted only for the fact that the alleged father was not excluded, then a likelihood that accounted for his phenotype, then a likelihood that accounted for all three phenotypes—the mother's, the child's, and the alleged father's. Why stop there? Why not consider a likelihood that accounts for all three phenotypes plus either the alleged father's denial, or some other piece of evidence introduced in court?

Shafer and Tversky point out that, outside of the realm of planned statistical experiments, problems do not come with the evidence on which likelihoods are to be based distinguished from the evidence on which prior probabilities are to be based.[22] This partitioning of the evidence must be done deliberately by the person who constructs the Bayesian argument. The standard terminology encourages a pretense that the partitioning is determined by timing: likelihoods are based on new evidence, priors on old evidence. But this is only pretense. It is clear in our example that it is pretense because the blood tests are unlikely to be the last evidence we obtain, no matter whose shoes we are in.

G. The Relevance of Likelihood

Several authors, including Professor Lempert[23] and Professor Kaye,[24] have advanced likelihood as a general tool for explicating and correcting legal doctrine. Professor Lempert, for example, explains judgments of relevance in terms of likelihood ratios. Given evidence is logically relevant, he explains, unless its likelihood ratio is close to one. Such evidence may be excluded, nonetheless, from a jury's consideration if the jury is ill-equipped to estimate this likelihood.

This reliance on likelihood is misguided because it overlooks the constructive nature of likelihood. A cogent likelihood argument does establish the relevance of the evidence it uses, but if we have not succeeded in constructing a cogent likelihood argument, then there is no content in talk about whether the likelihood ratio is close to one, and we are free to consider other ways of establishing the relevance of the evidence.

Professor Lempert's references to difficulties that a judge or jury might have in estimating a likelihood suggest that a likelihood, as opposed to a prior probability, has objective reality. Even in the case of blood test evidence, this is not necessarily so. If we cannot identify a population from which we can say the observed phenotypes were randomly drawn, then our likelihoods are simply meaningless symbols.

We might defend a reliance on likelihood by appealing to Savage's personalistic interpretation, according to which an individual must always have both subjective prior probabilities and subjective likelihoods.[25] We could then say that given evidence is relevant if any reasonable juror can give it a likelihood ratio different from one. This approach leads, however, to all the difficulties associated with the personalistic pretense to all-encompassing probability opinions. Moreover, it leads to the paradoxical result that evidence is relevant if it is weak. When we know little, almost any likelihood ratio may seem reasonable.[26]

V. Is There a Theory of Argument Construction?

If Bayes's Theorem is not a general recipe for constructing probability arguments, how do we construct them? Can we develop a theory that tells how to do so? This question has gained prominence in recent years because computer scientists have begun trying to incorporate the ability to construct probability arguments into expert systems.[27] Their initial efforts have had limited success, but the effort itself represents an important challenge to statisticians and probabilists.

Genuine progress towards automating the construction of probability arguments will depend in part on progress in the construction of artificial associative memories.[28] Human probability judgment, poor as it is, depends on an ability to retrieve memory of situations that are fairly similar to a given situation. It also depends on an ability to evaluate the relevance of the instances retrieved and to adjudicate between conflicting instances and analogies. How can we mimic this human ability?

It is here that the courtroom metaphor may have its greatest impact. The adversary system, the system the law has evolved to deal with conflicting arguments, may turn out to be an essential element of automated probability argument.

NOTES

† © 1986 by Glenn Shafer.

* Professor of Statistics, University of Kansas School of Business. The author has benefited from conversations with Mikel Aickin, Paul Meir, and Howard Stratton. Research for the article has been partially supported by NSF grant IST-8405210.

[1] Personal communication with Lorraine Daston (1984).

[2] J. BERNOULLI, ARS CONJECTANDI (Basel 1713).

[3] Garber & Zabell, *On the Emergence of Probability*, 21 ARCHIVE FOR HIST. EXACT SCI. 33 (1979).

[4] Shafer & Tversky, *Languages and Designs for Probability Judgment*, 9 COGNITIVE SCI. 309 (1985).

[5] *See, e.g.*, Schum, *The Behavioral Richness of Cascaded Inference Models: Examples in Jurisprudence*, in 2 COGNITIVE THEORY 149 (1977).

[6] *See, e.g.*, H. SOLOMON, MEASUREMENT AND THE BURDEN OF EVIDENCE (1983).

[7] L. COHEN, THE PROBABLE AND THE PROVABLE (1977).

[8] Ekelöf, *Free Evaluation of Evidence*, 8 SCANDINAVIAN STUD. L. 45 (1964); *see also* Ekelöf, *Beweiswert*, in FESTSCHRIFT FÜR FRITZ BAUER 343 (W. Grunsky, R. Sturner, G. Walter, and M. Wolf eds. 1981); Ekelöf, *My Thoughts on Evidentiary Values*, in EVIDENTIARY VALUE: PHILOSOPHICAL, JUDICIAL AND PSYCHOLOGICAL ASPECTS OF A THEORY 8 (P. Gardenfors, B. Hansson, and N. Sahlin eds. 1983).

[9] Shafer, *Lindley's Paradox*, 77 J. AM. STATISTICAL A. 325 (1982).

[10] L. SAVAGE, THE FOUNDATIONS OF STATISTICS (1954).

[11] *See, e.g.*, M. FINKELSTEIN, QUANTITATIVE METHODS IN LAW (1978); R. BENDER & A. NACK, TATSACHENFESTSTELLUNG VOR GERICHT (1981).

[12] *See* Savage, *Elicitation of Personal Probabilities and Expectations*, 66 J. AM. STATISTICAL A. 783 (1971).

[13] *See* Shafer and Tversky, *supra* note 4.

[14] *See generally* G. SHAFER, A MATHEMATICAL THEORY OF EVIDENCE (1976); *see also* Shafer, *supra* note 9; Shafer, Belief Functions and Possibility Measures, to be printed in 1 THE ANALYSIS OF FUZZY INFORMATION (J. Bezdek ed. 1986).

[15] Freund, *Puzzle or Paradox?*, 19 AM. STATISTICIAN no. 4, 29, 44 (1965).

[16] See Shafer, *Conditional Probability*, 53 INT'L STATISTICAL REV. 261 (1985).
[17] See id.
[18] See Speed, *Discussion of* Conditional Probability, *by Glenn Shafer*, 53 INT'L STATISTICAL REV. 276 (1985).
[19] Essen-Möller, *Die Beweiskraft der Ähnlichkeit im Vaterschaftsnachweis, theoretische Grundlagen*, 68 MITT. ANTHROP. GES. 9, 598 (1938). For an excellent review, written primarily for statisticians, see Berry & Geisser, *Inference in Cases of Disputed Paternity* (1982) (Technical Report No. 404, available at the Department of Statistics at the University of Minnesota).
[20] See Lee, *Estimation of the Likelihood of Paternity*, in PATERNITY TESTING 28 (H. Polesky ed. 1975); *see also* Weiner, *Likelihood of Parentage*, in PATERNITY TESTING BY BLOOD GROUPING 125 (M. Sussman 2d ed. 1976).
[21] See Berry & Greisser, *supra* note 19, at 3-8.
[22] See Shafer & Tversky, *supra* note 4.
[23] Lempert, *Modeling Relevance*, 75 MICH. L. REV. 1021 (1977).
[24] Kaye, *Probability Theory Meets Res Ipsa Loquitur*, 77 MICH. L. REV. 1456 (1979).
[25] See Savage, *supra* note 10.
[26] I discuss this point in more technical detail elsewhere. See G. Shafer, *Comment*, 66 B.U. L. Rev. 629 (1986).
[27] *See, e.g.*, B. BUCHANAN & E. SHORTLIFFE, RULE-BASED EXPERT SYSTEMS: THE MYCIN EXPERIMENTS OF THE STANFORD HEURISTIC PROGRAMMING PROJECT (1984).
[28] *See generally* T. KOHONEN, SELF-ORGANIZATION AND ASSOCIATIVE MEMORY (1984).

Glenn Shafer,
Professor of Statistics,
University of Kansas School of Business.

DAVID A. SCHUM*

BEATING AND BOULTING AN ARGUMENT†

Anyone who reads scholarly works on evidence law will eventually encounter Sir Matthew Hale's assertion that our adversarial legal system—involving the parties in contention, their counsels, and the tribunal—"beats and boults out the truth."[1] Professor Shafer argues, essentially, that it also takes a fair amount of "beating and boulting" to construct a defensible and useful probabilistic representation for many frequently encountered evidential ingredients of an inferential problem, regardless of context. It is certainly true that no evidence comes to us with established "relevance" or "weight"; such attributes have to be developed on the basis of careful argument, the construction of which is a creative act. In the process, we will not always be sure when we have constructed the most useful or defensible argument. Shafer tells us that the necessity of this "beating and boulting" process is commonly unrecognized in many applications of probability, some of which involve Bayes's Theorem. In such applications, it may appear that the person formulating an inference problem supposes that the relevance or weight of the evidence, or both, have objective reality when, in fact, they must be established according to judgmental arguments in which there is always some element of arbitrariness. Shafer advocates a widespread use of "courtroom metaphors" in constructing the arguments that underlie the assessment and combination of probabilities in inference tasks.

Since I agree wholeheartedly as to the usefulness of "courtroom metaphors" in analyzing probabilistic inference, I will confine my remarks to a few quibbles and some additional examples of the argument structural difficulties Shafer mentions.

Arguments, Probability Judgments, and "Thought Experiments"

Shafer tells us that none of the common interpretations of probability emphasizes the constructive character of probability judgment; all interpretations fail to acknowledge the roles played by human judgment and creativity. To make interpretations of probability sensitive to these judgmental ingredients, it would be helpful to know something about how people actually construct probability arguments as part of their judgmental process. Shafer reports a suggestion he and psychologist Amos Tversky have recently offered about how people may approach the assessment of probabilities. They argue that Bayesian probability judgments, for example, are made by matching an actual problem to an imagined list of canonical examples involving well-defined random sampling or other operations in which there are known probabilities. Another way of specifying this matching operation is to say it involves a "thought experiment" in which the evidential ingredients of an actual problem are imagined to have been generated by random sampling operations from well-defined reference classes. An articulation of such a thought experiment might indeed offer a reasonable argument for the adjudged probability, whatever it is. It is also likely that parts of such arguments may be more persuasive than others. However appealing such an argument is, though, we should take care to note that it rests upon an imaginative process and has no factual basis.

Professor Shafer provides a good example of a Bayesian thought experiment involving disputed parentage.[2] This example reveals the subjective quality of all ingredients of Bayes's Theorem. Some Bayesian statistical scholars allege that prior probabilities are subjective but that likelihoods based on frequentistic or actuarial information are objective. Shafer's example illustrates how having an actuarial basis for likelihood assessments does not make them any less subjective. He argues that different possible thought experiments arise when we contemplate different and possibly finer distinctions in our data; which distinctions we choose to consider is a matter of subjective judgment. One problem here is that our distinctions may easily outrun our data. We wish to take more things into account, but can only do so less convincingly. Each distinction we make defines a new dimension for partitioning our data. If we add enough dimensions we may soon discover that most of the cells in our partition have no data points unless we have a prodigious amount of data that are, in fact, "measured" on all the dimensions we wish to consider.

Now suppose we consider events, so common in trials, for which there are no actuarial records to support probability assessments. We could easily get a person's probability judgments about these events, and would, without hesitation, label such judgments "entirely subjective." If asked how this person made the judgment, we might say that it was based on intuition, educated by appropriate background knowledge. One problem, of course, is that the probability judgment required may involve distinctions that exceed this person's background knowledge, just as the actuarially based assessments just discussed did. But Shafer notes yet another problem. We must not assume that a person's background knowledge is necessarily well defined; some elements of this background may be better defined than others.

So when a person makes a probability judgment, we would be safer saying it was based on whatever background knowledge the person was able to retrieve.

Concerning background knowledge, which may include specific prior evidence or previously encountered evidence in the problem at hand, I find it useful to draw an analogy between probability judgment and the background or "contrast" phenomena one observes in human color and brightness perception.[3] It is well established that the perceived color or brightness of an object depends, in part, on the characteristics of the background against which the object is presented. Some features of this background are more salient than others in inducing contrast enhancement. Similar, often subtle, contrast effects are at work in probability judgments. One such subtlety involves what may be termed "reciprocal" contrast. As the color and brightness of a background may influence the perceived color and brightness of an object, so does the object itself influence the perceived color and brightness of elements of the background. Similarly, the perceived value of elements of an evidentiary "background" may change in light of some new or current evidence. Thus, I agree with Shafer about the need for care in discussing how background knowledge influences probability judgment; we cannot simply assume that this background is always well defined or the elements in it equally salient.

Shafer's comments about thought experiments, in which actual problems are matched with canonical examples, are much better as a guide for students of probability in thinking about the usefulness of alternative probability systems than as a descriptive statement about how people "actually" behave when asked to assess a probability. In fact, it is difficult to imagine any untutored person actually doing such matching, particularly in situations involving the complex masses of evidence so frequently encountered in trials. It is also difficult to imagine what empirical research operations it would take to make us reasonably confident that such matching behavior was taking place. I believe people do have *feelings* of uncertainty, doubt, and related attributes. When prompted, they can express those feelings in any of a variety of ways involving betting, sampling, or other metaphors. Often such methods of expression are chosen by experimenters, and may be quite unnatural to the person whose judgments are being sampled. Thus, thought experiments and canonical examples, like likelihoods and other probabilistic ingredients, have a fictional quality, but they may still be very useful fictions on occasion.

On the Acquisition of Evidence and the Representation of "Knowledge"

Shafer's works on probabilistic reasoning are appealing to many behavioral scientists interested in human reasoning skills. Those who have not heeded Shafer's work have often reached insupportable conclusions about human reasoning skills. Shafer's comments about the importance of how evidence is acquired contain a message for all of us interested in inferential behavior. Shafer employs a conundrum called "Freund's Puzzle" to illustrate the necessity, in the Bayesian conditioning process, of specifying the "ground rules" by which the conditioning events have arisen.

There are many such puzzles, some of which convey the same message as Shafer's favorite. My favorite is the "Prisoner's Puzzle."[4] I state the message conveyed by such examples slightly differently than Professor Shafer does, but the messages themselves are quite similar. Commonly, the expression $P(H|D)$ is read, "the probability of H conditional on 'knowledge' of D." In the conventional probability system $P(H|D)$ is *defined* as $P(H$ and $D)|P(D)$. The trouble is in the difficulty of determining from some statement describing D exactly what "knowledge" has been conveyed. Superficially similar statements may appear to provide the same knowledge even when they do not. Apparently minor changes in the wording of D can make all the difference. There is certainly nothing about the conditioning process itself that will tell us which alternative statement of D actually represents the matter of concern. Here is a simple example. More than one Bayesian study has been muddled by failure to distinguish between two conditioning events like the following: E = the event that defendant ran away from the house in which the crime was committed, and E^*_i = the event that Witness W_i testifies that the defendant ran away from the house in which the crime was committed. To equate E^*_i and E is to assume that W_i is a perfectly credible witness.

I believe there are antidotes to at least some of the difficulties Shafer attributes to Bayesian formulations.[5] The antidote comes in the form of meticulous structuring of the reasoning linkage between what is taken as evidence or knowledge and what is to be inferred. This is one reason I take Wigmore so seriously. He has given probabilists at least the beginnings of a systematic way of clarifying what we perceive these linkages to be. Careful structuring of an argument from evidence to major hypotheses or facts in issue uncovers what one believes are major sources of uncertainty; in the process, the quality of the evidence can be revealed.

There is a message here for behavioral scientists who stake their claims about human inferential competence on comparisons between judgments of probabilites and calculations of these probabilites using conventional canons. As Shafer notes, the conditioning process underlying these calculations is quite arbitrary and we might well inquire why anyone should be expected to behave in accordance with them. As he further notes, however, there are instances in which this conditioning process can be put on a more solid basis. These instances involve construction of a probability argument in which the ground rules underlying the acquisition of evidence can be specified. The "courtroom metaphor," with its focus on the careful structuring of argument from evidence to hypotheses or facts in issue, assists in the process of specifying these ground rules. I add only that these ground rules may be remarkably complex, even for apparently "simple" evidentiary situations.

On the Quality of Likelihoods and Likelihood Ratios

Shafer applies some Bayesian thought experiments to inferential problems concerning disputed parentage. In the process, he constructs various likelihoods corresponding to various arguments we could make about the importance of certain kinds of evidence bearing on conjectured parentage. Some of these thought experiments and the arguments they suggest turn out to be more useful than others. As we know, the likelihood and likelihoods ratio ingredients of Bayes's Theorem concern, within the conventional probability calculus, the weight, value, or importance of our evidence. Shafer's message here is that establishing evidentiary weight by these ingredients is a subjective process, and that different assessments of evidentiary weight are possible depending on how we construct our arguments that underlie the assessments. In short, there is no objectively unique "weight" to assign any evidence.

I was surprised, and a bit disappointed, to discover that Shafer mentions almost nothing about his own belief function system. This system certainly has much to say about the process of assessing the weight of evidence, so I will add a bit of Shaferian wisdom to a Bayesian thought experiment, which involves hearsay evidence.

At the risk of appearing immodest, I believe I have spent at least as much time as anyone alive constructing and pondering likelihoods and likelihoods ratios. I have done so for every species and combination of evidence I could lay my hands on. Subject to the constraints Shafer brings out in his Bayesian thought experiments, the formal process of constructing and analyzing likelihoods and their ratios can be very informative about the process of establishing the value of evidence. But in studying these representations, one encounters other difficulties, some of which involve evidential problems

Shafer mentions in his other works.[6] I will discuss a difficulty concerning instances in which we might reasonably say that we have a *lack of belief* rather than *disbelief*.

Hearsay evidence comes to us from a chain of sources and, in many instances, we have contact with only one of these sources. For example, A allegedly tells B about some event, and then B tells us what A is supposed to have said. Often we have no contact with A. Even longer hearsay chains are possible, as we all discover watching television newscasts. We may hear something like the following: the newscaster (A) says "our correspondent (B) reports that a Pentagon source (unidentified) told her (B) that . . . ." Here we have three sources—A, B, and an unidentified person. In evaluating the significance of the substance of this report we must consider the credibility of all three sources. Though we might have some opinion about the credibility of A and B, we probably have none at all about the unnamed source since we have no evidence about this source other than that he or she works at the Pentagon.

It is possible, though tedious, to construct Bayesian thought experiments for hearsay evidence. Obviously, the chain of reasoning involved is catenated or cascaded. Some of the likelihood ingredients we encounter in the process concern the credibility of sources in this hearsay chain. They appear as conditional probabilities of the general form, $h = P(E^*|E)$ and $f = P(E^*|E^c)$, where E^* is the report from a source or witness that event E occurred. In most cases, there are many other conditioning events besides E and E^c (not-E) to account for. In our hearsay example, there are h and f ingredients for each of the three sources. Now, most of us would imagine that hearsay evidence coming from a chain of sources, one of which is unidentified or about which we know nothing at all, would have little inferential value. This hearsay evidence might suggest some new possibility, but still have little weight. In a Bayesian formulation for this hearsay evidence we can effectively destroy its value by setting $h = f$ for any one of the sources. This says we believe the source's report E^* is just as likely if E—the event reported—occurred as if E did not occur.

There is a problem with a Bayesian formulation for this hypothetical. Consider the h and f values for our unidentified (Pentagon) source about whom we know nothing. How do we indicate that we "know nothing" about this source? We say to ourselves, "Since I know nothing about this source, the newscaster's report can usually have little, if any value, and so I ought to set $h = f$ for the unnamed source. This will ensure that the newscaster's report will have no value." The trouble is that all sets of equal h and f values are equally effective in destroying the report's value. It happens that different pairs of equal h and f values have decidedly different meaning. Suppose we set h = f = 0.99. This says the source has a distinct bias *in favor of* reporting E; this source will report E* with high probability regardless of whether E occurred. Suppose, instead, we decide to set h = f = 0.05; here the source apparently has a distinct bias *against* reporting E*, since we believe his or her report has low probability regardless of whether E occurred. Thus the inconsistency becomes apparent. Setting h and f equal corresponds with some particular knowledge we claim to have about the behavior of this unnamed source, but, in fact, we have no knowledge at all about this source. Since we have no evidence about this unnamed source, we have no basis for a belief about the credibility of this source. I can see no way to capture this entirely plausible scenario within Bayesian formulations.

In my own contribution to this book,[7] I acknowledged the debt owed Glenn Shafer for his insights about the inferential tasks of concern to us all. His paper for this symposium increases our debt. I now find myself describing my own work in Shafer's terms—as "thought experiments" involving particular arguments. I acknowledge that any argument I might construct for some purpose has been selected from among many other possible arguments, some of which do not necessarily find ready expression in "Bayesian" thought experiments. We must all undertake the process of "beating and boulting" persuasive and useful probabilistic representations for evidential concepts. This process, which rests upon our imagination, is difficult and always arbitrary to some degree. These are Shafer's most valuable contributions to this symposium.

NOTES

† © 1986 by David A. Schum.

* Professor of Systems Engineering, George Mason University.

[1] SIR MATTHEW HALE, THE HISTORY OF THE COMMON LAW OF ENGLAND 164 (C. Gray ed. 1971) (1739).

[2] We need not consider only Bayesian thought experiments. We could, with equal facility, consider thought experiments, like those of Shafer-Dempster, that involve different scales of canonical examples.

[3] Schum, *Contrast Effects in Inference: On the Conditioning of Current Evidence by Prior Evidence*, 18 ORGANIZATIONAL BEHAV. & HUM. PERFORMANCE 217, 217-19 (1977).

[4] P. PFEIFFER & D. SCHUM, INTRODUCTION TO APPLIED PROBABILITY 89 (1973).

[5] These difficulties may well relate to any system of probability, not just Bayesian formulations.

[6] G. Shafer, A Mathematical Theory of Evidence § 7 (1976).
[7] Schum, *Probability and the Processes of Discovery, Proof, and Choice, infra* at 213.

David A. Schum,
Professor of Systems Engineering,
George Mason University.

DAVID A. SCHUM*

PROBABILITY AND THE PROCESSES OF DISCOVERY, PROOF, AND CHOICE†

I. Human Inference "On Stage"

It does not seem inappropriate to think of a pattern of inferential behavior, as it develops in legal contexts and elsewhere, as a stage play in which the plot is often very complex and is frequently made up as the play proceeds. Inferential "plays" often have many actors in different roles; in law-related matters there are, at least, investigators, advocates, and factfinders. On occasion, several or many of these actors may appear on stage at the same time. In some inferential plays there are discernable acts and scenes involving the processes of "discovery," "proof," and "choice." When matters are resolved in a court of law, these acts and scenes are played out in a more or less well-ordered sequence. In other inferential contexts (in medicine and science, for example), two or more of these acts or scenes may be "on stage" at the same time or they may recur in cycles. Some inferential plays concern events in the past; others concern events that may happen in the future. The latter cases involve particularly difficult problems because the world continues to change while we are trying to understand it well enough to make sensible predictions or inferences about the future.

It is fair to ask who or what should give *direction* to the complex forms and combinations of human inferential activity in such plays. In jurisprudence, procedural and other rules provide at least some direction for these complex patterns of inferential behavior. However, not being entirely content with the direction afforded by experience, some scholars in jurisprudence have sought additional direction from the works of scholars of inference, probability, and related matters. Some of these legal scholars may now regret having advertised for directorial assistance from external sources. Who would have predicted that evidence scholars would be asked to read about "likelihood ratios," "belief functions," and "inductive" or even "fuzzy" probabilities? Perhaps Wigmore was wrong to have openly solicited the assistance of logicians and others in developing canons of reasoning for instances in which conclusions must be drawn from a large mass of fallible evidence.[1]

Strong reaction against initial efforts by some evidence scholars to obtain such direction was not effective in preventing further efforts by others. For example, Finklestein and Fairley[2] actually obtained the assistance of the

Reverend Bayes for direction in one inferential scene in which the finders of fact must (apparently) aggregate probabilities. Speaking for the *producers* of inferential plays in jurisprudence, i.e., society, Professor Lawrence Tribe[3] argued that scenes under such direction, even if provided by a man of the cloth, would offend public standards of decency. In spite of Tribe's arguments against the formalization of evidentiary processes, a few evidence scholars[4] continued, quite openly and shamelessly, to suggest that Bayesian canons could be helpful in providing direction in at least some inferential scenes of interest in jurisprudence.

Since these earlier times, we have seen opened a Pandora's Box from which a swarm of possible systems for the "direction" of inferential plays in jurisprudence has emerged. The current "probability debates" among scholars of evidence concern, in part, which of these possibilities merit serious consideration. There is no shortage of possibilities; for the first time we have a variety of well-articulated formal systems for representing various attributes of inference and inference-related tasks. The major "schools" of inferential "direction" currently being discussed include:

(1) The Pascal/Bayes School of Probability and Uncertainty,
(2) The Bacon/Mill/Cohen School of Inductive Probability,
(3) The Shafer/Dempster School of Non-additive Beliefs,
(4) The Zadeh School of Fuzzy Probability and Inference, and
(5) The Scandinavian School of Evidentiary Value.

Assessing the potential contribution of these "schools" of inferential direction is no small task. Each seems to have its own language; as some will be quick to point out, it is often quite impossible to make translations from one language into another. Given such a difficulty, it is natural to inquire about the extent to which ideas from these different schools actually involve the same inferential matters. Do these different schools concern different attributes of human inference or do they look at the same attributes in different ways? Might ideas from these schools provide some direction for certain inferential acts/scenes but not for others? Answers to such questions require examination of the entire performance (i.e., all the acts and scenes) of inferential plays as well as consideration of the inferential roles played by the various actors who "have their entrances and their exits" during the performance.

Having some curiosity about such matters and considerable interest in the current "probability debates," I thought it might be useful to examine these various views of probability and inference when we do consider the entire performance of an inferential play and the behavior of the various actors in it. These remarks concern some of the things I observed when I performed such an examination. Being identified with the Pascal/Bayes school, I cannot claim to have been completely unbiased when I began this task. Perhaps I ought to begin by telling you a bit about what my expectations were when I began this examination.

A. *Not A Reward But A Summons*

The date October 15, 1978, is quite significant for me. Prior to this date it is fair to say that I was overflowing with confidence about the ability of the Pascal/Bayes school of probability and inference to capture for study and analysis a wide array of the evidential subtleties apparent in the task or "act" of factfinding. I had even written a piece[5] on how Bayesian likelihood-ratio formulations for the "weight of evidence" are "behaviorally rich" in the sense that they capture a wide assortment of evidential subtleties of long-standing interest in jurisprudence. By the way, it was on or about October 15, 1978, when I received in the mail a copy of Jonathan Cohen's book *The Probable And The Provable*,[6] which I had been asked to comment upon for the University of Michigan Law Review.[7]

So, when this book arrived, I quite expected to be rewarded by the discovery that someone else had observed the "richness" of Bayesian canons in representing various evidential and other inferential subtleties as they appear in reasoning from evidence in a trial at law. Never was human expectation more seriously misguided. What I received instead was a summons to answer the multiple-count charge Cohen was presenting against the entire Pascal/Bayes school. (To this day, I remain unsure about whether Cohen actually considers these charges against us to involve criminal or civil misconduct.) In the act of preparing a defense against these charges, which I did later present,[8] I had to admit that I would face some difficulties. One charge concerned faulty advertising: propagation of the idea that the Pascal/Bayes school was the only one having adequate credentials for directing inferential plays. Other embarrassing charges involved the apparent innocence within the Pascal/Bayes school of the importance of evidentiary completeness; still others involved assorted difficulties associated with the manner in which probabilities are defined and combined on the Pascal/Bayes view. As far as inferential plays in jurisprudence are concerned, Cohen argued that experience-given direction for these plays conforms not to the canons of Pascal/Bayes, but to canons stemming from Bacon and Mill.

Upon reading *The Probable And The Provable*, my general expectations about the Pascal/Bayes school were somewhat softened and so, a short time later, when I began to read Glenn Shafer's book *A Mathematical Theory Of Evidence*,[9] I was not altogether surprised to learn that someone else was bringing action against Pascal, Bayes & Co. Some of Shafer's charges sounded to me quite similar to those mentioned by Cohen; but he added some new ones as well. As I read Shafer's work, I was especially taken by his acknowledgment of the importance of an ingredient frequently overlooked in much formal and empirical research on human inference: human imagination. From evidence we imagine or discern new and more differentiated possibilities; in turn, we imagine or discern additional categories of potentially relevant evidence and perceive the significance of evidence in different ways.

A third person bringing action against the Pascal/Bayes school has other concerns; he is Lotfi Zadeh, a person who has achieved an international reputation for his very precise thoughts about imprecise or "fuzzy" matters. Since 1965, Zadeh has been arguing that all of us must cope with concepts having fuzzy rather than distinct boundaries and that our discourse about these concepts is necessarily imprecise.[10] This is certainly true in many inference and decision tasks. Zadeh shows us how to grade our imprecision in rather precise ways and how to aggregate or combine these gradings across combinations of imprecise statements. In the process, he cautions us against confusing "imprecision" and "uncertainty" and argues that we should not expect rules for combining assessments of our imprecision to be the same as rules for combining assessments of our uncertainty. In particular, he argues that the rules of Pascal/Bayes will not be satisfactory for combining our assessments of imprecision.

Not all of the "actions" brought against Pascal, Bayes & Co. appear to be of equal severity; one somewhat less severe than those mentioned above is in the form of a "class-action" suit brought against the Pascal/Bayes school by a collection of Scandinavian thinkers.[11] This suit alleges that applications of the canons of Pascal, Bayes & Co. have not always addressed appropriate issues arising in the act or scene we might label "proof." In particular, it is argued that instead of focusing on the probability of some hypothesis, given evidence, we should be concerned about the probability that the evidence proves this hypothesis. It is alleged that, in applications of Bayesian canons, there has been a failure to acknowledge that sources of evidence are "evidentiary mechanisms" which may or may not be working properly and that, in the act of proving major facts-in-issue, we have to give attention to whether they are working. What we must guard against is saying that we have proved a certain point on the basis of "correct" evidence when, in fact, the evidence was delivered "by accident" from an improperly-working evidentiary mechanism.

B. *Other "Actions" And Some Cognitive Dissonance*

I do not wish to suggest that these four "actions" brought against the Pascal/Bayes school are the only ones; there are others that might be mentioned. Nor would it be fair to suggest that all current actions are ones being brought against Pascal, Bayes & Co. For example, both Zadeh[12] and the Scandinavians[13] have brought actions against the Shafer/Dempster school, partly on grounds involving the manner in which dissonant evidence is treated in this school. Speaking of dissonance, I will now tell you about my state of mind as I began observing how these various schools of probability and inference might best be placed in providing "directorial" assistance in inferential "plays" of interest in jurisprudence and in other contexts.

On the one hand, I continue to view as impressive the wide assortment of evidentiary subtleties that can be captured for study and analysis using Bayesian canons. Further, such analyses have resulted in conclusions about these subtleties which, to my knowledge, no one has yet found foolish or unreasonable. No criticism of Pascal, Bayes & Co. has caused me to be any less enthusiastic about the richness of Bayesian canons in capturing certain inferential attributes. On the other hand, the canons of Pascal, Bayes & Co. either are entirely silent or say very little about an equally wide assortment of other characteristics that are also important in inference tasks. The other views I have mentioned certainly do capture many interesting attributes of human inference.

So, I cannot ignore the accomplishments apparent in the works of persons representing the other views I have mentioned; neither can I ignore the results of Bayesian analyses. I attempted to resolve my own cognitive dissonance about such matters by deciding to ponder the following question: What attributes of inference are best represented by each of the schools I have mentioned? I began such deliberation with a strong expectation that no existing view can represent all of them. Human inference, as it appears in various contexts including those of interest in jurisprudence, is simply too rich an activity for us to expect any one system of inference and probability to cover entirely. Other related expectations seemed rather natural. Some views may appear to trade depth of coverage of important evidential and inferential matters for breadth of coverage of these matters; for other schools, the opposite trade-off is apparently made. Beginning with these essential expectations, I asked: What can each of the aforementioned schools do best in providing direction for the acts and scenes of importance throughout the entire performance of an inferential play? Let me share with you some possible answers to this question.

II. The Performance of an Inferential Play in Jurisprudence

It is common to hear a courtroom trial described as a "drama." However, the trial itself appears to be just one act in a much longer play having other acts and scenes and many actors, each one playing discernable inferential roles. The following is an attempt to characterize the acts and scenes, as well as the actors and their roles, in inferential plays of interest to evidence scholars. Like any general characterization, this one will suffer to some extent when matched against specific instances you might easily identify. Perhaps I have left out a scene or two; my general ordering of the acts and scenes may not seem appropriate to all instances of civil or criminal litigation; and you may easily believe that I have taken too much liberty in my characterization of the events in any act or scene. I do admit to having interest in human inference as it occurs in contexts other than jurisprudence; thus, my descriptions of certain acts and scenes may carry some excess meaning as far as they concern matters of interest within jurisprudence. I believe it both interesting and profitable, however, to compare patterns of inferential activity across various contexts in which such activity occurs. Many precise human judgments are comparative in nature; perhaps characterizations of inferential behavior in one context may be sharpened by comparing them with characterizations of similar behavior in others.

A. *The Acts And Scenes*

Figure 1 below shows the prominent acts and some of the scenes in inferential dramas of interest in jurisprudence. In this diagram, time moves to the right and there appear to be three major acts which occur in the following order: (I) Discovery, (II) Proof, and (III) Deliberation and Choices. Also observe that I have identified three scenes in Act I: (1) Generation, (2) Elimination, and (3) Argument Structuring. Particular scenes in Acts II and III might be identified but I have not put them in since they are, for the most part, unnecessary for my present purposes. For example, some of the scenes in Act II might correspond to the (antiphonal) order of evidence presentation as commonly enforced by the adversarial system of our Anglo-American courts. There is certainly room for argument about how I have placed some of scenes. In particular, some will wonder why I have placed "elimination" and "argument-structuring" in Act I rather than in Act II. As we proceed I will tell you about the basis for my choice in this matter.

Also shown in this figure are the general points at which various actors appear and, for some of them, a *very* rough account of how long they remain on the stage. Finally, this diagram gives an equally rough account of points at which major inferential roles become necessary in the various acts and scenes. On this matter there is also much room for controversy; part of the difficulty is that two of these roles are not easy to distinguish.

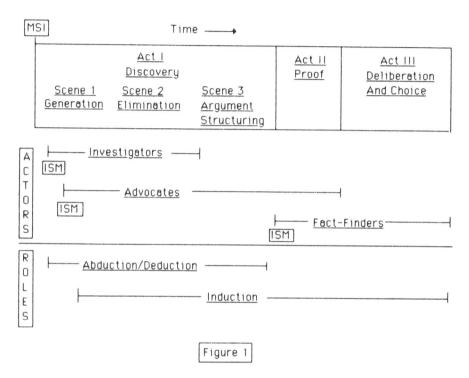

Figure 1

1. Act I: Discovery

In their recent work, Binder and Bergman[14] identify a specific point at which inferential plays in jurisprudence begin; they refer to such a point as a "moment of substantive importance" (MSI in Figure 1). At such a moment, events occur that may provide the basis for a subsequent criminal or civil trial: two cars collide at an intersection, a convenience store is robbed and the clerk shot, or a payment is not made at an agreed-upon time. Investigators and/or advocates appear on stage and begin asking questions about what happened. In the process, of course, they become interested in events leading up to the MSI as inferential problems are recognized and the play begins.

(a) *Scene 1: Generation of Possibilities and Evidence.* It may only rarely happen that an inferential problem springs forth in well-posed form in the sense that it is immediately obvious what hypotheses should be entertained and what relevant evidentiary tests of these hypotheses are necessary. In fact, an often distinguishing feature of the beginning of this initial scene is the lack of possibilities/hypotheses and of evidence. So, the process of generating possibilities and evidence begins, a task undertaken by investigators of some sort and by advocates when they are called upon. In this process questions are asked of persons likely to have some knowledge of events at and surrounding the MSI, observations are made by the investigators and advocates themselves, and information begins to flow. On occasion, there may be witnesses in a position to observe the events at the MSI; there may, for example, have been an eye-witness to the robbery and shooting at the convenience store. Often, testimony from such witnesses is potential direct evidence on major facts-in-issue in a subsequent trial. Knowledge obtained about events occurring before and after the MSI becomes potential circumstantial evidence. This corresponds with Wigmore's[15] categorization of evidence as being concomitant (at the MSI), prospectant (before the MSI), or retrospectant (after the MSI).

It is not an easy matter to specify the precise point at which investigators or advocates begin to entertain or generate hypotheses, guesses, or conjectures about what happened at the MSI. Almost certainly, the immediacy of the generation of possibilities depends upon the available information as well as on the knowledge and prior experience of the investigators or advocates. Often, however, the initial hypotheses or possibilities entertained are vague or undifferentiated. For example, early information may suggest only the possibility that the person robbing the convenience store was a white male under 30 years of age. Since evidence, as it arrives, may exhibit any pattern of conflict or contradiction, several diffuse or undifferentiated possibilities may be entertained. The hope, of course, is that further evidence will suggest more refined or more specific hypotheses.

As this process continues, the skillful investigator or advocate will begin to put existing hypotheses to use in generating additional classes of potentially relevant evidence. New questions are asked and further evidence is obtained. Perhaps this new evidence will allow the generation of more specific or refined hypotheses; it may even be suggestive of novel hypotheses not previously imagined or discerned. Binder and Bergman remind us that new evidence can also be generated or discovered through careful analysis of existing evidence.[16] Often we simultaneously have hypotheses in search of evidence and evidence in search of hypotheses. In either search a premium is placed upon the imaginative skills of investigators and advocates. The relevance and significance of existing evidence may change, in light of newly generated possibilities, at the same time that new possibilities are being generated from evidence. In short, this process of generating hypotheses and evidence involves intellectual reasoning processes in which imagination or creativity is a most important ingredient.

A major trouble, of course, is that there is usually more than one plausible hypothesis or theory to explain a given set of "facts." In a negligence suit, for example, plaintiff's advocate will put a face on events at and surrounding the MSI that may be quite different from the one put on by defendant's advocate. Perhaps it is not always apparent which of several possible explanatory faces will best suit the interest of one's client. The advocate who expects a successful day in court must give attention to a variety of possible explanations for existing evidence, including those likely to be advanced by his or her adversary at trial. Furthermore, success at trial may also depend on an advocate's assessment of the probative force of evidence on various alternative hypotheses or theories. Binder and Bergman argue throughout their work that an advocate is certainly not a passive collector of evidence; successful advocacy requires considerable imagination in generating hypotheses and evidence and then in constructing persuasive argument from the evidence to the hypothesis to be defended.

(b) *Scene 2: Elimination of Possibilities.* As a result of the generation process in Scene 1, there may be several more or less specific hypotheses or possibilities about events at and surrounding the MSI that are being considered. Not all of these possibilities may be equally attractive and so we suppose that there is some effort by investigators or advocates to eliminate those possibilities that seem less interesting than others, i.e., those apparently unsupported by available evidence. In a criminal matter, an investigator knows that he or she cannot bring all suspects to trial, only those specifically chargeable with the offense. In a civil suit, the contending advocates attempt to focus on points or issues about which there is actual disagreement and then to offer that explanation of the existing evidence on these points that seems most congenial to the interests of their clients.

At this point, concern may develop about the completeness or sufficiency of coverage of evidence judged to be relevant on material issues. If we are to eliminate some plausible hypothesis or possibility, we ought to have specific evidentiary grounds for doing so. Prematurely discarding some possibility or hypothesis may cause embarrassment in the later light of a more complete assessment of evidence. At the same time, there is equal concern about the extent to which available evidence covers matters judged relevant in discriminating among those hypotheses retained for consideration. Presumably, all hypotheses actively considered should be put to the same evidentiary tests. If evidence on some favored hypothesis fails on completeness grounds, this will certainly be noted, either by one's opposition or in preliminary hearings such as those before a grand jury.

So, comfort in discarding and retaining hypotheses is, to some extent, dependent upon the completeness of our coverage of evidence on relevant matters. It may not always be possible to determine what constitutes completeness of evidentiary coverage. Another origin of comfort or discomfort, apparent in Scene 2 or even earlier in Scene 1, involves the credibility of sources of evidence we have. Here is a witness who reports observing a certain event that could provide a strong argument for belief in some particular hypothesis rather than in others being considered. The very foundation for any argument from this event, however, rests upon the credibility of the witness who reports its occurrence. It is both important and natural to inquire about the strength of this credibility-related foundation. The literature in evidence law certainly reflects the wide array of subtleties apparent in the task of assessing the credibility of sources of evidence, whatever these sources may be. We can neither eliminate nor advance some hypothesis based on the report of a certain event unless the source of this report is worthy of belief.

(c) *Scene 3: Argument Structuring.* Here is a scene which may, in fact, have taken place at the same time as our earlier scenes. One advocate anticipates the necessity for defense of his or her arguments in a subsequent confrontation at trial with the opposing advocate. A very natural expectation is that one's arguments will be dissected or decomposed by an opponent, often in minute detail. So, in the process of assessing the potential weakness and strength in one's overall argument, the advocate gives careful attention to marshalling the evidence and structuring his arguments based on the available evidence. Knowledge of the opponent's evidence and its likely use is certainly a key element in this process. Thus, there is continuing interest in evidentiary completeness issues. Perhaps there is a need for additional witnesses (even forensic "experts") who will provide ancillary evidence to strengthen some linkage in an argument or its foundation. Such needs must be continually assessed as the arguments on both sides develop.

In this scene there is also concern about the apparent probative value of evidence items, taken individually and in the aggregate. As Binder and Bergman note,[17] consideration of individual evidence items is necessary in order to show exactly what point or points any given item will actually prove. Such consideration is also necessary to assess the probative value of an entire collection of assembled evidence on hypotheses or possibilities an advocate has chosen to defend. It is in such considerations that the "catenated" (Wigmore's term) or "cascaded" (in modern jargon) nature of argument from most evidence to hypotheses becomes apparent.[18] In a cascaded or catenated argument from evidence to hypotheses there are at least two reasoning stages interposed between evidence and major hypotheses or facts-in-issue. Apparently, there are just two evidentiary situations which are not catenated in this sense. One involves direct evidence about a major fact-in-issue; the other involves the unlikely instance in which a *perfectly* credible source delivers circumstantial evidence one reasoning stage removed from major facts-in-issue. Any careful structuring of arguments from evidence to hypotheses must acknowledge the existence of chains of reasoning from evidence to hypotheses, as Wigmore so masterfully illustrated.

There are two basic forms of argument structuring which, for want of better terms, I shall call *temporal* and *relational*. Each of these structural forms serves a very useful purpose and it is difficult to see how an advocate could function at trial without careful attention to each form. On the one hand, an advocate may wish to tell a coherent story or offer a believable scenario of events before, at, and following some MSI. Such a scenario rests upon a carefully developed evidence chronology. An explanatory scenario based on evidence arranged in chronological order has a temporal structure which, among other things, may assist in establishing causal connections. But telling a good story may not be enough; as noted previously, one's story or scenario will be dissected by an opponent and so one must give attention to the nature of the chain of reasoning from any evidence items, alone or in combination, to major hypotheses. In short, one has concern about the manner in which evidence items are related to interim or to final hypotheses and to each other. One of the major virtues of relational structuring is that it facilitates the tracking of dependencies among evidence items and among elements of one's overall argument. For "dependency" substitute the word "subtlety"; the literature in evidence law is a treasure house of evidentiary subtleties. Some of these subtleties involve evidence in one chain of reasoning while others involve evidence items in different chains.

This scene involving argument structuring seems to represent a transition between the act of "discovery" and the act of "proof." Careful argument structuring, either temporal or relational, may reveal certain evidentiary or other deficiencies calling for further generation of evidence or possibilities. At the same time, such structuring forms the basis for a subsequent defense of arguments on chosen hypotheses when the "moment of proof" arrives. We would be hard pressed to identify a scene in which greater demands are placed on the actors in it. The behaviorists among us may say that the process of argument structuring is the process of building "cognitive models" of the inferential problem at hand, and that there is more than one possible "model" for any given problem. That there is more than one possible model is one of the reasons why there is a trial in the first place.

But how does one develop skill in the structuring of argument? This must certainly have been one of Wigmore's concerns as he advertised for help in the task of evaluating masses of evidence. We all know of his work on a charting system for facilitating relational structuring.[19] That such a system has practical as well as pedagogical value is being demonstrated by William Twining and Terry Anderson.[20] Ward Edwards has been particularly concerned about methods for facilitating the development of explanatory scenarios in a variety of contexts, including those in which we have the very difficult task of inferring a MSI that may occur in the future.[21] More will be said about argument structuring skills when we later consider the actors and their inferential roles.

2. Act II: Proof

We now come to the drama of the courtroom. Trials may be viewed as a test of the strength of arguments advanced by advocates for the parties in contention. Perhaps strength of argument is not the only thing being tested; everyone hears of instances in which advocates compete in appealing to the sympathies and other emotions of factfinders. Though there are some constraints imposed upon the discovery act of the play, more numerous experience-given constraints bind the adversarial factfinding at trial. Some of these constraints regulate the manner in which arguments may be presented and the kinds of evidence admissible. Some concern which party has the burden of proof; others involve the rights of the various participants. A certain division of labor is enforced: the court will consider points of law and the jury will consider matters of fact. Since this act of proof is typically open to the public, an attempt is made to offer scenes that appear to conform to societal values and expectations.

But the advocates involved in this contest do have some degree of freedom in the strategies they use to present both their arguments and the evidence believed to support these arguments. Similarly, they have a choice of strategies for countering opposing arguments and evidence. There are many subtle aspects involved in a choice among various tactics.[22] Thus, we must not overlook the fact that advocates have decisional as well as inferential roles to play. This is where the level of concern about argument structuring in the preceding act may make all the difference. If an advocate is unsure about just what inferential role some item of evidence can play, how can he/she convince a factfinder of the importance of this evidence? An interesting question concerns the extent to which an advocate is able to convey to factfinders his/her "cognitive model" of the problem at hand. As noted, there are many intriguing and inferentially valuable subtleties lurking in evidence. The literature in evidence law is quite revealing of the number of cases in which the outcome hinged on evidentiary subtleties. These subtleties can play no role if they are unrecognized by the advocate; nor can they play a role if advocates cannot convince others of their importance.

During this act of proof various instructions are given to the fact-finding body. One of concern to us is an instruction about their role in the next act, the act of deliberation and choice. The instruction addresses the standards of proof they must adopt in reaching a conclusion. As we know, there is a value-related element in these instructions: the required standard of proof is higher in criminal cases than in civil cases. This reflects the increased severity of possible penalties to defendants in criminal cases. Upon completion of these instructions, the final act begins.

3. Act III: Deliberation And Choice

This final act has an interesting characteristic: whether in a jury trial or in a bench trial, it takes place behind a closed curtain. The final process of deliberation upon the overall weight of the evidence and reaching conclusions or verdicts is not open to public scrutiny. This being so, how are we ever to characterize what happens in it?

A few things about this final act are apparent. First, the inferential tasks encountered throughout this entire play are undertaken as the basis for a choice of action affecting the parties in contention; they are not undertaken simply for the purpose of satisfying curiosity about the behavior of often interesting persons. So it is unavoidable that value-related issues are encountered, particularly in this final act, when conclusions are reached and judgments are made. How value-related ingredients affect the process of inference in this final act is not easy to specify. Another thing we know is that final conclusions must be stated in unequivocal or unambiguous terms; attempts to "hedge" stated conclusions using numerical or verbal qualifiers are discouraged. Finally, conclusions and judgments are often reached on the basis of deliberation by a group of presumably impartial or disinterested persons. Assessing the likely effects of group dynamics on this final inferential and choice process is interesting on its own.

B. *The Actors and Their Inferential Roles*

In discussing theories of probability and their bearing upon inference in any natural context such as jurisprudence, it seems reasonable to ask whose inferences we are talking about. In the present view of inference in jurisprudence, the behavior of actors in at least three important classes should be considered. The required inferential behavior of actors in these three classes is clearly different; for example, factfinders are usually not asked to generate possibilities or evidence, and investigators or advocates are not asked to draw final conclusions on the basis of evidence assembled by the opposing parties. One reason to be concerned about the various actors and their inferential roles is that otherwise we may expect more of a theory of probability than it is able to deliver. For example, consider the final choice process confronting factfinders as they deliberate upon a verdict and its consequences. As we know, there are considerations of probability and of value in any choice made in the face of uncertainty. We should not expect a theory of probability to capture subtleties in value judgments. However, we might easily expect a theory of probability to be congenial in some way to the process of combining probability and value ingredients in the act of making a choice.

1. Inferential Roles and Their Placement in the Acts and Scenes

In Figure 1 above, I have placed three kinds of inferential reasoning that have been identified and, in a very rough way, I have identified intervals during an inferential play in jurisprudence in which these three forms of reasoning seem to be of special importance. What I might have done, to be on the safe side, is simply to suppose that all of the mentioned forms of reasoning may occur in any of the acts and scenes. In identifying these forms of reasoning, I am aware of various points of controversy which bear upon how I have placed them in the acts and scenes of Figure 1. One major point concerns my inclusion of induction in Act I on "discovery." Other points concern my specific placement of the "elimination" and "argument structuring" scenes in Act I rather than in Act II on "proof."

First consider Scene 1 (generation) of Act I (discovery). By what reasoning process, if any, are hypotheses or possibilities generated? In the view of some persons, the process of generating hypotheses involves imagination, insight, or originality and is thus a matter for psychologists and not philosophers or logicians.[23] Apparently, we cannot turn some formal or logical crank and have useful hypotheses emerge. It has been argued that new ideas, in the form of novel hypotheses, are not generated either by deductive or inductive reasoning. In one view, imaginative or creative thought appears to involve yet another form of reasoning, one which C.S. Peirce gave the label "abductive."[24] A certain item or collection of evidence may resist explanation by hypotheses being entertained; we ponder the matter and can often generate a new hypothesis that seems to explain evidence refractory under our existing hypotheses. To be retained as a useful possibility, this new hypothesis must also explain other evidence we have. In addition, we must put this new hypothesis to work in *deducing* new categories of evidence potentially relevant, not only on our new hypothesis, but on others as well. In comparing the three forms of reasoning, we may paraphrase Hanson[25] and assert that deductive reasoning shows that something *must be* true, inductive reasoning shows that something *is likely to be* true, but abductive reasoning shows that something *may possibly be* true.

So, abductive reasoning seems to begin at least in a "bottom-up" process, from evidence to hypotheses. But inductive inference is often characterized in exactly these same terms.[26] The question is whether abduction differs, if at all, from induction. If they are different, we might simply characterize discovery as the creative process of generating hypotheses from which relevant evidentiary tests are subsequently deduced, and then characterize proof as the inductive process of grading the likelihood of our hypotheses in light of evidence we obtain. It seems that any such characterization would be inappropriate if it implied that there were no room for imaginative thought in devising and conducting evidentiary tests of hypotheses. This is one reason for the overlap of "abduction/deduction" and "induction" in Figure 1. Anyone who has ever designed an experiment comes to recognize that there is often room for considerable ingenuity in the design of methods for data collection and analysis that will allow valid inferential tests of hypotheses being entertained.

Salmon's argument that discovery is a process "involving frequent interplay between unfettered imaginative creativity and critical evaluation"[27] is another reason for the overlap of all three reasoning processes in Act I. The following is a brief look at arguments concerning the distinction between abduction and induction. Hacking claims there is no sharp distinction between them.[28] However, Levi argues that, since abduction only concerns the *expansion* of possibilities which may or may not turn out to be useful (i.e., having "informational virtue"), we must at least apply different standards for grading the adequacy of abductive and inductive reasoning.[29] Levi argues that avoidance of error is therefore a consideration relevant to induction, but not to abduction.[30] Hacking goes on to argue that there has never been a system of probability congenial to the process of theorizing or abduction; probability metrics seem to apply in inductive inference and decisions under uncertainty, but not in situations in which we attempt to grade the degree of our confidence in theories.[31] Jonathan Cohen tells us that early writers such as Bacon may have failed to distinguish between induction as a method for discovery and induction as a method for proof.[32] Cohen adds that Mill later made this distinction; Mill argued in the process that his inductive methods, if not methods of discovery, were nevertheless methods of proof.[33]

There is substantial current interest in "abductive" reasoning and at least one entire work is available on the topic.[34] In this work, Eco and Sebeok present a valuable and entertaining collection of papers written by persons who, though they represent a variety of different disciplines, have a common interest. Their common interest is in exploring the connection between "abduction" as defined and investigated by Peirce and the reasoning "programmed" into Sherlock Holmes by Sir Arthur Conan Doyle. Their common view is that Holmes was given particularly acute powers of abductive reasoning, i.e., Holmes was especially adept at generating novel and often surprising possibilities from evidence and at deducing ingenious evidentiary tests of these possibilities. There is also considerable interest in abductive reasoning among our colleagues whose research in artificial intelligence involves development of so-called "expert systems."[35] Some of these systems are advertised as having the capability of generating possibilities.

Now consider the "elimination" and "argument structuring" scenes in Act I. Here are my reasons for placing these scenes within "discovery" rather than in "proof" in plays of interest within jurisprudence. First, if induction belongs in the act of "discovery," it seems that "eliminative" and not "enumerative" induction characterizes the process of narrowing the range of the possibilities or hypotheses we take seriously or the points about which there is disagreement. The elimination of concern to us seems to rest on "variation of circumstances"[36] or the incorporation of various kinds of evidence. Surely there is room for considerable imagination in the identification and ordering of our evidentiary tests. My placing elimination within "discovery" seems entirely consistent with at least two modern conceptions of the discovery process in jurisprudence.[37]

If the persons involved in an inferential problem had only themselves to convince, I might be tempted to place "argument structuring" in Act II rather than in Act I. But we have a trial involving advocates for parties in contention; careful structuring of one's argument as well as anticipating counterargument may, as mentioned earlier, stimulate further discovery efforts to generate additional possibilities or additional categories of directly relevant or indirectly relevant (ancillary) evidence.

Concerning Act II of inferential plays in jurisprudence, it might be said that the proof is in the pudding and the pudding is the trial. I have no hesitation about including induction as the major inferential role in this act; this is what Wigmore noted.[38] The exact nature of the induction in this act seems to be a point of contention among some of the views of probability we are considering. According to Cohen's view, it might be seen as the penultimate stage in a process of elimination that began much earlier on. The ultimate or final stage of this process might be thought to occur in Act III when factfinders deliberate on all of the evidence (or at least all of it they can remember). I also have no hesitation about extending induction as the major inferential role in Act III.

2. The Actors

Interesting things are currently being said about some of the actors in our play; not all of them are flattering and some may even be true. Before we consider the actors and some of the things said of them, there is one general point that ought to be mentioned. You may notice in Figure 1 above that the appearance on stage of each of the three classes of actors is marked with the letters ISM; these letters stand for "initial state of mind." As we all know, one of the issues on which views of probability differ concerns how we ought to represent a person's expectations—his "initial state of mind" or "prior beliefs"—before any specific evidence is considered in the problem at hand. An interesting question concerns what probabilistic representation seems to best characterize these initial mental states for the various actors. Perhaps it is true that investigators or advocates typically begin with different patterns of initial beliefs, expectations, or prior beliefs than do factfinders.

Depending upon the case at hand, the first persons we observe on stage following the MSI are investigators and/or advocates. We might also include as actors the various other persons directly involved in the matter such as the injured parties themselves or other witnesses subsequently interviewed. It may be argued that there are distinct inferential and decisional components of the observational and reporting behavior of any human source of evidence.[39] But we already have enough actors to consider in our efforts to sort out the contributions of the alternative systems of inference and probability under consideration.

Let us first consider the investigators and their initial states of mind. There may be many kinds of investigators and they may differ in their inferential behavior as much as Inspector Lastrade differed from Sherlock Holmes. Holmes, as we know, was an investigator and not an advocate or a factfinder. The work of Eco and Sebeok contains a marvelously thorough account of the various reasons why Holmes was such a good investigator; most of them involve, in one way or another, his abductive reasoning skills at generating possibilities and appropriate evidentiary tests of them. Among other things, Conan Doyle gave him the capacity to ask good questions, to appreciate the importance of evidentiary details and subtleties, and to remain open-minded. However, there are some who believe that many of Holmes's arguments would not have stood up if and when his cases had come to trial.[40]

What can be said about the initial state of mind of the investigator or advocate who arrives shortly after the MSI? Because there is very little evidence and few, if any, identified possibilities or hypotheses at this point, it is certainly hard to see how they could disbelieve anything or that they have disproven anything. Perhaps their initial states of mind are best characterized by saying that they have a lack of belief or a lack of proof, since they certainly have a lack of evidence. Characterizing their initial state of mind as "ignorance" seems a bit harsh; perhaps "innocence" is better. As mentioned earlier they may begin to entertain only vague or undifferentiated possibilities which they hope to make more specific or more refined in light of evidence they will observe. Some, like Holmes, may be adept at "eliminating the impossible so that the residuum, however improbable, will contain the truth."[41]

Let us consider the behavior of the advocates during their performance in Acts I and II. It is quite interesting to compare what is said about the inferential behavior of advocates with what is frequently said about the inferential behavior of persons who may serve as factfinders. Twining,[42] Twining and Anderson,[43] and Binder and Bergman[44] have, in various ways, criticized the kind of training advocates typically receive in preparation for their inferential roles during the three scenes we have identified in Act I on discovery. But at no point do any of these persons assail the basic inferential competence of even the beginning student of advocacy. As we shall see, factfinders do not fare nearly so well in other kinds of evaluation. Binder and Bergman have concern about the "generation" process, noting that "students can be excused for graduating from law school thinking that facts are like starving trout, ready to be reeled in at the drop of a question or two."[45] The need for generating and carefully evaluating alternative hypotheses is emphasized throughout their work. In addition, Binder and Bergman discuss the need for both the "temporal" and "relational" forms of argument structuring identified above.[46] Like Twining and Anderson, they give particular attention to the process of identifying generalizations which license or warrant reasoning steps and, thus, form the "glue" holding arguments together.[47]

In their efforts to enhance the training of advocates in the process of marshalling masses (or messes) of evidence and forming persuasive arguments therefrom, Twining and Anderson ask us not to overlook the basis for "relational" structuring provided many years ago by Wigmore. "Wigmorean" argument structuring is patently hierarchical in nature, and encourages the layout of argument steps or stages in considerable detail. By such a process one specifies reasoning linkages among evidence items, interim hypotheses, and final hypotheses, along with the generalizations that appear to license or warrant these linkages. Wigmore's analysis of the various uses of evidence at trial[48] and Toulmin's fine-grain analysis[49] of the formation and defense of a reasoning linkage are very similar. Relational structuring, in addition to facilitating the tracking of evidentiary dependencies, has also been extremely helpful in attempts to add to our formal understanding of evidentiary subtleties and the manner in which evidence can be usefully categorized.[50]

A major trouble is that important temporal relationships among events of interest are frequently suppressed in relational structuring, a process that tends to be "timeless." As noted, it is often important for advocates to be able to provide a coherent story about what happened before, at, and following a MSI. This requires an account of the order in which events seem to have occurred and an interpretation of the inferential significance of this ordering. Opposing advocates may, of course, disagree about the order in which events actually occurred and will, in any case, disagree about the interpretation or significance of the events. Ward Edwards has given considerable thought to the construction and use of explanatory "scenarios" in complex inference problems.[51] Noting that there are several logical forms of explanation, Edwards alerts us to the importance of scenario development not only in the act of "proof," but also in the act of "discovery." A scenario is, in essence, a complex hypothesis which can be put to use in generating new evidence necessary to fill in gaps in an explanation of what happened. In addition, Edwards provides us with some very good examples of how there is often probative value lurking in the sequence in which events occur. It is not exactly clear what the relationship between temporal and relational structuring is or should be; however, Edwards notes that at least local temporal factors can be incorporated in relational structuring.

Of the behavior of the advocates during Act II on proof, it might be said that we have, as in *Hamlet*, a play within a play. In Act II the advocates themselves put on a play whose audience consists of factfinders, those persons who form the third category of actors in our play. In this play within a play, the advocates resort to tactics, histrionics, and even inferential reasoning in telling a story of behalf of their clients. As we know, advocates even have some choice about who their audience will be.

We come, finally, to the factfinders, the actors in whose hands rest the final outcomes of our inferential plays in jurisprudence. Here we either have a collection of "ordinary citizens" as jurors or a judge if it is a bench trial. Let us suppose that we have a jury whose members represent a cross-section of our society and whose participation in Acts II and III is due to no particular set of intellectual or other credentials. Here we have an interesting situation; the other actors received at least some training for their roles and have had varying amounts of experience in them. But now we have a collection of persons who will determine how the play comes out but who, on the surface at least, appear to be the least qualified of any of the actors in the play. What confidence can be placed in such a process? The answer you will get to this question depends upon whom you ask.

In the view of some psychologists who study human inferential and decisional performance, people appear to be ill-equipped to perform these tasks. As one well-known account of research in this area has it, current research paints a "blemished portrait" of human inferential and decisional capabilities.[52] Allegedly, we are all prone to inadequacies of various sorts; the words "suboptimal," "biased," and even "irrational" or "incoherent" creep into many current assessments of human capabilities. What most of these studies assail is our basic intellectual competence rather than other factors such as the adequacy of the training we receive in preparation for inference-drawing or decision-making tasks. This stands in sharp contrast to the assessments of other actors we have already considered. Taken seriously, the conclusions reached in many of these studies suggest that we are ill-advised to assign to ordinary people the crucial fact-finding roles in our inferential plays.

Many of the studies upon which such depressing conclusions rest have involved the use of abstract and unfamiliar tasks; more hopeful conclusions have been reached in other studies in which there has been an attempt to make inference tasks similar, in various ways, to those a factfinder would encounter at trial.[53] There has been no scarcity of criticism of research on human inference based on abstract and unfamiliar tasks. Cohen has argued against the conclusions reached on the basis of such studies and has gone even further in suggesting that no empirical study of human inference can result in a conclusion that subjects are "irrational," "suboptimal," or "biased."[54] Von Winterdeldt and Edwards[55] and Phillips[56] mention the various ways in which subjects in these studies are typically put at a disadvantage.

Perhaps we ought to assign crucial fact-finding roles to experienced logicians or probabilists; surely they have some knowledge of the process of weighing evidence and drawing conclusions from it. However, it has been said that ten logicians placed end to end would never reach a conclusion. One reason for this is the lack of a settled conception of what the weight of evidence actually means. Indeed, this is one of the most important points of disagreement among the views of probability and inference we will examine.

What can be said of the initial states of mind, expectations, or prior beliefs of factfinders before they are presented with specific evidence in Act II on proof? Should we think of their ISM in the same way we thought of the ISM of investigators and advocates at the point they came on stage? There appear to be several reasons why we should not equate the initial states of mind of investigators/advocates in Act I with those of factfinders in Act II. In the first place, factfinders are given specific instructions regarding what they should presume; the prior presumption of innocence is one example. Whether or not such instructions are ever obeyed is an interesting question. The point is that factfinders rarely begin their inferential tasks in Act II with a complete lack of evidence; they see the defendant, hear the charges, and may be told, if it is a felony trial, that there has been a grand jury hearing in which the defendant was indicted. Some factfinders may even take such information as strong evidence against defendants. Finally, the essentials of the major issues at trial are explained to them, so we cannot say that they begin their inferential tasks in ignorance or innocence of what possibilities exist. If we characterize the ISM of factfinders in Act II as one of "lack of belief" we may be mistaken; perhaps we would come closer to the mark if we simply say that they may lack proof but not beliefs.

After the "play within a play" is over and the factfinders have heard or seen the evidence presented, they go off by themselves to deliberate about a verdict and, in some cases, about the magnitude of the consequences. The process of deliberation is subject to any or all of the influences so carefully outlined and studied by Hastie, Penrod, and Pennington.[57] Our present interest lies in characterizing the inference process in jury deliberations and in noting certain possible influences on them. As mentioned, factfinders are instructed about required standards of proof for the case at hand. Opinions differ about the probabilistic equivalents of "beyond a reasonable doubt," "balance of probabilities," as well as about "the presumption of innocence" in Act II. We have already mentioned the role of value considerations in setting these standards of proof.

An obvious thing one could say about the inferential tasks performed by factfinders is that they are often extremely difficult. Some tasks involve multiple count charges, multiple defendants, complicated legal issues, large masses of evidence, and carelessly explained court instructions. On occasion the issues at trial may be subtle concerning the points of law involved and, occasionally, there may be many defendants on trial. As I write this paper, there is a murder trial taking place in Washington, D.C., in which there are ten defendants and several advocates, each lawyer representing various subsets of defendants. How does a factfinder make inferences in such difficult situations? One suggestion is that, in many inferential situations, people attempt to simplify an inference-drawing task by turning it into a decision-making task.[58] We may all do this from time to time; in the face of many uncertainties, we may simply choose to believe some things and not others, effectively suppressing our uncertainty in the process. In fact, procedural rules prohibit expressions of uncertainty anyway; as noted above, the jury must produce an unequivocal judgment that is not to be hedged in terms of numerical or verbal qualifiers.

With all of this a preamble, let us see what each of the five views of probability and inference has to say about the actors and their roles in the acts and scenes of the inferential plays of concern to us.

III. Theories of Probability in the Acts and Scenes of an Inferential Play

My present task is to specify how the various views of probability and inference capture important evidential and inferential attributes that emerge as the play unfolds. As I proceed, I will attempt to present each view in its best possible light; my belief is that each one has something significant to offer. There are issues on which these views take entirely different stands; part of my present task is to bring out some of these issues so that discussion about the alternative views may be more precisely focused. Here is an acknowledgment I must make before I begin. My failure to mention a particular view of probability in connection with a certain issue does not necessarily mean that this view has nothing whatever to say about the issue. Such omission may, of course, simply reflect my own imperfect understanding of the view in question and its implications (ones that may be obvious to others). In addition, it seems fair to say that all of the views we are to examine are still in the process of emergence and that their champions will continue to extend coverage of these views to more attributes of inference and decision processes. The last word has not yet been said about the oldest system, Pascal/Bayes; we certainly hope and expect to hear more from the developers of the newer views being examined.

A. *The Shafer-Dempster School of Non-additive Beliefs*

I begin my assessment by considering the work of Glenn Shafer on a system for expressing our beliefs or credal states in light of evidence. I do so because I believe that Shafer has responded to many evidentiary and inferential concerns that arise in the earliest acts and scenes of our inferential plays. Though I should not like to go on record as saying that Shafer's theory of probabilistic reasoning is also a theory of discovery, much less a theory of generation or abduction, it does seem apparent that he has given concern to the state of our beliefs under evidentiary conditions that do occur during various stages of the process of discovery. The reader of Shafer's major work[59] cannot fail to notice the large number of examples he uses involving Sherlock Holmes the investigator in "The Case of The Burgled Sweetshop." It appears that Shafer considers both the unfettered imagination and the critical evaluation that Salmon, as we noted, said were characteristics of the discovery process.

Shafer, as well as Cohen and others, has various concerns about the suitability of the Pascal/Bayes system in epistemic situations, those in which we attempt to grade the subjective degree of our beliefs in hypotheses or possibilities in light of evidence. Shafer's major expressed concerns have to do with the additivity property of Pascalian probabilities and the fact that probabilities on this scale range in interpretation from "disbelief" to "belief" (or as Cohen would say, from "disproof" to "proof"). The various properties of Pascalian probabilities combine to render this system frequently unattractive in epistemic situations; they preclude our expressing various credal states which may well be experienced by actors in any of the acts and scenes of an inferential play. For example, we may wish to distinguish between "disbelief" and "lack of belief," or between "disproof" and "lack of proof"; we noted that such distinctions may be particularly necessary in the early scenes of an inferential play in which we have few possibilities and very little evidence. If the ISM of an investigator or advocate be one of "ignorance" or "innocence," where such credal states reflect lack of evidence, then we have no obvious way of grading such states within the Pascal/Bayes system. Shafer provides us with a cogent argument about why a uniform or flat prior probability distribution across possibilities is not a representation for "ignorance."[60]

Shafer asks us to consider a particular way of grading the support that the evidence offers possibilities we are considering; the manner in which we are allowed to assign evidentiary support can lead, in his system, to non-additive beliefs about these possibilities. Shafer's system is also behaviorally rich, but in ways other than ones we can observe in other systems. The following properties are particularly important in our present assessment.

(1) In Shafer's system we are given license and means to express a belief that a certain evidence item or body of evidence can provide support for one specific hypothesis but none at all for others. This takes us back to a distinction, made years ago, between "mixed" and "pure" evidence.[61] Evidence is said to be mixed if it supports or points toward H to one degree and supports or points toward not-H to another. Pure evidence, on the other hand, may offer some support for H, but none at all for not-H. A bit later on, I will discuss how the Pascal/Bayes system effectively treats all inconclusive evidence as mixed by requiring us to believe that such evidence is to some degree consistent with the truth of every hypothesis being entertained. In Shafer's view, the distinction between pure and mixed evidence is preserved and we are free to say that evidence can provide no support at all to certain hypotheses or collections of them. The result of this freedom to withhold evidentiary support, by leaving it uncommitted in various ways, is a non-additive system for assigning numerically expressed beliefs in possibilities or hypotheses.

(2) For Shafer, the weight of evidence on some hypothesis or collection of them is related to the support the evidence provides for the hypothesis or hypotheses. Shafer even offers a conjecture about the precise nature of this relationship and the conditions under which this relationship may hold.[62] One feature of this relationship is that evidence offering zero support to a certain hypothesis also has zero weight on this hypothesis. Since we are free to withhold support from any hypothesis for an item or body of evidence, we can say that this evidence provides zero support for or has zero weight on this hypothesis. It will also be true that, for this evidence, we will have zero belief in this hypothesis.

But the role of zero in Shafer's system is entirely different than it is in the Pascal/Bayes system. In the latter, a possibility once assigned zero probability cannot be resuscitated in light of new evidence, however compelling it is; the reason is that zero here means "disbelief," "disproved," "impossible," or "almost surely impossible" (for so-called "null" events). In Shafer's system, zero means "lack of support" or "lack of belief." Our lack of belief in H can be revised in light of new evidence to subsequent graduations of positive belief in H. Thus, zero in Shafer's system is not the "absorbing barrier" that it is in the Pascal/Bayes system.

(3) An element of richness in Shafer's system involves the array of credal or belief-related measures that can be captured from an initial assignment of evidentiary support, provided that such assignment is made in accordance with certain rules. Considering a certain item or body of evidence, a total

support of 1.0 can be allocated (additively) across the 2^n possible subsets of n hypotheses, with the constraint that the subset containing none of them (the "empty" set) receives zero support. One measure derivable from such support assignment is a measure called degree of belief or degree of support (Bel). For some non-empty subset A of hypotheses and some item or body of evidence, Bel(A) indicates the support this evidence seems to offer A as well as to any subsets of A. Thus, Bel is a credal state representing the aggregate of all support this item/body of evidence provides for A as well as for other sets which entail A.

But, since we are allowed to withhold evidentiary support from A or from any other subset by leaving some of it uncommitted, it will not always happen that Bel(A) and Bel(not-A) sum to one. Suppose we withheld some support from hypotheses in A; perhaps, in light of other evidence, we might wish to commit some or all of this presently uncommitted support to hypotheses in A. A measure on A called an *upper probability* or *plausibility* (P*) can also be determined from our initial evidentiary support assignment; this measure shows the total amount of support that could subsequently be committed to hypotheses in A pending other evidence. Thus, the interval [Bel(A), P*(A)] represents a hedge against future evidence that may cause us to commit further support to A.

An initial evidentiary support assignment accomplished according to Shafer's rules leads to the capture of the credal state *doubt*. In Shafer's system, the degree to which you doubt A [Dou(A)] is equivalent to the degree to which you believe not-A, i.e., Dou(A) = Bel(not-A). Finally, *ignorance* for Shafer means complete lack of evidence. If we have no evidence at all bearing upon H or not-H, we would be free to assign Bel(H) = Bel(not-H) = 0, leaving the entire potential evidentiary support of 1.0 uncommitted between them.

(4) In Shafer's view, the appropriate way in which to combine beliefs based on "distinct" bodies of evidence is called "Dempster's rule"; hence, I have referred to this system as the Shafer-Dempster system. Using this rule, a new aggregate belief function called an "orthogonal sum" can be determined from the belief functions for each of the distinct bodies of evidence. Shafer argues that use of this rule allows a true pooling of evidence. As mentioned earlier, there is some controversy surrounding Dempster's rule, discussion of which requires mention of the distinction between consonant and dissonant evidence.

Consider two items of evidence E_1 and E_2. If they both favor or support H, they are said to be consonant; but if one favors or supports H and the other favors or supports not-H, they are said to be dissonant. The trouble is that application of Dempster's rule to dissonant evidence can result in non-zero evidentiary support being allocated to the inconsistent (and empty) conjunction (H and not-H); this is in violation of Shafer's rules for evidentiary support assignment. Such a situation represents an apparent conflict in credal states in which evidence appears to support both sides of a contradiction. The remedy proposed by Shafer is to disregard whatever support was allocated to H and not-H in applications of Dempster's rule. He seems quite troubled over such dissonance in evidence,[63] since evidentiary conflicts and contradictions abound in many situations, particularly in trials at law.[64]

(5) I have left until last a discussion of certain features of the Shafer-Dempster system that seem to make clear the connection of this system with various belief structures or credal states during the process of discovery and, possibly, during other acts in our inferential play. The set of possibilities or hypotheses you entertain at any moment is called by Shafer your "frame of discernment"; possibilities you entertain always reflect your present state of knowledge and any assumptions you make. Shortly after the MSI, you the investigator or advocate may have only a *coarse* frame of discernment; it is coarse in the sense that it reflects only vague, general, or undifferentiated possibilities. Consider again the robbery and shooting at the convenience store; early evidence may bear only upon whether the subject was male or female, white or non-white, and over or under 30 years of age. So at this early moment, your frame of discernment may consist of the eight undifferentiated possibilities represented by the possible combinations of just these three attributes. Shafer stresses that the structuring of a frame is a creative act that embodies impressions, hunches, guesses, and assumptions, as well as evidence.[65]

However, with additional evidence, you may make any number of alterations to your frame of discernment. This evidence may allow you to *refine* your frame by identifying more specific or differentiated possibilities. For example, new evidence might cast suspicion upon two particular white males under 30 years of age. In addition or instead, new evidence or new assumptions might allow you to *abridge* your frame by eliminating certain distinctions and possibilities. For example, the new evidence might convince you that the subject was, indeed, male and so you eliminate four of your earlier possibilities involving female subjects. But this process may work the other way around. Suppose you had begun under the assumption that the subject was male and either white or non-white and over or under 30. New evidence or revised insight may cast doubt on your initial assumption that the subject was male; so you decide to entertain the possibility that the subject was female and *enlarge* your frame of discernment on this basis.

Now, it may be expected that the support or weight provided by an item/body of evidence to each possibility in a frame of discernment depends upon how these possibilities are defined. Further, this evidentiary support might appear to be different as a result of refinements, abridgments, or enlargements of this frame. Shafer devotes considerable attention to the manner in which evidentiary support/weight can change or resist change when we restructure a frame of discernment in various ways. So, newly discovered evidence may cause us to alter a frame of discernment; in turn, alterations of this frame may cause us to think of the evidence in different ways.

One important evidentiary matter affected by this dynamic discernment process concerns possible interactions between two or more bodies of evidence. Shafer tells us that the combination of two or more bodies of evidence according to Dempster's rule requires that the bodies of evidence be entirely distinct and that the frame of discernment, over which the evidence combination takes place, should discern the "relevant interaction" between these bodies of evidence.[66] There are, of course, many subtle but inferentially valuable ways in which items/bodies of evidence can interact or be related. Nearly every theory of probability acknowledges the fact that the inferential weight or value of one item/body of evidence depends not only upon your considered possibilities, but also upon what other evidence you have. In Shafer's system, the apparent evidentiary weight or support provided by one body of evidence can interact with that provided by another. Shafer notes that you may recognize such evidentiary interactions at one level of discernment of your frame of possibilities but not at another. Perhaps, for example, a "coarse" frame of discernment will obscure evidentiary interactions that only become apparent when we refine some of the possibilities.

In discussing the discernment of evidentiary interactions, Shafer touches upon a very important practical matter in many inferential contexts, namely, how much evidentiary detail should we record and preserve. If we record too much detail, we can overwhelm ourselves with information; if we record too little, we may fail to discern certain inferentially valuable relationships between items/bodies of evidence. Shafer tells us that a frame will discern relevant interactions between two bodies of evidence if the details in one body it fails to discern are independent of the details in the other body it fails to discern.[67]

Consider the final scene in Act I that I have labeled "argument structuring." In this scene the advocates marshall evidence (perhaps finding the need to generate more of it) and attempt to structure defensible arguments favoring conclusions they believe will be consistent with their clients' interests. Such structuring seems to involve temporal considerations based upon evidence chronologies and relational considerations frequently based upon a microscopic analysis of hierarchical evidence-hypothesis and evidence-evidence relationships. Study of such structuring processes reveals a very wide array of evidentiary subtleties, some of which I will discuss in connection with other views of probability and inference.

Shafer's system of belief functions is quite silent about these structuring processes and the evidentiary subtleties they reveal. As just noted, Shafer's system is rich in allowing us to capture a variety of commonly encountered credal states. However, it is not obvious how Shafer's system lends itself to the fine-grain evidentiary analyses that argument structuring requires. Others have noted this same problem with Shafer's system.[68] Many of Shafer's concepts are vague and may not be very helpful in answering specific questions about evidentiary subtleties; examples here include "evidentiary support," "distinct bodies of evidence," and "relevant interactions." In other views of probabilistic reasoning, you the actor are given quite specific answers to questions about evidentiary subtleties and how they may be incorporated in the process of assessing the probative value of evidence. In Shafer's system, as thus far articulated, the actors themselves bear the entire burden of determining what factors influence evidentiary support, what characteristics determine distinct bodies of evidence, and what forms relevant evidentiary interactions can take. The structuring of argument is a species of task-decomposition; other views of probability offer suggestions about how this decomposition might take place and answer specific questions about evidentiary and inferential subtleties that arise in the process.

In fairness, however, there are hints in Shafer's work that argument-structuring matters might receive future attention; one such hint concerns how a belief function might be "discounted" to reflect uncertainties we may have about the sources of evidence.[69] In relational structuring you essentially lay bare your recognized uncertainties about how observable evidence is linked to major facts-in-issue. Each reasoning step you identify exposes a source of potential uncertainty. Some of these uncertainties are associated with the force or weight of an argument that could be made *if* the events reported in the evidence did, in fact, occur. Other uncertainties concern the strength of the foundation for this argument and are rooted in the credibility of the sources reporting these events. In some cases, the strength of an argument based on the assumed occurrence of an event must be discounted depending upon what is known about the sources of information about this event. In other cases, however, what is known about the sources can enhance the value of their reports to us.

I end my comments on the Shafer-Dempster system by passing to the factfinders and their inferential roles in the acts of proof and of deliberation and choice. As noted earlier, factfinders in a trial at law are not normally involved in the discovery process; carefully articulated possibilities are presented to them along with carefully selected evidence. Their task is to "weigh" the evidence on these possibilities and render a verdict or make a choice from among these possibilities. It seems fair to say that every probability system we are considering aspires at least partially to describing the way people "actually" combine and draw conclusions from evidence. Shafer is no exception, since he tells us that Dempster's rule of combination permits descriptions of probabilistic reasoning that are more faithful than other systems to the way human beings actually think.[70] Though I entertain the possibility that he is correct, I have seen no empirical evidence on the matter. There is ample evidence that people do not reason in accordance with the Pascal/Bayes system; many purely "descriptive" models of human inference behavior have been proposed, but none, to my knowledge, have the properties of belief functions.

One thing that does seem certain is that credal states represented by "lack of belief," "doubt," and "ignorance" are not states peculiar to investigators and advocates in discovery roles. These same states may well characterize attributes of beliefs held by factfinders at various points in their evaluation of evidence. Shafer has recently introduced the idea of "canonical examples" to help provide an intuitive basis for his method of assigning evidentiary support.[71] Some canonical examples play the same role that betting analogies are intended to play in applying the Pascal/Bayes system. One trouble, of course, is that it may be difficult to see how possibly non-additive beliefs about hypotheses could be combined with assessments of value in the act of making a choice. As we know, expected-utility choice models stemming from the Von Neuman-Morgenstern axioms assume additive beliefs about those hypothesized states which act, in concert with possible actions, to produce the consequences whose relative value we need to assess. But there are rules for choice which do not rest upon the axioms forming the basis for expected-utility maximization; perhaps some of these systems will be found to be congenial to the incorporation of non-additive beliefs.

B. *The Bacon/Mill/Cohen School of Inductive Probability*

I believe that Cohen's system of inductive probability is responsive to certain elements of the discovery process as I have described it; the exception involves the act of generating possibilities and evidence. In addition, this system has many things to tell us about the act of proof and, possibly, the act of deliberation and choice. Much will hinge upon how we construe the nature of the inductively inferential roles undertaken by actors in these final two acts. It seems fair to say that, although Cohen did not foment the "probability debates" in jurisprudence, he has certainly provided additional dimensions and considerable focus for these debates. Before 1977 and the appearance of *The Probable And The Provable*, there was ongoing debate about whether elements of the Pascal/Bayes system had anything to offer to the study of evidentiary processes in jurisprudence.[72] Cohen introduced a specific alternative probability system, one he argued was more consistent with existing experience-given rules concerning the presentation, evaluation, and aggregation of evidence in trials at law. Debate now continues on how adequate Cohen's alternative system actually is in representing these experience-given rules.[73]

Cohen's expressed concerns about the applicability of the Pascal/Bayes system in jurisprudence are somewhat more extensive and specific than Shafer's concerns. This is understandable, in part because Shafer did not direct his comments toward jurisprudence in the same way as Cohen. Cohen noted difficulties involving the Pascalian additivity property and its special case in the negation rule, and in the interpretation of limiting probability values within the Pascal/Bayes system. Cohen also pointed out difficulties associated with the manner in which probabilities for conjunctions are determined in this system. He compiled a list of six "anomalies and paradoxes" that he claims will emerge in any Pascalian account of judicial probability.[74] I have already spoken my piece on these difficulties[75] and will not do so again except as they arise in my present attempt to place Cohen's inductive probability system within the acts and scenes I have identified.

Consider the scene in Act I on discovery that I have labeled "elimination." In anticipation of a trial, unsupported possibilities must be eliminated and attention focused on the ones that can survive this process. I believe a case can easily be made that the narrowing or elimination of possibilities that surely takes place on the part of investigators/advocates during discovery is a process that continues during the acts of proof and, possibly, in the act of deliberation and choice. Since Cohen's system of inductive probability is rooted in eliminative induction, it seems natural to suppose that this system can offer guidance wherever such inductive processes occur.

In most cases that will eventually come to trial there are more or less well-defined points or issues that have to be satisfied by the plaintiff or the prosecution. Each of these issues suggests possible general categories of relevant evidentiary tests whose details depend upon the particular case at hand. The general evidentiary categories of means, motive, and opportunity are typically associated, for example, with felony cases; other lists of relevant points are developed for situations involving negligence, fraud, breach of contract, or other familiar actions. Cohen argues that we should have a means for grading the extent to which possibilities, hypotheses, or theories survive the relevant evidentiary test appropriate to the matters at issue. He further argues that the system of Pascal, Bayes & Co. is incapable of this task since it seems to ignore a rather fundamental consideration, namely, *how many* relevant evidentiary tests does a possibility seem to survive?

Thus arises our concern about the completeness of coverage or sufficiency of evidence on matters at issue. I argued earlier that concern about completeness of evidentiary coverage must surely arise on the part of investigator or advocate during the process of discovery. However compelling is the motive and means evidence against a certain suspect, a case against this person will fail unless it also includes evidence of opportunity. So, in the early stages of Act I, certain possibilities can often be eliminated on the basis of the most elemental tests; Suspect X cannot have committed the robbery in Des Moines since believable evidence places him in Duluth at the time of the crime. Cases which tax the ability of investigators/advocates are those in which several or many possibilities continue to remain viable in the face of evidentiary tests during the process of discovery. In such cases, identification of a single possibility may well hinge on any one of a very large number of subtleties associated with the evidence and the sources from which it comes.

As Cohen tells us,[76] the grading of possibilities, hypotheses, or theories in science rests upon an experimental process in which the circumstances are varied. As Bacon and Mill argued, we gather support for one theory over another by subjecting them to an ever-wider range of evidentiary tests. The weight of evidence favors the theory that best resists being invalidated by relevant evidentiary tests we construct. Further, our evidence is incomplete to the extent that we overlook potentially relevant evidentiary tests. Another way of stating this situation is to say that, among other things, science searches for "invariances," possibilities that seem to remain true in spite of the best efforts to invalidate them. In legal matters investigators and advocates attempt to generate or discern possibilities that will hold up in the face of the different categories of evidence relevant to the matters at issue. As I mentioned earlier, there is ample room for imagination and ingenuity both in the construction of relevant evidentiary tests in particular cases and in the selection of a strategy for performing them.

Cohen argues that probabilities, construed as chances, do not provide a suitable metric for grading the extent to which possibilities resist our attempts to invalidate them. His "inductive" probability system is an ordinal means for doing so. In this system the weight of evidence on some possibility is related to the amount of evidence and is indicated by the number of relevant evidentiary tests this possibility survives. We might recall the heuristic device of Sherlock Holmes: eliminate the impossible and the residuum, however improbable, must contain the truth. Cohen asks us not to confuse high probability, in the sense of Pascal/Bayes, and strong evidentiary weight, since the Pascal/Bayes system says nothing about how many relevant evidentiary tests we have employed in the process of elimination or about how completely the tests cover those that might have been employed. In short, the Pascal/Bayes system says nothing about the *completeness* of our evidence.

Now consider the "argument structuring" scene in Act I on discovery. I shall argue that Cohen and Pascal, Bayes & Co. have more to say about this challenging process than any of the other systems we are examining; however, Cohen's views are quite different from those of Pascal, Bayes & Co. First, consider "relational" structuring, in which arguments from evidence to major facts-in-issue are meticulously laid out. As noted earlier, Wigmore and several of his heirs (e.g., Twining, Anderson, Edwards, Schum) have appreciated the catenated, hierarchical, or cascaded nature of the structure of such arguments and the evidentiary subtleties such a structure reveals. Apparently, so does Cohen.[77] Perhaps better than any other matter, the structuring of argument lays bare the difference between Cohen's metrics for inductive proof and evidentiary weight and the Pascal/Bayes metrics for uncertainty.

Consider the very simplest case of a catenated or hierarchical argument structure, or, if you prefer, "an inference on an inference." Witness W_i reports the occurrence of event E; we label the testimony from W_i as E^*_i to distinguish it from event E itself. The necessity for such distinction was obvious to Wigmore and others.[78] Now, suppose that event E, if true, is relevant circumstantial evidence on some hypothesis H. So our chain of reasoning from evidence E^*_i to hypothesis H has two links, one from E^*_i to E and one from E to H. The latter involves the strength of the argument from E, if true, to H; the former involves the foundation for this argument and it rests on the credibility of Witness W_i. We need to supply some specific basis for each of these two reasoning steps. This basis seems to come in two parts. The first part consists of a general sanction for such reasoning and it has been called alternatively a "generalization,"[79] a "warrant," or a "license."[80] The second part consists of specific ancillary evidence which, when available, serves to back up a generalization, warrant, or license.

First consider the reasoning step from E^*_i to E. Can we conclude that E occurred from W_i's report of its occurrence? Cohen tells us that we can compile a list of relevant variables, the satisfaction of which will supply us with an appropriate license or warrant for inferring E from E^*_i. These variables might include: (i) observational capability, (ii) lack of bias, (iii) a reputation for veracity, (iv) testimony consistent with other information, and (v) composure during testimony. These are common-sense generalizations of the sort our legal system credits all of us with comprehending. Each of these points is, of course, a potentially relevant evidentiary test; in establishing the credibility of W_i, we perform such tests by gathering specific ancillary evidence, if available, on each of these points. If W_i passes these tests, then we are licensed to infer E based upon W_i's testimony E^*_i. In other words, what we believe true in general should also be true in this particular instance involving W_i. An inductive probability will grade the extent to which the credibility of W_i survives this series of tests.

Now consider the second reasoning step from E to H and suppose we can come up with another list of points which would sanction the inferring of H from E. This list would include entirely different relevant variables and possible evidentiary tests that they suggest. Another inductive probability grades the number of these points which seem to be satisfied in the present instance involving E and H. So now the question is: How can we combine the inductive probabilities at each of these two stages in order to grade the strength of an inference from E^*_i to H? Cohen says, quite simply, we cannot do so. Each of these stages of proof involves entirely different relevant variables or points, different evidentiary tests, and different standards for satisfaction on any of them. From a measurement point of view, we could not do so anyway, since inductive probabilities have only ordering properties and cannot be algebraically combined. The result is that inference up a reasoning chain, in Cohen's view, is rather an "all or none" affair; if we cannot, to our satisfaction, believe E based on E^*_i, the reasoning chain is completely severed at this point. The Pascal/Bayes system acknowledges all of the points made above about inference on inference, but takes an entirely different view about the aggregation of probabilities over the stages of inference.

Cohen has given attention to many of the evidential and inferential subtleties that are brought out in relational structuring. Among these subtleties are inferential transitivity and other order-related effects in reasoning chains, as well as evidential corroboration and convergence.[81] In fact, Cohen should be credited with stimulating current discussion about some of these important evidentiary subtleties, as I have acknowledged elsewhere.[82] Cohen's further concern about the causal relevance of evidentiary tests leads me to suspect that he would also see the importance of what I have called "temporal" structuring (that based on evidence chronologies). As I noted earlier, the order in which events occur is itself a source of probative value; Cohen's system of inductive probability acknowledges the frequent importance of careful ordering of evidentiary tests.

Let us now consider inductive probability and the final acts of proof and of deliberation and choice. We can begin by noting that Wigmore, in construing the inductive nature of judicial proof, makes reference to Mill's methods for (eliminative) induction.[83] He argued that they could play at least a subordinate role in our attempts to determine the certainty we have that "the alternative chosen is the right one out of all those conceivable."[84] So it might be thought that eliminative induction characterizes at least some aspects of the fact-finding process. Cohen relates the development and ordering of relevant variables to Mill's methods.[85] As stated earlier, Cohen argues that experience-given rules for the fact-finding process seem to conform to inductive rather than Pascalian probability. First consider the ISM of factfinders. Cohen argues that a small or near-zero prior probability on "guilt" does not capture the prior presumption of innocence as stated prescriptively by courts in criminal cases. Inductive probabilities have certain properties in common with Shafer's Bel, one of which is that zero does not indicate disproof but lack of proof (for Shafer, zero indicates lack of belief rather than disbelief). Further, an inductive probability of zero on guilt, unlike zero probability in the Pascal/Bayes system, can be revised upward in light of evidence, i.e., we can go from a state of lack of proof to various gradations of proof in the same way we can go from lack of belief to gradations of belief.

So, the fact-finding process may be, at least in part, eliminative in nature. Advocates present their respective arguments on the matters at issue and factfinders complete the process of elimination by deciding for one side or the other. Cohen credits the ordinary persons who perform the role of factfinding, whatever it involves, with the cognitive competence necessary to do the job.[86] If we construe the fact-finding process to involve a Pascalian balancing of probabilities, current research on heuristics and biases suggests we may have placed the important fact-finding responsibility in the wrong hands.[87] I must remark at this point that identification with Pascal, Bayes & Co. does not imply agreement with the stated implications of research on heuristics and biases. At least two of the authors in this book (Edwards and Schum) have spoken out against these stated implications as well as the adequacy of the research upon which they are based.

Let us now consider the final act, the act of deliberation and choice. Here we must consider Cohen's criteria for acceptance of hypotheses or possibilities and how these criteria concern the forensic standards of proof courts prescribe for the factfinders in reaching their verdicts. We have already discussed Cohen's advice that we not confuse a high value of Pascalian probability with strong weight of evidence, on the grounds that such probabilities are insensitive to completeness of evidentiary coverage. So, the forensic standard "beyond a reasonable doubt" in criminal cases seems to Cohen not to be necessarily implied by a high Bayesian posterior probability. He argues that "beyond a reasonable doubt" means that the evidence you have just heard and/or seen leaves no lingering reasons for you to doubt; in short, the prosecution has covered all its bases with regard to means, motive, and opportunity. Since inductive probabilities grade the number and completeness of evidentiary tests that have failed to invalidate a possibility or hypothesis, Cohen argues that they come much closer to the mark in representing "beyond a reasonable doubt."[88] In civil cases, where the standard is proof "on the balance of probabilities" or on a "preponderance of the evidence," Cohen argues that plaintiff wins any point on the preponderance of evidence if there is *more* believable evidence in his/her favor, something Pascalian probability does not indicate.[89]

Finally, we must consider the choice process and the articulation of inference and value assessment. In science, inference is part of a knowledge-acquisition process that does not always lead to a specific choice from among alternative courses of action. So, in the realm of science there is no particular reason to be concerned about how probabilities and values might best be combined in making a choice. But in legal matters factfinders do have choices to make, often in the face of drastic consequences to the parties in contention. As noted earlier, there are identified and respected methods for combining Pascalian probabilities and value assessments in the process of choice. Cohen's ordinal inductive probabilities fare no better in such combination than do Shafer's non-additive Bels. But Cohen still has a very important argument concerning the choice process that might go something like this: A high posterior probability on a certain possibility, in combination with a certain pattern of value assessments, may seem to force a particular choice. But unless this high posterior probability also indicates completeness of evidentiary coverage, our choice may seem quite unpleasant in light of other evidence we did not consider.

C. *The Pascal/Bayes School of Probability and Uncertainty*

Almost everyone agrees that, kept in its proper place, the Pascal/Bayes system of "conventional" probability is an enormously flexible and rich system for allowing us to express our uncertainties about a wide variety of matters of interest. The trouble seems to be one of determining "the proper place" for this system. We have just examined two recent views about areas in which the Pascal/Bayes system seems to be deficient. Shafer argues that it will not always do in grading our beliefs in epistemic situations and Cohen argues that it grades the wrong matters and overlooks important ingredients in the process of inductive reasoning, particularly in trials at law. Their points in criticism of Pascal, Bayes & Co. are well-taken and the alternative systems they propose deserve the careful attention of anyone interested in human reasoning on the basis of incomplete, inconclusive, and unreliable evidence.

In spite of my appreciation for what we have learned from Shafer, Cohen, and others yet to be considered, I believe that there is a definite place for Pascal, Bayes & Co. in our continuing efforts to better understand the process of inference and its evidentiary basis, regardless of the context in which inferences are necessary. I believe there to be evidential and inferential subtleties whose capture for study and analysis requires the degree of flexibility and richness evident in conventional probability. In fact, the Pascal/Bayes system may be superior to other extant systems in its ability to capture the many evidential subtleties that are commonplace in jurisprudence. While recognizing certain deficiencies in the Pascal/Bayes system, we can ill afford to ignore what this system can tell us about evidence.

Regarding Act I of inferential plays in jurisprudence, I make no claim that the Pascalian system in general, or Bayes's rule in particular, is informative about the process by which hypotheses and evidence are generated. Nor do I suppose that Bayes's rule represents the process of eliminating possibilities on the basis of evidentiary tests of the sort that Cohen seems to have in mind, i.e., those which can "falsify" or "invalidate" possibilities. Conventional probabilities are just one means of expressing the extent of our uncertainty about matters of interest, and Bayes's rule is one representation of how we might combine various uncertainties in the process of determining the relative likeliness of hypotheses or possibilities we take seriously. Though it is certainly true that investigators and advocates experience uncertainty throughout the discovery process, I shall begin by focusing on the uncertainties faced by the advocates in the scene I have labeled "argument structuring."

As we noted earlier, Binder and Bergman tell us that advocates need to analyze evidence on an item-by-item basis for two essential reasons: (i) to be able to show exactly what element each item proves, and (ii) to be able to evaluate the overall strength of the evidence. Wigmore's methods for relational structuring help to accomplish at least the first of these two objectives. What seems to be involved in the process of argument structuring is a careful consideration of the properties of evidence items we have when they are considered alone or in combination. Some evidence will be directly relevant in the sense that a reasoning chain can be set up from the evidence to major facts-in-issue. Other evidence will have indirect relevancy since it is, essentially, evidence about evidence. Indirectly relevant or ancillary evidence serves to strengthen or weaken a reasoning linkage. Argument structuring surely involves consideration of the many evidential subtleties as revealed in the literature on evidence law. Some of these subtleties concern events reported in the evidence; others concern the sources from which the evidence comes. It would seem that skill in advocacy depends in part upon knowing what questions one might ask of evidence.

Certain ingredients in Bayes's rule, when expanded to incorporate what are judged to be the reasoning linkages in an argument from evidence to major hypotheses or facts-in-issue, suggest appropriate and interesting questions to ask of evidence and can supply some answers about the probative weight of the evidence on major facts-in-issue. This point separates Shafer, Cohen, and the Bayesians; the concept of evidentiary or probative weight is entirely different in each of these systems. For Shafer, the weight of evidence on hypotheses is related to the *support* the evidence (pure or mixed) provides or fails to provide hypotheses; for Cohen, the weight of evidence and the *amount* of evidence are related. For the Bayesians, evidentiary or probative weight is graded by ingredients of Bayes's rule called likelihood ratios.[90] About the only weight-related matter on which these three systems appear to agree is that evidentiary or probative weight has vector-like properties, i.e., that evidence has force which gets applied in certain directions (toward certain hypotheses).

A likelihood ratio indicates the likelihood of an item (or items) of directly relevant evidence under one hypothesis or possibility relative to its (their) likelihood under another hypothesis or possibility. There is a constraint here which I mentioned earlier in connection with the distinction between pure and mixed evidence. For any evidence we say is inconclusive, we must suppose that it has non-zero likelihood under both hypotheses involved in a likelihood ratio. This simply says that we believe inconclusive evidence to be to some degree consistent with the truth of either hypothesis. Evidence supports the hypothesis under which its likelihood is greater and by the amount indicated in the ratio; it supports neither one if this ratio is 1.0. So, effectively, a likelihood ratio treats all inconclusive evidence as if it were mixed evidence; there is no likelihood ratio representation for pure evidence, that which offers support to one hypothesis and none at all to others. As noted by several persons, a likelihood ratio determination of probative weight seems entirely consistent with evidentiary relevancy as it is defined in Federal Rule of Evidence 401.[91]

My own earlier dissatisfaction with Bayes's rule as a canon for probabilistic inference took a different form and produced different results than did the expressed dissatisfactions of Shafer and Cohen. In its pristine form, Bayes's rule says virtually nothing about the intricacies of probabilistic inference nor about the array of commonly observed subtleties in evidence. In my attempts to make it say something about these intricacies and subtleties, I had to be better informed about evidence. So, I began, now over 15 years ago, to read about how scholars of evidence in jurisprudence think about their stock-in-trade. I was especially curious about how evidence is categorized in jurisprudence and how it is used in arguments in a trial at law; I had already come to Wigmore's conclusion that virtually all arguments from evidence are, when carefully analyzed, catenated, hierarchical, or cascaded in nature. I reasoned that Bayes's rule could be made responsive to the different forms and types of evidence when considered alone and in combination, and to the subtleties apparent in reasoning catenations. The major mechanism for doing so involves formulating appropriate likelihood ratio expressions for various species of evidence and for the manner in which they can become linked to major facts-in-issue and to each other. Each link identified in a chain of reasoning exposes a source of uncertainty. A likelihood ratio, appropriately formulated, allows you to incorporate every source of uncertainty you recognize and also prescribes how these uncertainties ought to be combined in the process of assessing the probative weight of the evidence setting up the reasoning chain.

Within the Pascal/Bayes system there is a single well-defined mechanism for trapping evidential and inferential subtleties; this mechanism involves the concepts of conditional independence and non-independence. An amazing array of subtleties in evidence and in the behavior of sources of evidence find expression by means of appropriate patterns of conditional non-independence of events involved in likelihood ratio expressions. Some of these subtleties appear in simple and uncomplicated reasoning chains; others appear when evidence combinations of various sorts are considered. Conditional non-independence is a good example of a simple concept that turns out to have profound consequences. In a recent paper Anne Martin and I reviewed our work on the trapping of evidential subtleties within the Pascal/Bayes system.[92] Here follows a different kind of summary of this work.

Table 1 below contains a listing of the forms and types of evidence and evidentiary situations to which we have given attention in our formal studies of the process of assessing the inferential or probative weight of evidence. This table represents a somewhat different classification of evidence than many of you will have encountered; however, it does not appear to be inconsistent with more conventional ways of categorizing evidence. Across the top of this table we distinguish between directly relevant and indirectly relevant (or ancillary) evidence. You may recall my earlier assertion that directly relevant evidence sets up a reasoning chain to major facts-in-issue.

	Directly Relevant Evidence		Indirectly Relevant Or Ancillary Evidence
	Direct	Circumstantial	
"Real" Or Demonstrative (+ or -)	*	*	*
Testimonial (+ or -)	*	*	*
Missing (e.g. silence, non-production)	*	*	*
Source Behavior (e.g. equivocation)	*	*	*
Accepted Facts	—	—	*

Table 1

Directly relevant evidence can either be direct or circumstantial in relation to some interim or final fact-in-issue. Along the side of this table appear the general forms of specific evidence and evidentiary "situations" which may also supply probative value. Specific evidence is conventionally thought of as being "testimonial" or "real" (i.e., demonstrative). The plus and minus under each of these forms of evidence indicates whether the evidence asserts the occurrence or non-occurrence of a certain event or events. The report of the non-occurrence of an event is sometimes called "negative" evidence.

But, in addition to specific evidence of a "real" or "testimonial" nature, there are other evidentiary situations which are sources of probative value. One involves missing evidence and commonly involves silence on the part of witnesses or the non-production of evidence. Another involves the behavior of sources of evidence. One of the most interesting features of Bayesian analysis is that you are encouraged to incorporate the fact that the observational and reporting behavior of a source of evidence may be, by itself, a possible origin of probative value. One good example involves equivocation (sandbagging or stonewalling, if you prefer), but there are many others as well. The bottom category simply acknowledges that judicially noticed evidence or an "accepted fact" seems usually to be introduced as ancillary evidence.

Likelihood ratio formulations for certain of these forms and types of directly relevant evidence have been particularly informative, even about especially difficult kinds of evidence involving hearsay and alibi testimony. But there is a wide variety of evidentiary subtleties to be considered when evidence items of these varying species are taken in combination. Following is a listing of subtleties we have examined for combinations of two or more evidence items.

(1) *Contradictory And Conflicting Evidence*. Dissonant evidence comes in two varieties: (i) two or more sources make some pattern of contradictory assertions about the same event(s), or (ii) two or more sources make assertions supporting different possibilities or hypotheses. Bayesian likelihood ratio formulations illustrate the difference between these two forms of dissonance and specify a means for resolving either form of dissonance. In either form of dissonance these formulations make clear that "majority rule" is *not* the mechanism by which evidentiary dissonance is resolved.

(2) *Corroborative And Convergent Evidence*. In one form of corroboration, two or more sources testify to the same event(s). Convergent evidence involves testimony or other reports about different events all of which point inferentially in the same direction. A particularly interesting form of convergence involves the possibility that one item of evidence can act to enhance the value of another.

(3) *Redundant And Non-redundant Evidence*. Bayesian formulations make clear the distinction between "corroborative" and "cumulative" redundancy, specify the exact locus of possible redundancy in evidence, and even supply means for measuring the extent of redundancy. Non-redundant evidence also has different forms which Bayesian analysis distinguishes.

My argument, in short, is that the Pascal/Bayes system can be informative about extremely specific evidentiary matters that arise in the process of argument structuring. Though this system stumbles in handling "ignorance," "doubt," "lack of belief or proof," and does not distinguish between "pure" and "mixed" evidence, here are just a few of the specific kinds of information that it gives us instead: (i) the incorporation of valuable ideas from the theory of signal detection in the study of the behavior of sources of evidence, and how their behavior influences the probative value of what these sources tell us; (ii) the incorporation of statistical communications theory in the study of redundancy and other properties of evidence; (iii) the justification for assigning extra probative value to sources who testify against preference; (iv) specific information about the nature of the interaction between source credibility and rareness of reported events in determining the value of testimony; (v) information about the effects of locating a weak link or a rare event at different points in a chain of reasoning; (vi) information about why a potential redundancy problem lurks behind every instance of contradiction; and (vii) an explanation why missing or equivocal evidence can often have more probative value than the known occurrence of the events for which evidence is missing or for which there is equivocation. Thus, I take strong exception to Shafer's statement that "Bayesian belief functions are awkward for the representation of evidence."[93] I believe that many inferential and evidentiary subtleties find ready expression within the Pascal/Bayes system, provided that appropriate argument structures underlying these subtleties are made clear.

Bayes's rule has been taken as a canon for the fact-finding process in our Act II on proof. Such use supposes that the inference task in factfinding is one in which subjectively expressed uncertainties are evaluated and combined by factfinders in their final assessment of the relative likeliness of major facts-in-issue in light of evidence. It can be argued that, though we have feelings of uncertainty, we do not have numbers floating about in our heads. Conventional "canonical examples" used with reference to the Pascal/Bayes system frequently involve either "chances" or "betting odds." We ask a person to convey the extent of his/her uncertainty by stating chances or by quoting odds. This has been unfortunate, in a way, since it appears to tie the Pascal/Bayes system irrevocably to classical or to relative frequency interpretations of probability and uncertainty. Many persons seeking to illustrate the "subjective" or "personal" interpretations of conventional probability have used such betting or chance analogies.

Feelings of uncertainty, as well as degrees of belief, are examples of subjective dimensions called "prothetic" dimensions;[94] such dimensions indicate "how much" (e.g., how loud) rather than "what kind" (e.g., what color). We commonly talk about having more or less uncertainty but not about having different kinds of uncertainty. Now, there are very many dependable studies of human sensory and perceptual judgments of intensity involving the use of numerical scales in which it has never been thought necessary to invent "canonical examples" to motivate judgments using these scales. The literature in the area of sensory psychophysics supplies a wealth of examples.[95] In short, conventional probability scales and their possible transformations such as odds or log odds scales do not need canonical examples to make them useful in studies of human probability assessments. Some writers on the "subjective" or "personal" view of probability seem to have brought this problem on their own heads.

Finally, it would be smug to say that the Pascal/Bayes system has exactly what it takes to make it congenial to the process of choice; after all, expected utility choice models assume the same properties possessed by conventional probabilities. But not everyone is convinced that these models have the last word on canons for choice. Our next view has something to say about these and other matters.

D. *The Lotfi Zadeh School of Fuzzy Probability and Inference*

After we listen to the evidence in a certain felony trial, you ask me what probability I would assign to defendant's innocence of the charged offense, on the evidence we have just heard. I tell you that this probability is 0.87; you then ask me how I managed to arrive at this precise number. I hedge a bit and say in response: "Well, for me, this probability could actually be anywhere between, say, 0.70 and 0.95; 0.87 seemed to be a reasonable best estimate." Nor would Lotfi Zadeh, or any of his army of followers, believe that I was capable of such estimative precision in my original statement. He would go even further in suggesting that the interval of probabilities I gave you is not an appropriate way in which to hedge on the matter at issue here, which Zadeh argues to be my *imprecision*. So I back off completely from the use of numbers and say: "On the evidence we have just heard, I believe it *very probable* that defendant is innocent of the charge." What I have now given you is an example of what Zadeh calls a "fuzzy" probability. What is interesting, and in many ways remarkable, is that Zadeh has given us a formal means for assessing and combining such "fuzzy" judgments in inference, decision, and in many other situations. Zadeh's original work[96] on fuzzy sets, those sets whose membership boundaries are indistinct or "fuzzy," has now generated an enormous amount of published research; in many ways, Zadeh's original paper is still the best tutorial on the subject.

Zadeh argues that in our judgments and discourse, we repeatedly employ what he calls *linguistic variables*. As this name implies, a linguistic variable is a word which can take on several possible states. For example, the word "height" is a linguistic variable if it takes on values such as "very short," "short," "not short," "tall," "very tall." Consider "age" as another linguistic variable; its values might be "very young," "young," "not young," "old," "very old." Zadeh shows us how to give numerical expression to such "fuzzy" statements and how to combine them in discourse involving connectives such as "and," "or," and "not." So, for example, we can quantify a statement such as: "Mary is tall but not very young." The common-sense reasoning that we are credited with having by our own legal system appears to involve fuzzy statements. Here is an example Zadeh has used of fuzzy reasoning: "If a car which is offered for sale is cheap and old, then it is probably not in good shape."[97] This statement involves several linguistic variables: "price," "age," "condition" (of the car), and the one that here concerns us, "probability."

The foundation for classical mathematical systems, including the Pascal/Bayes system of probability, is provided by two-valued logic: a statement can either be true or false, or a particular element either is or is not a member of some set of elements. There are no intermediate truth values or no gradations of set membership. The trouble, says Zadeh, is that most of our everyday reasoning concerns propositions, events, or sets whose boundaries are fuzzy; for example, the sets "tall males," "cheap cars," "old persons" are all fuzzy in nature. What we have to do is to grade the *degree* of set membership of each element of concern to us; we do so by using numbers in the interval (0,1), where zero indicates that the element is definitely not in the set, one indicates that it is definitely in the set, and numbers between zero and one indicate various gradations of set membership. Consider the fuzzy set "tall males" and four particular males whose heights in feet and inches are: 4'10", 5'8", 6'1", and 6'10". Most of us would say that the 4'10" male is not tall and so we would say that this person is "0.0 tall"; the 6'10" male is definitely tall or "1.0 tall." We might say that the 5'8" person is "0.4 tall" and that the 6'1" person is "0.7 tall." In short, the numbers between zero and one are used here to indicate our necessary imprecision in assigning elements to sets or categories whose boundaries are indistinct.

So, each element of a fuzzy set gets assigned to it a certain grade of membership in the set of concern. In handling various combinations of ordinary sets, events, or statements, we make use of the operations of unions, intersections, and complementations as these connectives are defined in the conventional theory of sets. As Zadeh tells us, a particular algebra is necessary for these connectives in order to represent combinations of fuzzy sets. Another important ingredient in Zadeh's system is a "fuzzy quantifier"; such quantifiers are expressed by words such as "most," "several," "least," "usually," "frequently," "more," "very," and so on. Zadeh shows us how we can, in fuzzy logic, draw fuzzy conclusions from fuzzy premises involving fuzzily quantified statements. Let us return to the notion of a fuzzy probability and its place in our current discussion of the acts and scenes in inferential plays in jurisprudence.

It seems safe to say that all of the actors in our inferential plays in jurisprudence utter many of their "lines" in the form of fuzzy statements; the ones that concern us are statements involving such concepts as "probability," "uncertainty," "belief," "support," and "possibility." In the act of discovery, for example, an investigator might say "it is *quite possible* that X did the deed"; here "quite" is a fuzzy quantifier. A factfinder might think, as I finally did in the example above, "it is *very probable* that the defendant is innocent on the evidence I just heard." Zadeh argues that the underlying probabilities in many situations are not known with sufficient precision and that our imprecision is not to be graded by probability intervals such as the one I used in the example that introduced the present discussion. Here is a brief example of a fuzzy analysis of this same example.

When pressed away from my unbelievably precise probability statement and from my hedge on it involving probability interval, I simply said: "I believe it *very probable* that defendant is, on the evidence, innocent." This is a fuzzily quantified statement. Such a statement can be given numerical form using the fuzzy concept of a "possibility." What I wish to grade is the "possibility" that various probability values between zero and one mean "very probable." I grade the possibility of any probability being "very probable" by assigning to it a number between zero and one. In doing so, I might say that any probability value less than 0.7, for me, has zero possibility of representing "very probable"; neither does any probability value greater than 0.95, to which I would assign a more extreme quantifier such as "very, very probable" or "extremely probable." So, for me, probability values in the interval 0.70 and 0.95 have some non-zero degree of "possibility" that they are "very probable." Further, suppose I say that 0.87 has the greatest possibility of being "very probable" and that possibility declines systematically on either side of 0.87. Such a pattern of judgments on my part leads to a precise graph of numerically expressed possibilities for any probability between zero and one. Using the concept of a "level" set, I can identify the range of probability values that has at least any specified degree of possibility of being "very probable." Such an interval, Zadeh argues, is a more adequate representation of my fuzziness concerning what I mean by the statement "very probable."

So, we have now discussed three judgmental intervals: Shafer's interval [Bel(A), P*(A)], the interval of subjective Pascalian probabilities I used as a hedge, and now a Zadeh "possibility" interval. These three intervals do not indicate the same thing. If you recall, the Shafer interval represents a hedge against further evidence in allocating support provided by a present body of evidence to a set A of possibilities. The Pascalian interval is sometimes referred to as a "second-order" probability. The interval I estimated (0.7 to 0.95) I used to convey my uncertainty about the uncertainty reflected in my precise estimate of 0.87, which you argued was too precise. The Zadeh possibility interval, just discussed, is a direct hedge, not on my uncertainty, but on my imprecision.

Zadeh's system, like those of Shafer, Cohen, and Bayes, is not a theory of the generation of hypotheses or evidence. Though there is a great degree of similarity between the connectives for conjunction and disjunction in Cohen's and Zadeh's systems, I cannot say that Zadeh's system is as specific a theory of possibility or hypothesis elimination as is Cohen's. In my view, the best and most specific examples of fuzzy probability within the context of jurisprudence occur in the process of argument structuring when advocates attempt to be specific in stating "generalizations" or "warrants" for particular reasoning stages in an argument from evidence to major hypotheses or facts-in-issue. Here is a specific example selected from the discussion of generalizations by Binder and Bergman.[98]

Consider the following four variations of a generalization of the sort encountered in legal argument.

"People who purchase guns subsequently used in a robbery:

(1) almost always are the robber,"
(2) usually are the robber,"
(3) sometimes are the robber,"
(4) rarely are the robber."

The underlined words in these generalizations are fuzzy quantifiers in the way Zadeh describes. Each of these generalizations can be expressed as a fuzzy conditional probability. Take (2) for example: we may say that the probability of a person's being a robber, given that he/she purchased the gun used in the robbery, is "usually," where "usually" is a fuzzy number in the interval of numbers between zero and one. The same applies to any of the other fuzzily quantified generalizations.

In a court trial, in the act of proof, every effort seems to be made to "de-fuzzify" the testimony of witnesses and other forms of evidence. Ordinary witnesses must state what they observed as unequivocally as possible. In spite of this, factfinders may encounter fuzzy aspects of evidence as well as instructions provided by the court. The forensic standards of proof, "beyond a reasonable doubt" and "preponderance of evidence," can easily be construed as fuzzy statements. Add to this the fuzzy or inexact way in which advocates and the court may interpret the evidence for factfinders. So it can be argued that probability and inference theories resting upon precise mathematical foundations are not particularly suited to represent the factfinding process.

So Zadeh can argue that the hierarchical, catenated, or cascaded arguments so common in jurisprudence are examples of fuzzy inference. Examples, such as the one just given, provide a stick with which Zadeh can beat Bayesians, including the present author, who apply the precision of Pascal, Bayes, & Co. to the analysis of arguments whose ingredients are acknowledged to be fuzzy. Though I have not seen any application of Zadeh's fuzzy logic and probability to cascaded inference of the degree of complexity apparent in Wigmorean relational structuring, I see no reason why such applications could not be made.

Finally, consider deliberation and choice process. It can easily be supposed that we have fuzzy values in the same way we have fuzzy probabilities. Conventional choice models, such as expected utility maximization, rest upon more or less precisely judged probability and value-related ingredients. Consider the consequences of acquitting a guilty defendant or convicting an innocent defendant; not only do these consequences have many attributes, but they involve judgments on these attributes that are seldom precise. There have been recent attempts to develop fuzzy decision models which incorporate both fuzzy probability and fuzzy value or utility assessments.[99]

E. *The Scandinavian School of Evidentiary Value*

The commonly understood purpose of applying Bayes's rule to the process of factfinding is to determine the (posterior) probability of major hypotheses or facts-in-issue in light of the evidence. Another view, currently held by a collection of Scandinavian evidence scholars (including, by the way, one renegade Englishman), is that this Bayesian objective is misguided. They argue that, instead of determining the probability of some hypothesis, conditional on evidence, we should be determining the probability that the evidence *proves* this hypothesis. In this view there is a particular focus on the value of the evidence rather than on the probability of hypotheses; therefore, this view and its consequences have been labeled the "evidentiary value model" [EVM]. A recent collection of papers provides a good look at the essentials of the EVM.[100]

The basis for the EVM seems to have come from the ideas of the Swedish jurist Per Olof Ekelöf. In a recent summary of the development of his views,[101] Ekelöf tells us of the growth of his interest in what a Swedish judge must focus upon and how this seems to concern the evidence itself rather than the "themes" or hypotheses to be proven. Obviously, Ekelöf has many of the same insights about the fact-finding process as did Wigmore. Ekelöf distinguishes between directly relevant evidence and "auxiliary" (ancillary) evidence; he notes that an "item" of evidence is an evidentiary fact together with "auxiliary" facts attached to it; he discusses how the influence of auxiliary facts depends on "commonplace generalizations"; he notes the existence of chains of evidence; and he discusses various subtleties associated with evidence items when they are taken in combination.[102]

In the EVM, evidence has value to the extent that some hypothesis, fact-in-issue, or "theme" can be validly deduced from it; thus, it is argued, we should focus on grading the extent to which evidence "proves" some hypothesis or theme. But all evidence comes from sources having any gradation of credibility. Within the EVM, any source of evidence, be it human or otherwise, is viewed as an "evidentiary mechanism."[103] The basis for the EVM resides in our beliefs about the extent to which our evidentiary mechanisms are "working properly." Here is an example of how the EVM grades the probability that evidence proves some hypothesis or theme, based on what is known about the source of the evidence.

Consider a major fact-in-issue or "theme" H and a direct testimonial assertion about the truth of H made by Witness W_i. Here we label W_i's assertion as H^*_i to distinguish it from H itself. In short, W_i is an eyewitness to the fact-in-issue. Since H^*_i is direct evidence on H, there is no intermediate reasoning step between H^*_i and H; i.e., this is not a catenated or cascaded inference. The only inferential issue here is whether or not W_i is credible as a source of evidence. Applying Bayes's rule to this problem we would determine $P(H|H^*_i)$, the probability of theme H, in light of evidence H^*_i. Here, says the EVM, is where we go wrong. Witness W_i is an evidentiary mechanism that may or may not be working properly; in fact, it is argued, W_i might have testified H^*_i without any real knowledge of the event(s) in H. As Freeling and Sahlin put it, "a witness in a court case who does not have any relevant information may nevertheless correctly testify to the guilt of the defendant."[104] So, in this view, we face the unpleasant possibility of convicting a defendant on the basis of correct evidence given "by accident" from an improperly working evidentiary mechanism.

Here is what the EVM suggests we should be considering instead of $P(H|H^*_i)$. We have two possibilities: A = the event that evidentiary mechanism W_i is "working properly," and not-A = the event that W_i is "not working properly." Since W_i has given direct evidence on H, event A entails H; i.e., if event A is true, then event H is true. What we have to grade, says the EVM, is the conditional probability $P(A|H^*_i)$, the probability that the mechanism is working properly, given evidence H^*_i. We should here recall Ekelof's comment that an evidentiary fact also consists of whatever "auxiliary" evidence is attached to it. So, in our case, H^*_i presumably also includes whatever information is provided for the court about Witness W_i. Now, within the EVM, $P(A|H^*_i)$ is thought to be an ordinary Pascalian conditional probability, i.e., $P(A|H^*_i) = P(A \text{ and } H^*_i)/P(H^*_i)$. Since event A implies or entails H, it must be true, within the Pascal/Bayes system, that $P(H|H^*_i)$ is greater than or equal to $P(A|H^*_i)$.

So, within the EVM, focus is upon conditional probabilities such as $P(A|H^*_i)$. Examples have been provided about the application of the EVM to situations involving various mixtures of pure and mixed evidence, as distinguished above, and for the combined evidence from two or more "evidentiary mechanisms."[105] It turns out that the EVM allows the same distinctions between pure and mixed evidence that Shafer's system allows. The affinity between the EVM and Shafer's system has not gone unnoticed and at least one person argues that ideas from EVM help to put Shafer's system on a better footing.[106] Our present concern is to place the EVM within the acts and scenes of inferential plays in jurisprudence. The EVM, of course, is not a representation for the scenes of generation and elimination in Act I on discovery. I should like to place it within the "argument structuring" scene in this act, but I cannot do so. On the surface, given Ekelof's and Wigmore's similar view, it could be expected that the EVM might allow some important additional insights into the complex process of hierarchical argument structuring. The trouble is that, at least in its present state of emergence, the EVM seems to ignore the hierarchical nature of argument from evidence. The basic idea of EVM seems to work well for direct evidence of the sort given in the example above. But I have not yet seen it applied to *circumstantial* evidence in which the evidence itself may be many steps logically removed from major facts-in-issue. Here is an example of the difficulty.

Suppose, instead of asserting H directly, Witness W_i had asserted the occurrence of an event that was just circumstantial on H. For example, W_i might have testified, not that he saw defendant commit the robbery, but simply that defendant was near the scene of the robbery at the time it was committed. If evidentiary mechanism W_i is working properly, i.e., even A is true, then we may justifiably conclude that defendant was at the scene. But his being at the scene does not mean that he committed the robbery.

It may be argued that the Bayesian likelihood ratio formulations for the weight of evidence do not allow for the representation of situations in which a witness has no real knowledge of the event(s) in his/her testimony and gives "correct" evidence "by accident." On the other hand, they do allow for the capturing of all of the specific grounds for impeaching and supporting witness credibility recognized by our Anglo-American courts.[107] They do so by making much more specific the array of observational sensitivity, motivational, and veracity grounds upon which the credibility of a source is impeached or defended; the vagueness of evidentiary mechanisms "not working properly" and "giving true evidence by accident" cannot capture the behavioral richness of the task of credibility assessment.

IV. Final Remarks and Some Possibly Fuzzy Conclusions

My remarks in this paper stem from a discussion I had with Peter Tillers over a year ago on the extent to which the "probability debates" are useful to persons with interests in evidence law. We both agreed that these debates are certainly enjoyable, at least to the participants. We also agreed at the time that "inference" is, after all, a rather "fuzzy" concept covering, as it does, a wide assortment of mental activity. We thought it might be true that the different views of probability, to some extent, concern different attributes of inferential activity. If so, we thought, someone ought to sort out the inferential and evidential attributes that do or do not intersect with the alternative theories of probability discussed in the debates. In subsequently examining attributes of inference and evidence, Tillers and I have taken somewhat different routes. In his mapping of the domains of inference, Tillers considers various configurations of inference problems as experienced by triers of fact and the extent to which theories of probability seem to bear upon these configurations.[108] I have taken a different tack by examining what alternative theories of probability and inference have to say about the activities of the various actors involved in the legal processes of discovery, proof, and deliberation and choice.

My present analysis of the contributions of alternative theories of probability and inference has not caused me to revise a certain conclusion I admit to having reached before I even began this analysis. My conclusion is that the inferential activity within jurisprudence represents an ideal locus for "airing out" alternative theories of probability and inference. The reason is that at least the major acts of discovery, proof, and deliberation and choice occur in a definite order; in no other context known to me is this necessarily true. In medicine, business, and in other contexts we may have all three of these acts on stage at the same time. When inferences involve possible future events, it is inevitable that the process of discovery is ongoing, whatever else is happening. In science, even though there may be a reduced "choice" element, we have at least discovery, deliberation, and proof processes going on simultaneously. Though I did experience some difficulty in deciding what scenes belonged in certain acts, I would have faced a preposterously difficult task in sorting out the five alternative systems in any other context. In many contexts, the application of probability models is looked upon with disfavor. One reason is that it is difficult to "extrude" mixtures of ongoing discovery, proof, and choice processes through any of the "templates" provided by the theories we have examined. Thus, for example, a Bayesian representation of some probabilistic reasoning task is not satisfying in situations in which discovery is still taking place and new or revised possibilities are being generated or discerned.

As I noted at the outset, each of the five views we have considered has its own particular language and/or concepts, and translations from one system to another are not always possible. In my discussion of these systems I have avoided the use of mathematics; thus, I may have purchased some readability at the expense of considerable precision. Different persons examining these five systems might well observe different distinctions among them or emphasize different points. For example, I have had very little to say about how numbers are used to indicate inference-related ingredients in these five systems. A useful discussion of the assumed scale properties of the numerical ingredients in each of these systems and a careful assessment of the behavioral implications of these scale properties would be a considerable task on its own.

Another point I did not emphasize concerns various formal relations that exist between certain of these systems. For example, Shafer has discussed the conditions under which both the Bayesian and Cohen systems seem to emerge as special cases within the Shafer-Dempster system.[109] In addition, I noted the similarity of the connectives in Cohen's and Zadeh's systems and the fact that the EVM calculation of the probability that an "evidentiary mechanism" is working is basically a Bayesian process. On the issue of "general" and "special" cases, an additional word seems necessary. It is frequently important in mathematics to have some set of expressions that cover "general" cases. For example, there is a general model of cascaded or hierarchical inference. Some years ago, Kelly and Barclay[110] formulated a general Bayesian algorithm for cascaded inference, provided only that (Pascalian) probability is being applied to discrete events. This algorithm turns out to be as mathematically impressive as it is uninformative; it answers no specific questions about the evidence upon which inferences are based. Such information has come from examination of very special cases of cascaded inference.

I have attempted to put each one of these alternative systems of probability and inference in its best light; I hope I have done so, in part because I believe that the system with which I have been associated has not always been presented to its best advantage in critical analyses. Here are some conclusions that seem to follow from the analysis I have made.

(1) The five systems we have discussed certainly do concern themselves with different attributes of inference when we examine all the acts and scenes of an inferential play. The Shafer-Dempster system captures possible credal states likely to be experienced during the act of discovery and about which other systems have little or nothing to say. Similarly, Cohen's inductive system grades the amount and completeness of coverage of evidence in the process of eliminating possibilities, and looks particularly interesting in those acts and scenes in which inferential roles can be construed as eliminative in nature. The Pascal/Bayes system can answer specific questions about specific forms and combinations of evidence and combines easily with assessment in the act of choice. Lotfi Zadeh reminds us that imprecision is omnipresent and suggests that there are ways to cope with it more straightforwardly.

(2) The various systems exhibit different patterns of apparent trade-offs involving breadth and depth of coverage of inferential attributes. The Shafer-Dempster system certainly exhibits breadth of coverage in capturing various possible credal states, but is found wanting in capturing the details of probative value assessment. My comment above about general and special cases was, by the way, not intended as a swipe at Shafer. Indeed, as I have noted, the Shafer-Dempster view captures certain details that elude some of its special cases. The Pascal/Bayes system looks better in later rather than in earlier scenes of our inferential plays, and offers depth of coverage in argument structuring. Cohen's system of inductive probability has a fair amount of breadth and depth. It can be argued that the processes of discov-

ery, proof, and deliberation represent different stages in a process of elimination. Cohen does consider the details of argument structuring even though his system does not allow for the aggregation of any uncertainties associated with identified reasoning stages. Zadeh's system certainly exhibits breadth since many of the inferential ingredients in any act and scene are fuzzy in nature.

(3) Any probability model billed as a "model for human inference" should specify whose inferences are of concern and what aspect of inference is at issue. The various actors in inferential plays of interest in jurisprudence have demonstrably different inferential roles and the theories we have been discussing bear upon these inferential roles to varying degrees. There may be a message here for behavioral researchers. We are told that people are poor at inference tasks because their probability assessments do not conform to the canons of Pascal, Bayes & Co. Perhaps Sherlock Holmes himself might have given non-Bayesian responses to the blue bus problem (he probably would have, since he appears to have been a Baconian). But he would more than compensate for this apparent weakness by discerning some shrewd possibilities and evidentiary tests of these hypotheses that might surpass the best efforts even of those who passed unfavorable judgment on him because of his performance on the blue bus problem. Let us suppose that it is finally recognized by psychologists that human inference is many-sided. We now have demonstrations of Bayes incoherence in probability revision tasks. If psychologists continue to focus upon human inadequacy, perhaps we can expect future demonstrations of Cohen incoherence in eliminative induction, Shafer/Dempster incoherence in evidentiary support combination, and Zadeh incoherence in coping with imprecision. Let us hope, however, that future behavioral research will adopt more imaginative objectives.

(4) There are certain things that are not, to my knowledge, said within any view of probability and inference I have discussed. One assertion not made goes something like this: "Reason with us and fewer errors will be made at trial." No sensible person would make such a statement absent information about what evidence was being considered and how it was evaluated. These alternative systems concern the *process* by which evidence is evaluated and combined as well as the array of considerations that appear important along the way. No probability system can guarantee the adequacy of the many judgmental ingredients any of these systems require. In addition, I have never seen or heard advocates of any of these systems argue in favor of the overt incorporation and combination of numbers on the part of factfinders or others at trial. To my knowledge, no one is contemplating the recommendation that hand calculators or computer terminals be issued to factfinders for

ready use in combining their own probability judgments in the act of reaching a conclusion. In one sense, the numerical aspects of these systems do not matter, but their arguments about evidentiary and inferential processes do. The use of numbers in particular examples simply facilitates the process of understanding what these arguments tell us.

(5) I have argued that the various systems of probability appear to bear somewhat differently upon the major acts of discovery, proof, and deliberation and choice. An interesting and difficult question concerns the relative importance of these acts. Suppose that there is one act whose importance clearly outweighs that of any other act. The view of probability most congenial to this act might then seem to be favored, at least because of its greater bearing upon the act having the greatest apparent importance. The trouble, of course, is that we would find it very difficult to settle on the order of precedence of these three acts in terms of their importance in human inference and decision tasks. Consequently, different persons may quite legitimately place varying importance on these acts, thus drawing different conclusions about the relative merit of the systems of inference and probability I have discussed. My present analysis assumes that these three major acts are of equal importance; at no point have I argued that one act is more important than the others. Thus, my placement of *any* of the systems of probability in certain acts and scenes should not, by itself, be taken as an indication of my perception of the merit of this system.

(6) It may be argued that this paper represents a defense of a platitude, namely, that human inference has many attributes and that no one system of inference and probability covers all of these attributes. My belief that we ought to attend carefully to what each of these views tells us about inference may seem to be a tepid conclusion drawn, perhaps, by someone who seeks to avoid public confrontation with colleagues and friends. The truth is that, if I actually believed that the "dutch-book" theorem says all that there is to be said about human inferential rationality, I would say so and endorse the Pascal/Bayes system; but this is not what I believe. If I believed that all inference of concern to jurists and others was basically eliminative in nature and that there is no virtue in studying how our uncertainties might be assessed and combined, I would willingly defer to Jonathan Cohen and suspend all further analyses of likelihood ratio expressions for evidentiary weight that I have planned. (If I ever decide to suspend this activity, it will be because I have finally had a fill of tedium and not because I believe that such analyses accomplish nothing.)

In her very considerate comments on this paper Lea Brilmayer took up the plight of the reader who may well be left hanging by my present analysis.

Though I have found a place in inferential plays where each of the systems of probability and inference has interesting things to say, I have not tied these systems of thought together in any systematic way. It would, as she suggested, be useful within jurisprudence to have some generalization or meta-theory that would be of assistance in linking these alternative systems and in showing the specific conditions under which each system might be useful. In response to her suggestion, I replied that, if it made sense to do so and if it could be done, such a meta-theory would probably result from the efforts of someone having the breadth and depth of view about human inference that Jonathan Cohen and Glenn Shafer have so clearly demonstrated.

NOTES

† © 1986 by David A. Schum.

* David A. Schum is Professor of Systems Engineering at George Mason University, Fairfax, Virginia. This paper was presented at the Boston University School of Law Symposium on Probability and Inference in the Law of Evidence, April 4-6, 1986.

[1] J. WIGMORE, THE SCIENCE OF JUDICIAL PROOF 8 (3d ed. 1937).

[2] Finkelstein & Fairley, *A Bayesian Approach to Identification Evidence*, 83 HARV. L. REV. 489 (1970).

[3] Tribe, *Trial by Mathematics: Precision and Ritual in the Legal Process*, 84 HARV. L. REV. 1329 (1971).

[4] *See, e.g.,* R. LEMPERT & S. SALTZBURG, A MODERN APPROACH TO EVIDENCE (1977); Lempert, *Modeling Relevance*, 75 MICH. L. REV. 1021 (1977).

[5] Schum, *On the Behavioral Richness of Cascaded Inference Models: Examples In Jurisprudence*, in 2 COGNITIVE THEORY 149-73 (J. Castellan, D. Pisoni & G. Potts eds. 1977).

[6] L. COHEN, THE PROBABLE AND THE PROVABLE (1977).

[7] Schum, *A Review of a Case Against Blaise Pascal and His Heirs*, 77 MICH. L. REV. 446 (1979).

[8] *Id.* at 446.

[9] G. SHAFER, A MATHEMATICAL THEORY OF EVIDENCE (1976).

[10] Zadeh, *Fuzzy Sets,* 8 INFORMATION & CONTROL 338 (1965).

[11] *See, e.g.,* EVIDENTIARY VALUE: PHILOSOPHICAL, JUDICIAL, AND PSYCHOLOGICAL ASPECTS OF A THEORY (P. Gardenfors, B. Hanson & N. Sahlin eds. 1983).

[12] L. Zadeh, *On The Validity Of Dempster's Rule Of Combination Of Evidence*, Memorandum No. UCB/ERL M79/24, Electronics Research Laboratory, University of California, Berkeley (1979).

[13] EVIDENTIARY VALUE: PHILOSOPHICAL, JUDICIAL, AND PSYCHOLOGICAL ASPECTS OF A THEORY, *supra* note 11.

[14] D. BINDER & P. BERGMAN, FACT INVESTIGATION: FROM HYPOTHESIS TO PROOF 191 (1984).

[15] J. WIGMORE, *supra* note 1, at 55-156.
[16] D. BINDER & P. BERGMAN, *supra* note 14, at 35.
[17] *Id.* at 90-91.
[18] J. WIGMORE, *supra* note 1, at 13; 1A J. WIGMORE, EVIDENCE 1113 (P. Tillers rev. ed. 1983); Schum, *Behavioral Richness, supra* note 5.
[19] J. WIGMORE, *supra* note 1, at 858-81.
[20] W. TWINING & T. ANDERSON, ANALYSIS OF EVIDENCE (forthcoming).
[21] D. VON WINTERFELDT & W. EDWARDS, DECISION ANALYSIS AND BEHAVIORAL RESEARCH (forthcoming).
[22] *See generally* R. KEETON, TRIAL TACTICS AND METHODS (1973).
[23] W. SALMON, THE FOUNDATIONS OF SCIENTIFIC INFERENCE 111-14 (1967).
[24] Peirce, *Abduction and Induction,* in PHILOSOPHICAL WRITINGS OF PEIRCE 151 (J. Buchler ed. 1955).
[25] N. HANSON, PATTERNS OF DISCOVERY 85 (1965).
[26] J. VENN, THE PRINCIPLES OF INDUCTIVE LOGIC (2d ed. 1907).
[27] W. SALMON, *supra* note 23, at 140.
[28] I. HACKING, THE EMERGENCE OF PROBABILITY 75 (1965).
[29] I. LEVI, THE ENTERPRISE OF KNOWLEDGE: AN ESSAY ON KNOWLEDGE, CREDAL PROBABILITY, AND CHANCE 43 (1980).
[30] *Id.* at 44.
[31] I. HACKING, *supra* note 28, at 75.
[32] L. COHEN, THE IMPLICATIONS OF INDUCTION 124-34 (1973).
[33] *Id.* at 128 n.2.
[34] THE SIGN OF THREE: DUPIN, HOLMES, PEIRCE (U. Eco & T. Sebeok eds. 1983).
[35] *E.g.,* Winograd, *Extended Inference Modes in Reasoning by Computer Systems,* in APPLICATIONS OF INDUCTIVE LOGIC 133 (L. Cohen & M. Hesse eds. 1980).
[36] L. COHEN, *supra* note 32, at 144.
[37] D. BINDER & P. BERGMAN, *supra* note 14, at 31; R. LEMPERT & S. SALTZBURG, A MODERN APPROACH TO EVIDENCE 109 (1977).
[38] J. WIGMORE, *supra* note 1, at 18.
[39] Schum, *Sorting out the Effects of Witness Sensitivity and Response Criterion Placement upon the Inferential Value of Testimonial Evidence,* 27 ORGANIZATIONAL BEHAV. & HUM. PERFORMANCE 153 (1981).
[40] Truzzi, *Sherlock Holmes: Applied Social Psychologist,* in THE SIGN OF THREE: DUPIN, HOLMES, PEIRCE 70 (U. Eco & T. Sebeok eds. 1983).
[41] W. BARING-GOULD, THE ANNOTATED SHERLOCK HOLMES 17 (1967).
[42] Twining, *Taking Facts Seriously,* 22 J. LEGAL EDUC. 22-42 (1984).
[43] W. TWINING & T. ANDERSON, *supra* note 20.
[44] D. BINDER & P. BERGMAN, *supra* note 14, at xvii.
[45] *Id.*
[46] *Id.* at xviii.
[47] *Id.*
[48] J. WIGMORE, *supra* note 1, at 31.
[49] S. TOULMIN, THE USES OF ARGUMENT (1964).

[50] Schum & Martin, *Formal and Empirical Research on Cascaded Inference in Jurisprudence,* 17 LAW & SOC'Y REV. 105 (1982).

[51] D. VON WINTERFELDT & W. EDWARDS, *supra* note 21.

[52] D. KAHNEMAN, P. SLOVIC & A. TVERSKY, JUDGMENT UNDER UNCERTAINTY: HEURISTICS AND BIASES (1982).

[53] *E.g.*, Schum & Martin, *supra* note 50, at 131-42.

[54] Cohen, *Can Human Irrationality Be Experimentally Demonstrated?*, 4 BEHAV. & BRAIN SCI. 317 (1981); Cohen, *On the Psychology of Prediction: Whose is the Fallacy?*, 8 COGNITION 89-92 (1980).

[55] D. VON WINTERFELDT & W. EDWARDS, *supra* note 21.

[56] Phillips, *Theoretical Perspectives on Heuristics and Biases in Probabilistic Thinking*, in ANALYZING AND AIDING DECISION PROCESSES (P. Humphreys, O. Svenson & A. Vari eds. 1983).

[57] R. HASTIE, S. PENROD & N. PENNINGTON, INSIDE THE JURY (1983).

[58] I have made no record of the number of ideas I have plagiarized from Ward Edwards; this is one I will acknowledge. As far as I know, Edwards has not yet said this in print, but we have discussed the matter on several occasions. One result of such discussions is that, when carefully considered, the act of turning an inference into a decision may not induce much simplification after all.

[59] G. SHAFER, A MATHEMATICAL THEORY OF EVIDENCE (1976).

[60] *Id.* at 208.

[61] I. HACKING, *supra* note 28, at 151-53.

[62] G. SHAFER, *supra* note 59, at 77-78.

[63] *Id.* at 223-26.

[64] I believe that some of the controversy here may be a tempest in a teapot. I constructed some examples of Shafer-Dempster and Bayesian "conflict-resolution" that were as qualitatively equivalent as I knew how to make them. The general results of this analysis convinced me that, at least under the conditions in my examples, these two systems do not say anything different about what happens when you combine conflicting evidence.

[65] G. SHAFER, *supra* note 59, at 285.

[66] *Id.* at 57.

[67] *Id.* at 178.

[68] Gordon & Shortliffe, *The Shafer-Dempster Theory of Evidence*, in RULE-BASED EXPERT SYSTEMS 292 (B. Buchana & E. Shortliffe eds. 1984).

[69] G. SHAFER, *supra* note 59, at 251-55.

[70] *Id.* at 27.

[71] Shafer, *Constructive Probability*, 48 SYNTHESES 1 (1981).

[72] *E.g.*, Finkelstein & Fairley, *supra* note 2; Tribe, *supra* note 3.

[73] *See, e.g.*, Kaye, *The Laws of Probability and the Laws of the Land*, 47 U. CHI. L. REV. 34-56 (1979).

[74] L. COHEN, *supra* note 6, at 49-120.

[75] Schum, *supra* note 7.

[76] L. COHEN, *supra* note 32, at 124-41.

[77] L. COHEN, THE PROBABLE AND THE PROVABLE 68-73 (1977).
[78] *See, e.g.*, J. KEYNES, A TREATISE ON PROBABILITY (1921); J. WIGMORE, *supra* note 1, at 13-15.
[79] D. BINDER & P. BERGMAN, *supra* note 14.
[80] S. TOULMIN, *supra* note 49, at 98.
[81] L. COHEN, *supra* note 6, at 93-115.
[82] D. Schum, A Bayesian Account of Transitivity and Other Order-Related Effects in Chains of Inferential Reasoning, Rice University Psychology Department Research Report #79-04 (1979).
[83] J. WIGMORE, *supra* note 1, at 38-43.
[84] *Id.* at 38.
[85] L. COHEN, *supra* note 6, at 245-46.
[86] Cohen, *Can Human Irrationality Be Experimentally Demonstrated?*, 4 BEHAV. & BRAIN SCI. 317 (1981).
[87] D. KAHNEMAN, P. SLOVIC & A. TVERSKY, *supra* note 52.
[88] L. COHEN, *supra* note 6, at 247-52.
[89] *Id.* at 252-56.
[90] I. GOOD, PROBABILITY AND THE WEIGHING OF EVIDENCE 63 (1950).
[91] R. LEMPERT & S. SALTZBURG, *supra* note 4, at 159; Lempert, *supra* note 4, at 1025; Schum, *supra* note 7, at 446-83.
[92] Schum & Martin, *supra* note 50, at 105-51.
[93] G. SHAFER, *supra* note 59, at 46.
[94] S. STEVENS, PSYCHOPHYSICS: INTRODUCTION TO ITS PERCEPTUAL, NEURAL, AND SOCIAL ASPECTS 12-13 (1975).
[95] *See id.*
[96] Zadeh, *supra* note 10.
[97] Zadeh, *A Theory Of Commonsense Knowledge*, Memorandum No. UCB/ERL M 83/26, Electronics Research Laboratory, University of California, Berkeley (1983).
[98] D. BINDER & P. BERGMAN, *supra* note 14, at 97.
[99] *E.g.*, Freeling, Fuzzy Sets and Decision Analysis, *IEEE Transactions on Systems, Man, and Cybernetics*, SMC-10, 1980.
[100] EVIDENTIARY VALUE: PHILOSOPHICAL, JUDICIAL, AND PSYCHOLOGICAL ASPECTS OF A THEORY, *supra* note 11.
[101] Ekelöf, *My Thoughts on Evidentiary Value*, in EVIDENTIARY VALUE: PHILOSOPHICAL, JUDICIAL, AND PSYCHOLOGICAL ASPECTS OF A THEORY, *supra* note 11, at 9-26.
[102] *Id.*
[103] *E.g.*, Freeling & Sahlin, *Combining Evidence*, in EVIDENTIARY VALUE: PHILOSOPHICAL, JUDICIAL, AND PSYCHOLOGICAL ASPECTS OF A THEORY, *supra* note 11, at 58.
[104] *Id.* at 59.
[105] *See id.*
[106] Levi, *Consonance, Dissonance, and Evidentiary Mechanisms*, in EVIDEN-

TIARY VALUE: PHILOSOPHICAL, JUDICIAL, AND PSYCHOLOGICAL ASPECTS OF A THEORY, *supra* note 11, at 27.

[107] Schum, *supra* note 5.
[108] *See generally* Tillers, *Mapping Inferential Domains*, *infra* at 277.
[109] G. SHAFER, *supra* note 59, at 44-46, 223-35.
[110] Kelly & Barclay, *A General Bayesian Model for Hierarchical Inference*, 10 ORGANIZATIONAL BEHAV. & HUM. PERFORMANCE 388-403 (1973).

David A. Schum,
Professor of Systems Engineering,
George Mason University.

WARD EDWARDS*

INSENSITIVITY, COMMITMENT, BELIEF, AND OTHER BAYESIAN VIRTUES, OR, WHO PUT THE SNAKE IN THE WARLORD'S BED?†

I need not comment on the ingenuity of Professor Schum's structuring of a legal proceeding as a drama, with acts, scenes, and characters.[1] His structuring is both right and clever. Indeed, I would extend his analogy by adding that different actors in such real-life dramas have different stakes in their outcomes, and that these stakes are, by the rules of law and quite in contrast to the rules of science, expected to influence how the reasoning process is handled. This is delicate ground. Perhaps 10 years from now, when the legal profession has adapted itself to the idea that its problems are little different from those faced by doctors, scientist, and others with explicit responsibility for inference and decision, we will all meet again to examine the tools for making values explicit. But explicitness about uncertainty is a topic quite large enough for this meeting.

The thrust of Professor Schum's paper is an attempt to be even-handed among Professors Shafer and Cohen, Bayes and neo-Bayesians, and Professor Zadeh. To do that, he argues that each position uniquely captures something of what goes on in the jurisprudential drama, and that none capture it all. No one can disagree with the latter point; any arguments would have to address the former, and in particular the assertion that points captured by the others had not already been captured by the neo-Bayesian view.

I will ignore the good words Professor Schum put in for Bayes and neo-Bayesians, and focus on the others mentioned. Schum's main point about Professor Shafer seems to be that Shafer captures the ideas of suspension of belief and its opposite, and that suspension of belief is a frequent element of the discovery process.

Is that believable? It is clearly true that Bayesians do not suspend belief in the sense of withholding probability; their probabilities always sum up to 1. It is also true that, as a commonsense idea, suspension of belief is valuable—so long as no action must be taken.

Yet middle-range probabilities are suspensions of belief. So, in a quite different sense, is the lack of full commitment to structurings of the data-generating process that Savage and many others have described. If I understood how Professor Shafer's thoughts contribute to our understanding of the data-generating process, I would commend and steal from him. That understanding is not resident in my head; I can only speculate about whether or not it resides in yours or his. David Schum has gone as far as anyone (except for Wigmore) by recognizing that such structures are typically complex, that often the same data can be represented in various different ways, and by looking at and classifying details of real data-generating processes.

The point that middle-range probabilites are suspensions of belief is perhaps best illustrated by considering the kinds of actions to which they lead. The most obvious point is that one is far more likely, from a strict Bayesian viewpoint, to accept the action that consists of gathering more evidence (if it is available) for middle-range probabilities than for extreme ones. Even that statement would require some careful hedging about the decision problem, the available acts, and so on. Yet it is true for a Bayesian that extreme probabilities (if well-calibrated) are far preferable to non-extreme ones, as decompositions of proper scoring rules easily show.

The universality of the multinomial model (or even the binomial one) in legal applications makes middle-range numbers the only form of suspension of belief conveniently available. As Schum points out in his discussion of Professor Zadeh's work, other numerical expressions of suspensions of belief, and indeed of degrees of suspension of belief, are possible and inevitable within the Bayesian viewpoint.

That brings me to a major omission from Schum's paper, and indeed from most discussion of Bayesian thinking in legal contexts. Every decision analyst takes for granted that sensitivity analysis is a central element of the analytic process. In its crudest and least interesting form, sensitivity analysis simply explores the importance of precision in number estimation by studying what happens when the numbers are varied. Much of David Schum's research has consisted of doing exactly that, and many of his insights have emerged from so doing. A still more important form of sensitivity analysis, however, consists of exploring the degree to which alternative formal structurings of the same inference or decision problem will lead to the same result. While Schum has done a lot of that too (though usually without labeling it as such) it appears to be missing from his contribution to this symposium.

In most decision-analytic contexts, sensitivity analysis typically leads to the comforting conclusion that the choice of an act is quite insensitive to the numbers that go into the process, and nearly as insensitive to model structure. Unfortunately, the one part of decision-analytic thinking to which that comforting conclusion cannot be expected to apply is inference. The reason for the difference is straightforward enough: virtually all of the other parts of decision analysis depend on what are technically known as linear models, or close relatives to them in some cases. Such models are generally highly insensitive for mathematical reasons about which I have published but will not review here. Unfortunately, the fundamentals of probability theory do not lead to linear models of inference, and it is easy to show that linear models of inference (which have been rather absurdly proposed and used in some artificial intelligence contexts) can often produce absurd results. We do not currently know what aspects of hierarchical Bayesian inference are highly sensitive to their numerical inputs and what aspects are not. Meanwhile, however, the safe thing to do is to expect that inference procedures will be sensitive to all their inputs. Perhaps I should add, however, that in most legal contexts of which I am aware, inference is translated into decision, and decision is the final output. The output of an inference process, therefore, should be appropriate but usually need not be precise; in general, decision processes are rather insensitive to their inputs.

Of course, sensitivity analysis, particularly when it is exploring sensitivity to model structure, is a form of suspension of belief, though not the same as the one Professor Shafer advocates. For as confirmed a Bayesian as I, it wouldn't be a bad metaphor to say that the inferential stages of a trial are an elaborate form of sensitivity analysis looking at the relation between the conclusion and the evidence bearing on it.

While multinomial formulations fit legal structures rather well, I need not remind anyone that many uncertainties are about means, and means have variances. I conclude that, while Professor Shafer does indeed call attention to the need for representation of partial and/or partially suspended belief, the Bayesians (a) did so earlier, and (b) have very attractive and sophisticated ways of meeting that need.

Professor Schum's pro-Cohen point is that Cohen emphasizes complete coverage of topics by evidence, while Bayesian formulations do not even have any representation of completeness of coverage. Originally, I was much impressed by that point. Now I have formed the opinion that, although the point is valid, there is considerably less to it than meets the eye. I consider now how the awareness of a need for comprehensive coverage came about and the ways in which Bayesian analyses do indeed respond to that need.

I think it unlikely that the awareness of a need for comprehensiveness of evidential coverage came full-blown to some primordial advocate trying to figure out how to persuade the warlord that the warlord's brother, next in line for the warlordship, may not have been the one who put the snake in his bed. Rather, it grew out of a series of such experiences, accumulated over time, and eventually was captured by such categorizations of evidence as means, motive, and opportunity. It may or may not be true that the notion of comprehensiveness of evidential coverage, now that we have it, can contribute to our structuring of new inferential problems. I suspect that it does, but without a model of how such problems are to be structured, that must remain just a suspicion.

Of course there are models from probability theory that recognize such a need for evidential completeness. The most obvious class of such models is called fault trees. Within a fault tree, the concurrent occurrence of several events is typically required for a catastrophe to happen; we then calculate the probability of the catastrophe by using the product rule of probability to combine the probabilities of the individual events. As I think about completeness from this point of view, all it seems to be saying is that Bayesians should not forget the product rule for this particular form of non-independence. While it could be argued that Bayesian discussion of non-independence probabilistic inference in legal contexts has made too little of this form, that merely implies inadequacies of performance in the structuring phase of the problem, rather than anything more fundamental.

Of the theories Professor Schum raised, the one that bothers me most is not Cohen's, but Zadeh's. There can be little question of the facts that (a) we do not typically carry three-decimal probabilities around in our heads, awaiting verbal expression on demand, (b) we much prefer verbal to numerical expressions of uncertainties (and of most other graded quantities) precisely because we feel that we can capture the degree of imprecision we feel much better with words than with numbers, and (c) the most obvious dodges, such as using ranges instead of point estimates, continue to leave us uncomfortable.

These issues seem to me to be more concerned with comfort than with capturing facts about evidence. It is also true that many of us dislike making decisions, and try to put them off as long as we can, even though we can have no reasonable expectation, in many such cases, that putting them off will ultimately improve them or make them easier. No one that I am aware of has proposed this as an argument against the normative merits of a utility-maximizing theory of decisionmaking.

It is crucial to understanding this issue to realize that every empirically derived number is inherently uncertain. A far subtler but equally important point is that probabilities, unlike most other numbers, in a sense communicate their own uncertainties. In our narrow focus on the task of getting the best numbers we can justify we sometimes lose sight of the real concern, which is not those numbers, but rather the occurrence or nonoccurrence of the events to which they refer. The numbers are only a convenient way of saying something about our knowledge of those occurrences. They are particularly convenient in that they explicitly quantify (whenever none of them is 0 or 1) our uncertainty over whether the events did (or will) occur. While I admit that such quantification can be uncomfortable, I also have come to believe that it is so valuable that I no longer feel the need to state a range when giving a probability estimate. Instead, I think hard about the probability estimate, and then give it.

Still, Professor Zadeh is right; it is uncomfortable to give such explicit numbers, even if you are very accustomed to doing so. It is hard to remember that they carry their own built-in information about precision—not precision of the numbers, but precision of the statements "A occurred" and "A didn't occur." But I do not believe that Professor Zadeh's procedures are more likely to succeed than Sherman Kent's attempt to address the same problem, though they are far more sophisticated.

Sherman Kent was a history professor at Yale who left to join the OSS; he eventually became the Dean of American intelligence analysts. Early in the 1960's, he argued for the use of probabilities to quantify uncertainties in intelligence analysis. Recognizing the discomfort that such numerical assessments produce in analysts, he proposed that the way out was to associate specific words with specific ranges of probabilities, and suggested what he hoped would be an appropriate set of words and of word-to-number conversions. However, intelligence analysts continued to use familiar words in ways that made the most sense to them. This may not have been as unfortunate as Kent probably thought it was. Later research has shown that expert analysts, inexpert about probabilities and unfamiliar with Kent's idea, associate ranges of probability with the specific words Kent proposed, and indeed with others that are used to communicate uncertainty, quite different both from one analyst to another and from Kent's ranges. This is a cautionary finding, in my view, for those who advocate communicating uncertainty by using words. All in all, this historical incident counts as a gallant failure. Kent identified a very real problem, that of explicit assessment of uncertainties in intelligence analysis. He recognized that probabilities were the only appropriate way out, and that the key problem in using probabilites in intelligence contexts, as in legal ones, is that people inexperienced in doing so are uncomfortable about making numerical assessments. He proposed a solution, but it didn't work. To this day, in both intelligence and legal contexts, the problem remains completely unsolved.

I do not think Professor Zadeh's solution will work any better than Kent's did. Once one understands the Zadehian system, verbal statements will come to have specific numerical implications, just as Kent proposed. Similarly, as those specific implications come to be understood, one will be equally reluctant to commit to them. I am confident that human ingenuity in finding verbal ways to equivocate will more than outdistance Zadeh's ingenuity in finding models that translate such verbal equivocations into numbers.

Just as lawyers need clerks, any holder of an explicit and highly controversial view needs one or more outriders, defending a position that is even more extreme. Thus the protagonist's position can be seen as moderate and kindly, in contrast to the stridency and intolerance expressed by outriders. In this comment I have been playing the outrider role for David Schum. It is probably safe to predict that some other day our roles will be reversed. But for today my setting, amidst distinguished lawyers greatly accustomed to advocacy, may perhaps excuse me for exploring just how far advocacy of my side of the argument can go.

The defense rests.

NOTES

† © 1986 by Ward Edwards.

* Director, Social Science Research Institute, and Professor of Psychology and of Industrial and Systems Engineering, University of Southern California.

[1] Schum, *Probability and the Processes of Discovery, Proof, and Choice*, supra at 213.

Ward Edwards,
Professor of Psychology and Industrial and
Systems Engineering and Director,
Social Science Research Institute,
University of Southern California.

PETER TILLERS*

MAPPING INFERENTIAL DOMAINS†

The nature of rational inference from evidence has been a major concern of twentieth century legal scholars of evidence. Throughout most of the twentieth century, however, the theoretical approach of Anglo-American legal scholars has been remarkably homogeneous. Their theoretical perspective has its roots in British rational empiricism and is part of what Professor William Twining has called the "rationalist tradition of evidence scholarship."[1] In this tradition great emphasis is put on the role of generalizations and inferential reasoning. This rationalist approach to evidence and inference has dominated legal scholarship for most of the twentieth century.[2]

The virtually unquestioned preeminence of the rationalist approach was broken by the rise of Bayesianism.[3] This "mathematicist" approach treats uncertainty in inference as a species of probability and emphasizes the importance of combining probability values in a logically consistent way. There are now a substantial number of prominent evidence scholars who subscribe to some version of Bayesianism. Moreover, Bayesianism, though vigorously assaulted at various times, seems to be gathering force in the law school world.[4]

In 1977 Dr. L. Jonathan Cohen, an Oxford logician and philosopher of science, published *The Probable and the Provable*. This book sets forth a "Baconian" theory of probability and inference.[5] Cohen's theory draws on the philosophical tradition of British rational empiricism, and has many affinities with legal scholarship in the rationalist tradition.[6] In particular, Cohen's theory of inference, like that of Wigmore and other legal scholars, emphasizes the importance of generalizations.[7] The parallels between Cohen's theory and rationalist theories are close enough to warrant calling the latter Baconian as well and, without imputing all of the wrinkles of Cohen's theory to the rationalist legal scholars, I shall sometimes do so.

Cohen's theory has attracted a great deal of attention in non-legal circles.[8] Although it initially had relatively little impact on legal scholarship,[9] interest in his theory among legal scholars is now intense and still growing. Although Cohen's complex theory has many idiosyncratic, controversial,[10] and problematic[11] features, Dr. Cohen has effectively put to rest any possibility that the "rationalist tradition of evidence scholarship" will be seen as a relic of the past. For example, Professor William Twining[12] has found support in

Cohen's theory for his own general thesis that a generalization-based theory of inference, while having its limitations, remains a powerful theoretical approach.

The rise of Bayesianism and Cohen's rejuvenation of the epistemological foundations of the rationalist tradition make it likely that the debates about Bayesianism and Baconianism will play a large part in legal theorizing about inference in the years to come. This attention to fundamentals is a refreshing change,[13] but the focus on Bayesianism and Baconianism also generates at least two dangers. First, there is the danger of theoretical myopia and narrow-mindedness. Although both Bayesianism and Baconianism are powerful theoretical perspectives, they are not the only important perspectives on inference, probability and related matters. Also important are, for example, Lofti Zadeh's theory of fuzzy probability, various kinds of frequentist and presupposition theories of probability, Glenn Shafer's theory of belief functions, and "scenario" theories of inference and forecasting. While the rise of Bayesian legal theorizing has made the theoretical diet of legal scholars less monotonous, that diet must become still more rich and varied.

If the papers presented in this book are any example, there may not be much reason any more to worry about theoretical narrow-mindedness. There is, however, a second danger—a more elusive but still important one. The *character* of the debates between some Bayesians and Dr. Cohen sometimes suggests that the disputants believe that diverse theories are necessarily mutually exclusive. To be sure, the participants in these controversies are largely concerned with the relative power, elegance, and coherence of their respective theories, but at times the tenor of the rhetoric suggests that some of them believe that theories of inference that rest on disparate logical matrices are necessarily mutually incompatible. For example, Professors Stephen Fienberg and Mark Schervish seem to embrace this assumption when they say,

> The subjective Bayesian approach to inference is a normative theory. It gives rules [to be followed by all who accept] the axioms of the theory . . . . To criticize a normative theory one must criticize either the axioms or the properties which follow from them. One cannot, however, accept the axioms without accepting the ensuing properties. Ultimately, the axioms themselves must either be accepted or rejected.[14]

It is a mistake to think that theories which ascribe different properties to rational inference are necessarily locked in mortal combat.[15] This paper attacks that mistake. It does so because the mistake rests on a false view of the functions and limitations of theorizing about inference. This paper is an effort to develop a better general perspective on theorizing—one that offers a contrast to the assumptions sometimes built into the claim that subjective Bayesianism is a normative theory, and one that also makes it clear that diverse models of inference are both permissible and necessary. Dr. Cohen once suggested that Bayesianism and Baconianism are complementary theories.[16] While I do not embrace all of the features of Dr. Cohen's theory,[17] I do embrace the general spirit of his suggestion and I widen its import.

I argue that an adequate portrait of rational inference requires the use of at least several logically disparate theories of inference. The need for these various models of inference can be recognized if it is understood that the trusted forms of conventional inference are the only appropriate benchmark for assessing the explanatory power of new theoretical models of inference. I further argue that theories of inference can only be considered rational models of inference to the extent that they capture the logical properties of these traditional forms of inference.[18] These considerations lead me to conclude that different theories have different domains, and that a central task of theorizing is the effort to delimit and describe those domains. This paper shows that (1) rational inference has a least four (and probably more) different types of logical properties, (2) there are at least four different kinds of logic in inference, (3) these four portraits of the logical properties of inference are complementary and not mutually exclusive, and (4) these different portraits or models describe different domains of inference.[19] The conclusion that rational inference incorporates at least four distinctive types of logic and theory supports the further conclusion that a *general* theory of inference must have a loose kind of "metatheoretical"[20] dimension. If a general theory of inference is possible at all, it is possible only as a theory that maps the domains and respective territories of various more discrete theories, each of which describes just one type of logic in inference.[21] The paper's other purpose, of equal importance, is to describe or suggest what the "foreign relations" are among those theories and of the domains in which each holds sway. (The possibility of concurrent theoretical jurisdiction is not ruled out.)

Part I will set forth the concept of a "nomological structure," and will describe four general types of nomological structures. Part II will use several types of nomological structures to analyze several hypothetical problems. I will show how attentiveness to the logical patterns and structures which are imprinted in evidence by human beings facilitates a meaningful, if provisional, assessment of the domains of existing models of inference. This sort of mental focus also identifies the existence of certain patterns of inferential reasoning which hitherto have received little attention. Part III will pull together the general implications of the entire argument for the possibility of a general theory of inference. There I will address the difficult question of how it is possible that a theory oriented toward the clarification of ordinary inference can improve the quality of ordinary inference. I will also discuss the related question of how a description of ordinary inference can have a prescriptive character. Finally, I will add a few words about the verification of theories of inference.

I. The Place of Theory in Evidence and Inference

A. *Theory as a Part of the Problem: The Concept of a Nomological Structure.*

All theories of inference recognize the importance of theory *for* inference. Extant theories of inference, however, do not pay enough attention to the importance of theory *in* inference. Theory can have a bearing on inference in two quite different ways. On one hand, there is the type of theory found in Cohen's account of inference. Viewed broadly, this type of theory consists of theoretical statements that establish a belief in a connection between two distinct sets of events or phenomena. On the other hand, theory can be a constituent of the thing about which an inference is sought to be made. In this article I use the phrase "nomological structure" to refer to theories of this second kind. My thesis is that theory is built into all evidence—that nomological structures are part of the fabric of evidence. More abstractly stated, my thesis is that inferential problems have a nomological structure that has ontological status. By this I mean only that theory is a part of the fabric of all things that are taken as evidence.

There are two ways in which theory is present in evidence—which is to say there are two ways in which theory is a constituent of a problem[22] of evidence and inference. First, a kind of logic or logical structure can be a property of the thing about which guesses are being made. Label this Theory I. Second, theory can be imported into a problem of evidence and inference by the way in which the evidence is seen and configured by the decision-maker. Call this Theory II.

The gist of my thesis about the actual place of Theory I in inference can be described by the "black box" metaphor.[23] A black box is a box whose inner workings cannot be directly seen. Nonetheless, an observer, by watching the outside of the box, can see what things go into the box—inputs, or antecedent conditions—and can also see what things come out of the box—outputs, or consequent conditions. The black box is a processing mechanism that acts upon inputs to change them into outputs. Of course, the quality of our predictions about the outputs of the box is often enhanced if we are able to remove the lid and peer into the box. To peer into the box, however, we need only to infer the internal structure of the box. Conversely, actual inspection of the innards of the box is not synonymous with peering into the box. An informative understanding of the innards of the box involves more than just an appreciation of the physical characteristics of the innards. A relevant understanding of the innards of the box requires a grasp of their logical structure. The aim is not simply to understand the physical characteristics of the innards, but to understand how those physical characteristics transform inputs into outputs.

Suppose that an observer tries to understand the behavior of the box without peering into it. This effort is not intrinsically futile. It is sometimes possible to predict the behavior of a black box by watching inputs and outputs and by trying to determine the relationships between them. Thus, for example, one might make thousands or millions of observations and then construct elaborate formulas describing the observed relationships between inputs and outputs. This external point of view, however, is often an inferior basis for making predictions and guesses about the behavior of the box. If the observer merely describes the relative frequency with which various inputs have produced various outputs, his predictions may be poor because they lack an understanding of the box's processing mechanism, its logic. The observer will be unable, on the basis of observed relative frequencies alone, to predict the behavior of the box in novel situations. Pure relative frequency statements simply fail to describe the significance which the box's processing mechanism assigns to novel antecedent conditions. The difficulty, of course, is that every situation is in some respect a novel one. But if we understand the logic that is incorporated within the physical characteristics of the box, we can predict the significance of differences in antecedent conditions.

As already noted, the observer may be able to construct various transformation rules describing the conversion of inputs into outputs without actually peering into the box. Indeed, observations of the relative frequency of inputs and outputs may suggest various transformation rules that fairly accurately predict outputs in some novel situations. If the observer does this, however, he is effectively "looking" into the innards of the box. The transformation rules that he constructs amount to a description of the innards of the box. More precisely stated, those rules describe the logical structure of the box's processing mechanism, which is all that we are interested in.[24]

As illustrated by the black box metaphor, my claim about the importance of Theory I in inference is uncontroversial. If one assumes that things actually exist, it cannot seriously be doubted that an understanding of processing structures generally enables us to make better predictions of outcomes and results. However, my thesis about the place of Theory II in inference is controversial, and a large part of the paper is devoted to its defense. Stated broadly, the thesis is that theory exists in every problem of evidence. The general import of this thesis may be expressed by saying that problems of evidence and inference do not present themselves ready-made. They are conceptually[25] constructed[26] and organized by human beings. The human brain is like a black box that ingests raw data and outputs information and evidence. The output bears the imprint of the logical structure of the human processing mechanism. In short, evidence itself is a kind of construct, and the person who assesses evidence also constructs it.

The general thesis of the existence of theory[27] in all evidence and inference can be stated more precisely. It can be phrased as a set of three claims about evidence: (1) evidence, considered as part of a problem of inference, is always configured by human beings; (2) evidence has logical properties, and, considered as part of a problem of inference, has a nomological structure; and (3) the logical properties of evidence are taken as real and actual to the extent that evidence is taken as real and actual, i.e., the nomological structure of evidence must be accorded ontological status to the extent that evidence is accorded ontological status.

The import of the first of these three discrete claims is apparent. The second claim asserts that conceptual shaping of evidence leaves evidence with certain logical properties—that theoretical shaping gives evidence itself a logical imprint. The third claim may seem obscure, but amounts only to the simple proposition that we must take the logical patterns in evidence seriously to the extent that we think that the evidence that we see is "the real thing" for all practical purposes.[28]

My thesis about the omnipresence of theory in evidence is the linchpin of my argument. Once this crucial premise is granted, the argument summarized next is fairly straightforward.

There are various theoretical matters present in inferential problems. These theoretical or nomological components of inferential problems are themselves potential sources of uncertainty. A useful model of a specific inferential problem must describe the kind of nomological matter involved. Inferential problems, however, are varied in character. Hence, there must be not one, but various models of inference.[29] These models of inference include Bayesian and Baconian models, but there must be still other models of inference. This is so simply because the nomological structures present in evidence often have non-Bayesian and non-Baconian properties. Hence, a description of inference solely in terms of Bayes's Theorem or generalizations alone grossly oversimplifies the nature of inference.[30]

B. *A Typology of Models of Uncertainty and Inference: Four Kinds of Nomological Structures in Inferential Processes*

One major purpose of devising a formal model of inference from evidence is to describe the logical structure of statements about uncertainty. Bayesianism and Baconianism both describe certain properties of inference that are often present in rational inferential activity. For example, subjective Bayesianism accurately describes those properties of rational inference that require a comparison of (1) the trier's subjective estimate of the probability of the existence of certain information when he assumes that the disputed factual proposition is true, with (2) his subjective estimate of the probability of the existence of that information when he assumes that the disputed factual proposition is false. Similarly, in other cases, an evaluation of the strength of a proposed inference rests on the decisionmaker's view of whether there are any generalizations supporting the proposed inference and, if so, whether they outweigh any opposing generalizations.

Baconian and Bayesian models both describe the structure of rational inference because (1) each model identifies some properties that are sometimes actually found in problems of inference and (2) those properties are subject to rational analysis and discussion. Analysis of these properties can facilitate the decisionmaker's understanding of his own view of the nature of the problem of inference facing him. Neither of these models, of course, guarantees that the decisionmaker will draw the "correct" inference since neither model guarantees that the decisionmaker will configure a problem of inference according to some "objective" benchmark. However, this limitation does not mean that these models do not describe the structure of rational inference.[31]

While both the Bayesian model and the Baconian model do say something about the logical structure of inference in certain situations, they fail to identify other important properties of inference that may be partly responsible for the uncertainty of a proposed inference. Still, these additional properties are susceptible of rational discussion, argument and analysis. If it is granted that evidence has properties other than those described and employed in Bayesian and Baconian theory, it follows as a matter of course that the logical properties and implications of evidence are not always described by Bayesiansism and generalization-based theories of inference. It is familiar learning among formal theorists that the logical implications of evidence are a function of their properties and, if those properties vary, so do their logical implications. The entire issue is whether evidence can have such properties and, if so, whether they are susceptible to rational analysis. I answer both of these questions affirmatively.

In many configurations of inferential problems there are types of nomological structures[32] that do not answer to the descriptions afforded either by Bayesian or Baconian theory. Specifically, I submit that the following four[33] nomological structures are involved, tacitly or expressly, in a great many inferential problems:

1. Nomological structures that are images of physical objects or of events or circumstances in space and time. An example is a thought in my mind that projects an image of an airplane with a specific shape in a particular place.

2. Nomological structures that are sets of principles and operations describing the transformation of events or things A into events or things B. This nomological structure includes sets of principles and operations that describe transformations in nature known as stochastic processes.[34] An example is the description "$F = MA$."

3. Nomological structures consisting of relative frequency statements and rules regulating extrapolation from such statements. Also included are rules defining the domains of relative frequencies, generalizations, and rules (explicit or implicit) governing their use and scope. An example is the statement "Ninety-five percent of all Swedes are not Catholics."

4. Nomological structures that interpret the meaning attached to spatio-temporal events. An example is the rules of grammar, when used to determine the probable meaning of a statement or sentence, and the rules of etiquette, when used to decipher the meaning of a person's social behavior.

In any given inferential problem the first three of these nomological structures are always potentially present, and may have a distinct bearing on the assessment of the probability of some fact or some proposition about a fact.[35]

No one of these three nomological structures can stand by itself; they are mutually dependent. This means that the specific content of each structure within this group of three structures is affected by the content of the other two. These three structures, however, are logically incommensurable; no one formula can describe the extent to which one nomological structure destroys or undermines the implications of either of the other two. Nonetheless, the interdependence among them is a theoretical necessity, even though as a practical matter it is not always possible to see the role that all three structures play in shaping a particular complex of evidence. Hence, it is not always essential or possible to use all three structures to describe a particular inferential problem.[36] Additionally, the relationships among the first three nomological structures apply to all four structures when the inferential problem at hand involves an assessment of the meaning attached to a spatio-temporal event.[37] In the discussion that follows I will try to support these various claims and theses by discussing examples of various kinds of problems of inference.

II. Thought Experiments About the Logical Properties of the Shapes and Shaping of Evidence

A. *The Interdependence of Theory and Evidence*

1. The Theory-Dependence of Facts and Evidence: How Theory Shapes the Things in the Mind's Eye

In this section, as well as throughout the entire paper, I advance the argument that evidence does not speak for itself but, rather, something in the nature of theory serves to shape and structure the evidence that we think we see. My thesis, if construed as a claim about the constitutive functions of theoretical perspectives, is not novel. There is a well-developed body of literature in the philosophy of science which asserts that facts and data are theory-dependent, that the observer's theoretical perspectives or interests define the types of phenomena that can legitimately count as meaningful data or possible facts.[38] Granted, I do extend the theory-dependence thesis by arguing not only that facts are dependent on theory, but also that theoretical perspectives serve to characterize and shape any data or evidence that are taken as genuine. Here again, however, my general thesis is not novel. This variant of the constitutive role of theory—the shaping of experience by reason—is, of course, found in Immanuel Kant's philosophy and, even more prominently, in the work of neo-Kantian philosophers such as Ernst Cassirer. In a roughly similar vein, Gestalt psychologists have emphasized the extent to which the human organism shapes and constitutes the things it sees in its environment. An emphasis on the power of the observer's perspective to shape experience also appears in the literature on theories of inference, probability, and statistics. Thus, for example, Professor Glenn Shafer has argued that data can be partitioned in various ways, and that how it is partitioned, dissected, or put together depends on one's theoretical "frame of discernment."[39] Roughly analogous arguments have been made for some time by theorists such as Willard van Quine. In short, the general thesis I advance and discuss in this section, the power of theory to shape evidence, is not at all novel; my view that evidence never speaks for itself is embraced by many scholars in a wide variety of disciplines.

Consider a simple illustration of the theory-dependence of data and information. Figure 1 on the next page gives an example of the configuration of

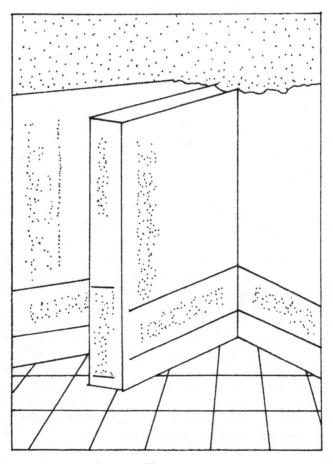

Figure 1

sense data into a spatial image. Figure 1 depicts what a hypothetical observer sees. The objects are rectangular, aren't they? Or are they? After all, if you measure the angles on the page, you will find that they are not right angles, and if you measure the lengths of the parallel sides, you will find that some of them are not of equal length. Nonetheless, while the observer may well think that Figure 1 accurately describes what he thinks he has seen (e.g., the observer might say that a photograph shows exactly what he saw), that same observer might also say that he saw rectangular objects. What happens, of course, is that the observer, either implicitly or explicitly, "straightens out" the angles and "evens up" the lengths of the parallel sides by making certain computations that allow for the effects of distance and perspective; he draws in his own mind rectangular objects that can take different positions in three-dimensional space.

My point, of course, is that the observer constructs the information he receives through his eyes into rectangular shapes in space, and that the information he receives through his eyes does not in itself show that the things he sees in space have a rectangular shape. This construction of a visual image is not a trivial act since, for example, it allows the observer to predict how he will see the objects from different perspectives and positions, something he cannot do if he believes that he only saw the shapes actually drawn in Figure 1. Moreover, Figure 1 also shows that the formation of the observer's belief in the existence of rectangular bodies in space involves a theoretical act. While it is true—as Gestalt psychologists have taught us— that the observer may not be able, by conscious thinking, to dispel his perception of a thing as having a certain shape, this sort of perceptual rigidity is not what concerns us. We are concerned with what the observer actually believes is out there, whatever his eyes may tell him. In fact, an observer in some situations will conclude that his eyes are deceiving him and that what is out there is not in fact what his eyes seem to suggest. In this event, the theory-dependence of information about spatial shapes becomes fully apparent. Thus, the accurate interpretation of the shape shown by a photograph often requires some explicit theorizing. Variations in perspective and focal length can easily distort the "true" shape of the thing recorded in a photographic image, and the effect of these variations can often be accurately assessed by precise mathematical calculations.[40]

2. Tacit Logic in Inference

As I have said, my thesis of the theory-dependence of evidence and information—of the manner in which theory affects the interpretation of information—is not original. Nonetheless, my analysis has several wrinkles, several emphases, that generate some important insights into the nature of inference and inferential reasoning. The first wrinkle is my emphasis on the extent to which the theoretical configuring of evidence is a tacit process rather than an explicit process. Thus, for example, suppose that the observer presented with the information in Figure 1 believes, without thinking about it very much, that he sees a rectangular object in space. To the extent that our general concern is with the quality of inferences, it is entirely appropriate to say that the hypothetical observer in this case has tacitly adopted a theoretically-generated inference of rectangular objects in space. This is an appropriate thing to say precisely because it is entirely possible for the observer to change his belief by thinking about its very foundations. The possibility of such a change of belief highlights the extent to which the observer's unexamined belief is theory-dependent.

Now the second wrinkle. As the discussion of Figure 1 suggests, the tacit character of the nomological shaping of information does not at all imply that tacit interpretations of evidence do not involve reason, theory, or logic. In fact, tacit interpretation often involves the use of theoretical processes that display a kind of logic. This emphasis on the logical character of tacit interpretive processes is one of the reasons why I have chosen to use the ungainly phrase nomological structures. The indirect allusion to the Greek notion of *nomos* serves as a reminder of the possibility that there is a kind of natural logic in things and information that at first sight simply seem to be there.

In sum, the shaping and structuring of experience by theory occurs at an implicit level as well as at an explicit and self-conscious level. In addition, this tacit shaping of evidence and experience can and should be seen as an activity that is conceptual or theoretical.

3. The Evidence-Dependence of Theory

The notion of theory-dependence of facts and evidence often seems to drift into or invite a sort of fact-skepticism since the thesis suggests an epistemological solipsism (the view that what we think we see or experience is entirely a function of our theoretical predispositions and beliefs). Nonetheless, epistemological solipsism or radical fact-skepticism is not a necessary implication of the thesis of the theory-dependence of facts. For my own part, I reject any form of radical fact-skepticism or epistemological solipsism. Indeed, I argue that theoretical structures are as much dependent on facts and evidence as facts and evidence are dependent on theoretical and nomological structures. The shaping of evidence by theory must have support in events or states of affairs that are taken as genuine data.

Assume once again that the problem at hand concerns the existence or non-existence at a particular point in time of a particular spatial event. For this purpose, examine the following diagram:

Figure 2

The solid lines in Figure 2 represent the shape of a thing in space that an observer sees or thinks he sees; the observer believes that light emanating from an object reveals the shape shown by the solid lines. The dotted lines, however, represent unseen portions of the object; something has obscured part of the observer's field of vision. (There are various ways in which this may happen even if there "really is" a rectangular shape out there. Thus, for example, the eyelashes of the observer may obscure part of his field of vision or the observer may have a blind spot in his eye.)

Figure 2 may be used to illustrate *both* (1) how theory or thought shapes facts and (2) how evidence and information shape theory. Consider first the theory-dependence of facts. It is quite possible that the observer, on seeing the shape shown by the solid lines, will decide that the shape of the entire thing he sees is a rectangle. If so, the observer has inferred that the shape of the unseen portion of the object is the shape shown by the dotted lines. This inference or extrapolation may have an explicit theoretical basis. It is quite possible that the observer has drawn that inference because he has decided that the solid portion of the shape indicates that the object is a man-made object. Typically, inanimate objects not shaped by human beings rarely have the property of being as smoothly rectilinear as the shape shown by the solid lines in Figure 2. On the basis of this reasoning (whatever we may think of its merits), he decides that the entire object is rectangular, perhaps because he believes that it is rare that man-made objects of the sort seen have both rectilinear shapes and other kinds of shapes. If the observer has reasoned this way, either explicitly or implicitly, he has used theory to constitute or infer a fact.

So once again we have an example of the theory-dependence of facts; with Figure 2, as with Figure 1, I have given an example of an image-generating nomological structure. Nonetheless, the example given by Figure 2 should also suggest that the formative power of theory is dependent on facts, information, data, and evidence. Brief reflection suggests that an observer who configures or sees a spatial shape in a particular way can be confident about the appropriateness of that particular configuration only if he believes he has some information on which to base his configuration. In my example, an observer has extrapolated the shape of the entire object from the shape of the visible portion of that object. The observer's uncertainty about the legitimacy of his extrapolation increases in proportion to his doubts about the accuracy of the information indicating the shape shown by the solid lines in Figure 2. Moreover, the extrapolation becomes entirely inappropriate in the observer's own eyes if it turns out that he believes that the information establishing the solid lines is non-existent. For example, if it turns out that the solid portion of the object is in fact highly irregular, the observer's inference about the unseen portion of the object is, in the observer's own theoretical or conceptual terms, entirely inappropriate and unfounded. Alternatively, the observer, either on further investigation or reflection, may decide that he was wrong in thinking that he did not "see" the portion of Figure 2 depicted by the dotted lines. Contrary to his earlier opinion, he may decide he did in fact see light emanating from the areas shown by the dotted lines. Depending on how he interprets this new information he may decide that the shape that he saw was not a rectangular one.

4. Some Realism about Evidence-Skepticism

It is of course true that the theory-dependence hypothesis can be applied to new information which the observer now accepts as legitimate. It can be rightly argued that the observer also constitutes the information or evidence that he now takes as genuine and reliable. Hence, it can also be argued that evidence lacks all independence and is merely the construct of the observer or factfinder. This argument proves too much, however, since it leads to an infinite regression. While any information can be analyzed as to how it has been put together by the observer's conceptual eyesight, an observer will not infer the existence of anything, whatever its shape, if he does not believe that he has some actual information to support that inference. It might, of course, be asserted that it is vain to suppose that "genuine" information can be found anywhere. This assertion, however, is gratuitous and in conflict with common experience. It is a rare person who will assert that an actual object has a particular shape if he believes he has no information except what he himself thinks. Most of us would consider such an observer demented.[41]

Inferences are coordinately determined by theory and evidence.[42] The shape of evidence is dependent on theoretical perspectives, but theoretical perspectives are also dependent on evidence. Hence, while evidence and information are, almost by definition, reports of a thing and are not the thing in itself, evidence may be so powerful in our minds that it overrides a theoretical perspective that stands at variance with it.[43] From this it follows that neither theory nor evidence has primacy as an abstract matter. Moreover, whether uncertainty about theory or uncertainty about the inferential implications of data has primacy depends altogether on the contours of a particular problem of evidence.

The paper now presents a series of thought experiments involving the question whether a particular person did or did not smile. My discussion will show that analysis of a factual question of this sort may depend primarily on the issue of how a certain bundle of information should be characterized and interpreted. Taken at the most general level, this discussion will illustrate the phenomenon of the theory-dependence of evidence. Conversely, however, my discussion of this hypothetical problem involving the smile will also illustrate the reverse form of this dependence, the evidence-dependence of theory and characterization. This analysis of the interactions between theory and information will show that the appropriateness of using particular models of inference such as Bayesianism and Baconianism is a function of the prominence of specific types of uncertainty in particular problems of evidence and inference. The remaining discussion in this section will examine in more detail the specific properties of particular theoretical approaches used in the interpretation of evidence. I will also examine the unique properties of a certain way of guessing about the aims and intentions of human beings.

B. *Interpretation of Evidence of Human Meaning: The Logic of Meaning and the Logic of Inferring Meaning.*

Problems of inference may require that guesses be made about human actors. Such guesses are uncertain due to our uncertainty about the aims, intentions, and meanings of human actors; since problems of inference require this same guessing, they possess the same uncertainty as the guesses themselves. The remaining discussion in this section focuses on this sort of uncertainty. My discussion of the properties of rational inferential reasoning about human behavior is broken down into three thematic topics: (1) problems of evidence demanding a mode of inferential reasoning that is not informatively portrayed either by Bayesian theory or by Baconian theories; (2) rational inference about human beings involving a story-based and coherentist mode of analysis and interpretation; and (3) the properties of a species of a coherentist mode of guessing uniquely applicable to guesses about human aims, intentions, and meanings. Taken as a whole, the argument in this part of the paper further substantiates and illustrates the general thesis of the paper about the nomological structuring of problems of evidence and inference.

1. Uncertainty about a Certain Smile: The Limits and Uses of Bayesian and Baconian Analysis

I shall now construct a problem of evidence and inference which cannot be adequately portrayed either by Bayesian or by Baconian models of inference. I shall also demonstrate that a problem of evidence and inference can be reconfigured or restructured so that it is amenable to either a Bayesian analysis or a Baconian analysis.

Please examine the following diagram:

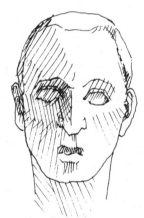

Figure 3

Figure 3 represents a face the observer saw at a particular time and place. The mouth, however, is not fully drawn since the observer did not see the entire mouth. The observer now wishes to determine the probability that the mouth was shaped thus:

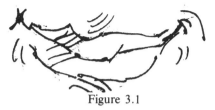

Figure 3.1

The observer believes it is also possible that the mouth was shaped thus:

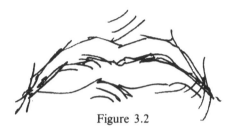

Figure 3.2

The observer may attack the factual question facing him in many different ways. Suppose he concentrates on the facial features shown in Figure 3. He decides the eyebrows are wrinkled and that the cheeks are dimpled. Moreover, he sees these facial features as indicating that the person shown in the diagram was in a humorous mood. On this assumption the observer then tries to assess the probability that the mouth had the shape shown in Figure 3.1. He frames the question in this particular way because he believes that the shape in Figure 3.1 is the shape of a smile.

With this sort of attack on the problem, the observer does not use Bayesian thought patterns in arriving at his subjective estimate of the probability of a smile. Instead, the observer imports the theoretical construct of humor into the problem. Bayesian analysis does not capture or portray the part of this inferential process that involves the appeal to the notion of humor. This is true even though the smile problem presents a situation in which the probability of the shape of the unseen portion of the mouth is in some degree affected by the shape of the visible parts of the face (dimpled cheeks, wrinkled eyebrows and the like). The intervention of the construct humor in the problem makes a decisive difference.

By using the notion of humor to resolve the problem, the observer resorts to a theoretical or nomological structure of his own making. In Bayesian terms, this observer is involved in thinking through the beliefs or theories that generate his subjective and relatively primitive assessments of the existence of the conditional relationships between evidence and factual hypotheses. Bayesian analysis, however, does not deal with the theoretical cause or basis of whatever assessments are taken as primitive; Bayesian analysis only deals with the coherent integration of the assessements that are taken as both primitive and subjective.[44] Hence, Bayesian analysis of the subjectivist stripe does not speak to the problem of inference as presented and structured here.[45]

It seems arguable that this problem retains a Bayesian dimension because the observer, even if drawing no direct and theoretical inference of a smile from the shape shown in Figure 3, seems to draw a theoretically unmediated inference of humor from evidence such as dimpled cheeks, furrowed eyebrows, and the like. In fact, however, this characterization of the inferential process is inaccurate. If the process of inference here is described more precisely, it becomes apparent that the problem that I have constructed excludes even the possibility of giving a Bayesian account of the first inferential step in the problem.

In the problem posed, the observer's first step is to decide how evidence and information about the shape of the cheeks and eyebrows is appropriately characterized and interpreted. His first step is *not* to decide whether the probability of the existence of humor is affected by the evidence of dimpled cheeks. The first step is rather to *examine* the shape of the cheeks and the eyebrows in order to decide how the cheek and eyebrow shapes he believes are reported by the evidence should be characterized. The initial description of the problem cavalierly referred to the shape of the cheeks as being dimpled. In fact, however, that description of the cheeks may already involve a characterization which is supportable only if the theoretical construct humor has already been built into the problem; only after the observer has imported the construct humor into the evidence can the cheeks appropriately be described as dimpled. This sort of inferential process is a theoretical process; it is equivalent to the process of thinking about theories in order to determine what connections should exist between two different things. Once again, however, Bayesianism does not portray the structure of this sort of theoretical process but takes instead an agnostic or indifferent attitude.[46]

It is possible, to be sure, that the *outcome* of my observer's analysis may be used and evaluated in a Bayesian fashion. Thus, for example, the observer may have other information that is relevant to the question of the shape of the mouth. In this event, it might be possible and fruitful to think of the observer as facing a Bayesian sort of problem, namely the problem of deciding how to adjust a prior estimate of probability on the basis of new information. Furthermore, it is also possible that the features of the face shown in Figure 3 are themselves subject to doubt and uncertainty. Thus, the features shown there must be treated as the report of, or evidence of, the shape and features of the visible portions of the face. In this situation as well, it is possible, though not indubitable, that a Bayesian description of at least part of the observer's inferential problem will be appropriate and informative.

The problem as posed, however, has been designed to exclude (among others) these two possibilities and complications. It has been implicity assumed that the observer himself believes that (1) he knows that the shape of the visible portion of the mouth is as is shown in Figure 3 or, alternatively, that he does not know how to question that belief, and that (2) the only pertinent information available to him is the information presented in Figure 3. This observer therefore cannot[47] describe the inferential problem that he sees himself as confronting by means of a Bayesian picture. The question for him[48] is whether a mood of humor entails a certain kind of facial expression.[49] The moral of this lesson is that whether Bayesian analysis helps to describe a problem of evidence and inference depends altogether on how the person faced with drawing an inference configures, structures, and sees the evidence and the problem facing him.

This relationship between the configuration of an inferential problem and the pertinence of specific models of inference can be illustrated further. I have described a problem of inference that cannot be informatively analyzed in Bayesian terms. But the problem might have been structured in a way that makes a Bayesian analysis informative. Thus, for example, had the observer faced the problem described below, at least part of the inferential project facing him could have been appropriately described in Bayesian terms. Suppose, as before, that the observer is confident that he did see the mouth-shape shown in Figure 3, but assume that this time he also saw some faint light images that appeared in his visual field at the places shown by the broken lines at either corner of the mouth on the following diagram:

Figure 3.3

In this situation a Bayesian analysis might be appropriate and descriptive of the character of the observer's inferential project. The observer thus might first form a preliminary estimate of the probability of a Figure 3.1 shape of the "unseen" portion of the mouth by considering the fully visible features of the face. With his "Bayesian prior" in hand, the observer might then estimate (1) the probability of the appearance of the light images described by the broken lines (in Figure 3.3) if the mouth did in fact have the shape shown by Figure 3.1, and (2) the probability of the appearance of those light images if the mouth did not in fact have that shape. The observer can then make a ratio out of these two estimates, turning them into the likelihood ratio,[50] and, using Bayes's Theorem, he can now calculate his own estimate of the probability that the shape of the mouth was in fact that shown by Figure 3.1.[51]

But my hypothetical observer in fact faced the problem based on Figure 3 rather than the problem depicted by Figure 3.3. As I have shown, this makes all the difference.

We are not substantially better able to think about the smile problem if we resort to a Baconian model of inference. Baconian theories are generalization-based theories of inference. Such a model is not responsive to the properties of the smile problem that occupy the observer's attention. Baconian theorists assert that inference consists of a transition from one factual proposition to another; they assert that the warrant for a connection between two factual propositions must be a generalization. The sort of inferential process involved in the smile problem, however, is not amenable to generalization-based treatment. The reason why is quite simple; the theoretical constructs being used are not generalizations and the theoretical processes being used involve more than arrays of generalizations.

Suppose a Baconian tries to portray the role that humor plays in the smile problem by referring to humor as disposition. He describes humor as an event that tends to be associated with smiling. If the observer talks about humor as a disposition that tends to be displayed in conduct such as smiling, the Baconian may advise him to express himself more clearly. The Baconian sees disposition or tendency as a shorthand expression for relative frequency.[52] To talk about the tendency of humor to be displayed by smiling is effectively to talk about the relative frequency of smiling and humor. The Baconian advises the observer to articulate as clearly as possible the generalization that the observer believes most accurately expresses the relative frequency of smiling and humor; he advises the observer to use that generalization as the basis for his ultimate inference. The Baconian argues that this analytical process amounts to rational analysis of inference, and that any observer who engages in rational inferential reasoning goes through this process.

Depending on the circumstances, the Baconian may have given poor advice about the path to clear-headed thinking about the smile problem. To be sure, the Baconian is right in saying that it is important to think about theoretical constructs that serve to explain the relationship between separate factual propositions. He is also right in saying that it is often helpful to talk explicitly about those theoretical constructs. The Baconian, however, advises us to use theoretical constructs that take the form of relative frequency statements. This advice may miss the mark, as humor can be an actual state of affairs as well as a theoretical construct imported into a problem by the observer. The character of humor as a real-world state of affairs is probably not described by any talk about relative frequencies. It is also not adequately described by talk in a more colloquial vein about dispositions, inclinations, or tendencies. Humor as an actual state has its own logical fabric. Thus, in the smile problem, *the observer's uncertainty about the existence of a smile may well depend on his uncertainty about the logical fabric of the smiler's humor.* If so, talk about relative frequencies will not illuminate the observer's analysis of the logical and theoretical patterns embedded in the smiler's humor. Talk about relative frequencies probably begs the question since the observer is interested in trying to analyze the mental processes that generate the relative frequency of smiling and humor.

In sum, generalization-based models of inference may miss the mark because guesses and inferences about facts sometimes involve guesses and inferences about the logical character and structure of real-world processes. The construct humor is, to be sure, an invention or creation of the observer, but it is at the same time a construct that the trier uses because he believes that it may accurately characterize the actual state of the actor's mind.[53] Nothing much hinges on whether I have fairly[54] attributed to Baconians as a class the view that tendencies or disposition must be interpreted as relative frequency statements. The important point is that models of inference may obscure the nature of inferential problems and thereby divert a factfinder from clear analysis of the question to be resolved.[55]

The occasional value of constructs such as humor or wit cannot seriously be doubted. If one understands how the witty mind of a particular person works, it is often possible to make some very good guesses about a person's behavior. Indeed, if one has no sense of a person's pattern of mental activity, one almost surely misses the person's jokes. In Section I, one of the nomological structures I describe converts, by means of logical operations, events of one type into events of another. Constructs such as humor are to some extent[56] nomological structures of this sort. For example, a factfinder who understands the complex logic that permeates the joking of another person—the topics that interest the joker, the twists in reasoning he uses to make a story funny, and the possible serious purposes of the joker—can make some pretty good predictions about what that person might say in humorous situations, such as at a party. In the language of the black box metaphor, the factfinder has some sense of the processing mechanism of this human box and can predict the humor outputs of this box when certain inputs, such as a particular type of party, are present.

To the extent that a problem of inference revolves around uncertainty about the character of an actor's mental processes, Baconian models of rational inference are unhelpful and uninformative. These sorts of inferential problems are not within the domain of generalization-based theories of inference. Of course, it is true that a factfinder will sometimes decide that there is no basis for gauging the specific character of an actor's mental processes. It may also be true that generalization-based analyses help to resolve problems of inference, even assuming that the problem involves someone who regulates his conduct by some sort of internal logic.[57] Neither of these concessions, however, impairs the force of my argument that generalization-based theories often do not address the problems generated by uncertainty about the practical logic used by human beings.

2. The Place of Stories in Inference: The Role of a Special Type of Theoretical Coherence in the Interpretation of Evidence.

The ranking of alternative theoretical constructs or generalizations plays a central role in Cohen's theory of inference, probability, and provability. Cohen asserts that the soundness of a factual inference may depend on the relative plausibility of different generalizations.[58] If he is right, Cohen has shown that the force of evidence depends to a great extent on the decision-maker's view of various theoretical constructs.

Cohen's account of the ranking of generalizations in vertical inference is similar to the approach many judges use when trying to decide whether or not the evidence in a case shows guilt beyond a reasonable doubt. Judges often approach this question by asking whether it is possible to imagine any reasonably plausible hypothesis indicative of innocence that is compatible with the evidence. In addition, judges often ask whether the hypothesis is the *only* plausible one compatible with the facts. Thus, when judges address questions concerning the weight of the evidence, they seem to be engaged in the ranking of alternative theoretical constructs. Thus, it appears that the

MAPPING INFERENTIAL DOMAINS 299

plausibility assigned to various theoretical constructs has a decisive bearing on the way in which some judges weigh the force of the evidence in the case.

But this analogy between Cohen's theory of inference and judicial approaches to questions of weight is imperfect.[59] While the judge is often searching for an explanation of the facts and the evidence, that explanation is usually not the kind of explanation to which Cohen alludes when he analyzes the role of generalizations in inferential chains. The judge who asks whether there is any plausible hypothesis compatible with the innocence of the accused is not simply asking what inferences are reasonably supported by the available evidence. Rather, he is asking whether one may read the facts compatibly with the claim of innocence. When addressing questions of weight in this last fashion, a judge faces a problem of factual interpretation[60] rather than a classic problem of inference.

The problem here is to describe more precisely the conceptual processes at work when the judge makes factual interpretations, and to weigh the relative plausibility of those interpretations. We can unpack these analytical processes if we provisionally assume that the judge, in ruminating about hypotheses, explanations, or interpretations, is effectively ruminating about "stories."[61]

Return now to the face problem. Suppose that the decisionmaker has heard the testimony of a witness. On the basis of that testimony the trier has concluded that the mouth had the following shape:

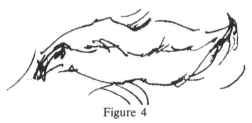

Figure 4

The question of fact facing the trier is whether the mouth forms a smile or something else. Suppose the question arises in a murder trial. S, a law student, is charged with killing T, one of his teachers. The prosecutor introduces evidence that convincingly proves that T's corpse has been found in a subterranean location. The evidence also shows that T was killed on June 1, 1980. The issue is whether S is the culprit. To support the claim that S killed T, the prosecutor introduces evidence that seems to suggest animus by S against the now largely-decomposed T. The prosecutor's evidence is the testimony of a witness who states that on April 1, 1980 T asked S a question and that T, apparently finding S's answer unsatisfactory, gave S a coin and told S to telephone his mother and tell her that he was coming home. The witness testifies that S responded to this ridicule by muttering, "You bloody bastard." S denies that he killed T and he offers R as a rebuttal witness. S wishes to undermine the claim that he harbored animus toward T. R testifies in the following way:

> I was a student in T's class and I sat next to S. On May 1, 1980, T told the following joke in class: "Arthur asked Harold, 'Who was that lady?' Harold answered, 'That was no lady. That was my wife.' " S thereupon turned to me and formed his mouth into the shape shown in Figure 4, which I have drawn for the benefit of the jury, and he said, "Hah! That's really very funny. This man T is no idiot, no?" I told him I thought the joke was idiotic and that T is indeed an idiot.

S introduces this evidence to show that it is unlikely that he harbored any lasting animus toward T since, S argues, it is unlikely that he would have laughed at a bad joke by T if he had harbored a grudge against T because of the incident on April 1, 1980. Obviously, however, the probative worth of this rebuttal evidence depends in part on whether the trier of fact regards the shape of the mouth shown in Figure 4 as a smile or as something else. The problem is not the shape of the mouth since that is taken as settled. (R was a very convincing witness.) The question is how that shape should be interpeted.

The trier can structure his interpretation of that problem in various ways. (How he structures his attack depends in part on how he structures the problem.) It is possible for the trier to structure his attack by thinking about various hypotheses and scenarios. In so doing he may take the following steps in the order listed:

> (1) The trier decides that the question he must resolve is a question about the meaning shown by the mouth in Figure 4. He decides that the shape of the mouth is as shown in Figure 4. He decides that the meaning shown by that shape can depend on various things other than the shape itself. (Provisionally, we shall label all sets of such other things as hypotheses or scenarios.)
> (2) The trier forms one hypothesis or scenario. The trier believes that this hypothesis, if true, suggests that he should read the shape in Figure 4 as a smile.
> (3) The trier forms an alternative hypothesis. The trier believes that this hypothesis, if true, suggests that he should characterize the shape of the mouth as being some other kind of expression, but one which is in any event not a smile.

(4) The trier contemplates whether there are yet other hypotheses or scenarios that have a bearing on his interpretation of the shape of the mouth but he cannot imagine any others.
(5) The trier evaluates the relative plausibility of the two scenarios or hypotheses. On the basis of this evaluation, he decides which of the two divergent characterizations of the mouth is more plausible or more likely true than not.

As detailed as this description of the trier's decision-making procedure may seem, it buries innumerable complexities and refinements. Nonetheless, the general pattern of the trier's procedure may be summarized still more succinctly. The trier's procedure consists, first, of forming two divergent hypotheses that produce divergent characterizations of the mouth and, second, of weighing the relative plausibility of the two hypotheses to resolve the question of whether or not S smiled.

Consider an example of this last way of attacking the question whether S smiled. The trier may look at the testimony of R about the events of May 1 and, putting aside the information about the shape of the mouth, he may decide that the remaining information—the word "Hah!" and the statement "That's really very funny, isn't it?"—supports the hypothesis that S's conduct on that occasion indicated humor. On the basis of this reasoning, the trier, to the extent that he finds the underlying hypothesis plausible, may well characterize the shape of the mouth as a smile. (I put aside the question of whether this reasoning makes the smile entirely redundant evidence.)

Of course, it is quite possible that this same trier entertains an alternative hypothesis. For example, the trier may hypothesize that S's testimony about matters other than the mouth shows that S's mood was one of derision because (1) the joke was indeed not very funny, and (2) genuine mirth is not usually expressed by the monosyllabic word "Hah!" On the basis of this sort of reasoning, the trier, to the extent that he has confidence in this second hypothesis, may characterize the shape of the mouth as a jeer or a snarl. Thus, how the trier finally chooses to characterize the meaning expressed by the shape of the mouth, and how confident he feels about that characterization, may well be a function of the trier's assessment of the relative strength of the two discordant hypotheses.[62]

My hypothetical trier's analysis does not easily fit within a Bayesian framework. This is not because the trier formulates the inferential problem as one involving the hypothesis of humor or derision. Bayesian analysis does describe the logical structure and implications of non-exhaustive divergent hypotheses. Bayesian analysis therefore is instructive if the trier structures the problem in a way that makes the question of the relative strength of the two discordant hypotheses the trier's focal concern. But Bayesian analysis offers a poor fit to the extent that the trier focuses his attention on the interpretation of the meaning (for example, humor) expressed by various verbal statements (for example, "Hah!"). This is because hypotheses such as humor and derision, as well as the factual hypotheses smile and jeer, are really *meaning-events*. Meaning-events are real world events. They are not, however, composed solely of the physical conduct by which meanings are expressed. Conduct is a meaning-event only if it is invested with meaning by an actor.

The real-world implications and consequences of meaning-events, however, depend in part on the logical fabric of the meaning of the actor. Meanings always involve and presuppose complex real-world mental activities having their own peculiar logic and pattern. These mental activities are performed by real people thinking real thoughts and expressing real meanings. The trier who uses the decision-making procedure involving alternative hypotheses about the events on May 1 plainly views those events as meaning-events. Moreover, he believes that the meaning of the shape of the mouth depends on the meaning of the other events on May 1. Hence, it is clear that the trier, when proceeding in the fashion described, must think through the logic underlying words such as "Hah!" in order to decide whether the probable consequence of such a meaning-event is the separate meaning-event of a smile or a jeer. Insofar as I can see, Bayesian analysis has little to say about how one tries to grasp the logic and reasoning underlying a meaning-event.[63]

In many cases, the trier must devote a great deal of attention to the logic underlying meaning-events. Extensive analysis may be required to remove the trier's doubts about an actor's intended meaning. This is not only due to the complexity of the practical logic of any individual person. It is also because the assessment of any individual's intent, meaning, or aim in a particular situation almost invariably requires an assessment of the interactions of a variety of persons and groups. Hence, it is no surprise that a trier may believe that his major inferential task is that of enumerating and classifying a wide range of practical logics at work among the persons and groups who play a part in the problem of evidence as he sees it. This trier's job is to divine both the logic regulating the conduct of various individuals and the logic governing real-world meaning-interactions among various actors, each of them having a separate practical logic. (The complexity of this sort of analysis boggles the mind.)

Consider, for example, the conversation between S and R. The trier wants to know what S meant by saying various things. A staggering variety of meaning-logics has a potential bearing on this simple question. For example, it helps to know the English language. It helps to know whether and how S modifies common usage of the English language. It helps to know what sort of wit and irony S normally uses. It helps to know how law students at that school normally talk with each other and what they normally mean when they use certain words. It may help to know that S and R are good friends. It may help to know that S is immature, and so on. Each of these helpful bits of information involves acquisition of knowledge about complex sets of rules that govern the behavior of people in various settings.

Step back from the details of the argument and consider the type of inference at work in the murder problem. We may think of the trier's mode of attacking the smile question as involving the trier's use of various stories to explain the evidence. The trier's focus is on the construction of a story rather than on the evidence because events such as the shape of the mouth are taken as settled. The trier is focusing on the question whether he can aptly characterize the shape of the mouth in a way that he sees as being consistent with a particular story about what various persons meant when they said various things. When trying out various stories about the events on May 1, the trier is attempting to tell stories that portray in different ways the kinds of logic that may be driving and motivating the actors in the play. The trier is searching, in effect, for an appropriate interpretive matrix, and he is trying out different matrices to see how well they fit particular sets of events that are taken as given but that can be variously read and interpreted.

It seems plain that it is not intrinsically irrational for a trier to approach a problem of evidence and inference in the way described here. It is conceivable that a Bayesian devotee might object to the trier's way of attacking the smile problem on the ground that the trier's emphasis on searching out an apt or fitting interpretive matrix impedes the rational analysis of inference. This is because, so the argument goes, rational inferential reasoning amounts to coherent integration of primitive probability estimates, and the trier's approach does not at all address the question of how various probability estimates should be summed up. The trier's approach focuses entirely on the question of what sort of an interpretive matrix should be used to define what the evidence actually is. The Bayesian analyst thus complains that Bayesian analysis is applicable only after certain facts are taken as genuine. The trier, however, spends all of his time worrying about how he should characterize the evidence, leaving Bayesian analysis out in the cold. This Bayesian analyst therefore argues that the trier should restructure the problem so that he can start thinking about redundancy, transitivity of doubt, and other matters of this sort. In support of this proposal, a Bayesian analyst can convincingly show that there are many different ways of portraying the problem that are amenable to Bayesian analysis.

The critique and proposal of such a Bayesian theorist are misconceived. The Bayesian straw-man I hypothesize—and let us not forget that the Wizard of Oz told us that straw-men do have brains—has things backward. The question is not whether inferential processes conform to rational patterns of inference. Rather, the question is how well models of inference describe or illuminate patterns of inference that quite clearly work well. (It would be hard to make sense out of the argument that an effective technique of information processing is irrational.) Thus, for example, I think that the sort of coherentist, story-based, and narrative method of evaluating evidence that I have thus far described is entirely rational because (1) this way of attacking a problem certainly involves thought, reasoning, and analysis, and (2) we often have the conviction that this way of evaluating evidence is a necessary and useful basis for finding facts.

The force of the claim about the rationality of coherentist thinking depends entirely on the force of the thought experiments used in this paper to support it. In support of this claim I have offered a thought experiment involving one species of coherentist analysis. Now consider another species which is equally non-Bayesian. This species of coherentist analysis also involves story-telling, but in this case the emphasis is not on constructing a narrative that offers a portrait of the logic of various actors. Rather, the emphasis is on trying to construct a story that irons out discrepancies between theory and evidence, and between different facts.

Probably everyone at some point has tried informally to make sense of some mass of evidence. In this more formal approach, however, the emphasis is on trying to construct stories and scenarios that serve to iron out discrepancies and conflicts among various items of evidence. There is scattered evidence that judges and jurors use this species of coherentist reasoning when trying to gauge the probability of a certain fact in issue. They ask whether other facts may be explained by a hypothesis that supports the fact in issue. When this type of coherentist approach is used discrepant evidence often is not treated as a report of a discrepant event, but rather as a fact. The trier sees his job as trying to find a plausible hypothesis or story that eliminates the apparently anomalous character of that fact. The story does its job of eliminating the anomaly by suggesting a different characterization for the thing that is taken as a fact. In that event, the jury may suppress uncertainty about the reliability of much evidence and it may, for example, largely disregard possible questions about the credibility of various witnesses.

In the murder problem involving S, for example, suppose the jury thinks that it is reasonably possible that the shape of the mouth in Figure 4 is a smile. The jury may then consider whether its interpretation is consistent with other evidence. If the jury believes, for whatever reason, that students' attitudes toward their teachers do not change quickly, it may decide to test the smile hypothesis by considering the evidence of the incident on April 1. If that incident can reasonably be construed as not suggesting any animus by S toward T, the jury might decide that the incident on April 1 reflected merely wry and good-natured humor by both S and T toward each other. The earlier incident then becomes not an expression of animus but an expression of good will.

The jury can, however, work this sort of process in reverse. For example, if it cannot bring itself to read the incident on April 1 as an expression of good will, the jury can rethink the incident on May 1 to see if it can be read in a way that resolves the inconsistency between the two incidents. With this idea in mind, the jury might decide that the report of the facial expression on May 1 was not in fact the report of a smile and that R has instead reported a jeer by S. In either case, the jury has such confidence in the durability of animus and humor that it is not willing to look at the evidence as indicative of conflicting attitudes. It is determined to describe what happened in a way that conforms with its belief that sentiments of animus and good will are stable.

Juries having a coherentist bent will occasionally be confronted with situations in which the matters that it treats as facts cannot plausibly be made to conform with its general hypotheses. In those situations, however, the trier may still be unwilling to abandon those hypotheses that seem to generate anomalies. A strategy is available to this trier that preserves the general hypothesis, but diminishes the consistency of the evidence. The trier may well decide (and rationally so!) to treat the fact as a report of a fact, a report that may of course be mistaken. If his commitment to a certain hypothesis is strong enough, the trier must conclude that one of the two conflicting reports is probably false.

For example, if the trier firmly believes that the April 1 incident shows S's animus toward T, he may then reconcile that belief with his hypothesis about the stability of the feelings of animus by reaching the conclusion that R's sketch of the shape of the mouth was probably incorrect.

Note, however, that the trier's willingness to abandon the effort to iron out every wrinkle in the evidence does not mean that the trier altogether abandons coherentist analysis. Note further that the trier has numerous strategies to preserve complete coherence in the smile problem. For example, instead of treating R's report as false, the trier may entertain the conjecture that the shape of S's mouth was attributable to S's nervousness.

These examples of the various ways that a trier can grapple with and resolve anomalies in the facts and the evidence shows both that a coherentist approach may be a rational way of attacking a mass of evidence and that a coherentist approach is not the only rational way of attacking evidence. A coherentist model of inference cannot be defended on the abstract ground that inferential thinking always has coherentist features. Which approach is the most rational depends to a large extent on the trier's presuppositions, his structuring of the problem as a whole, and on whether and how he revises his opinions as he receives new information. For example, it is entirely possible that the trier has some confidence in the hypothesis of the stability of feelings of animus, but that his commitment to that hypothesis is not absolute or very strong and that the acquisition of new information leads the trier to abandon it. Thus, it would not be at all surprising or irrational if a trier of fact, previously firmly committed to the idea of the durability of feelings, decided simply to abandon that belief in the face of the testimony of two witnesses who seem to be overpoweringly credible. Alternatively, the trier may simply admit that when there are large bodies of evidence it is practically inevitable that there will be wrinkles in the evidence that cannot be ironed out in any fashion, no matter how long one labors. He may conclude that as a practical matter it is impossible to find an explicit and noncontradictory explanation for every anomaly and inconsistency that he sees in a large mass of evidence.

We have seen some of the ways in which stories may be used in inference. We have seen that the properties of such story-based inferential reasoning are varied. Story is a rubric that refers to various sorts of theoretical operations. First, stories refer to efforts to reconstruct the practical logics of various actors. Second, they refer to coherentist efforts to devise or discover explanations for the evidence that tend to eliminate inconsistencies between different items of evidence. Third, all stories are inventions of the trier of fact, and the trier can invent an alternative story. Fourth, some story-based reasoning "changes" evidence while other such reasoning accepts evidence as factual. Fifth, there are other kinds of stories not emphasized in this article. For example, some stories are scenarios that try to describe possible sequences of events in space and time on the assumption that those events are causally or stochastically related. All of these ways of thinking are rational ways of thinking because evidence does not speak for itself, and problems of evidence and inference do not structure themselves. Human decisionmakers structure and configure both evidence and problems of evidence and inference. Depending on the initial configurations of evidence and problems of evidence, a human decisionmaker may find that telling various kinds of stories is an informative and rational way of both reinterpreting evidence and restructuring problems of evidence.

The thesis of the rationality of various kinds of story-telling, like the broader thesis of the rationality of various models of inference, may seem troubling because, as the discussion of various kinds of story-based attacks on the smile problem illustrates, different approaches to problems can and often do yield very different inferences. For example, I have shown that various kinds of stories can be told, that the trier can sometimes abandon coherentist story-telling altogether, and so on. These various ways of reading evidence, however, can generate very different conclusions. The importance of reliable factfinding naturally makes us wish to know which way of attacking a problem is the right one. The implication of my argument is plain; no one way is always best and, further, no abstract criteria specify which way of attacking a particular problem is the best way. It follows that the decisionmaker must decide which way is best.[64] Nonetheless, it does not follow that any way of thinking about a problem of evidence is as good as the next one and that we are free to pick and choose any way of thinking we please. It also does not follow that nothing is to be gained by the analysis of the logical structure of inference. Models of inference clarify a decisionmaker's thinking about inference by clarifying the logical implications of the properties that he ascribes to evidence. Hence, when the decisionmaker is able to identify the properties he ascribes to evidence, there is reason to believe that he will think more consistently about the implications of his own beliefs about the evidence that he sees. Moreover, the precision and care with which various properties of evidence are described by various models of inference may aid the decisionmaker in taking the preliminary but crucial step of identifying with precision the properties that he ascribes to particular complexes of evidence.

A serious question about the value of theorizing about inference remains because formal analysis of evidence and inference is limited by the "garbage in, garbage out" principle. The appropriateness of using particular models of inference depends on the way the decisionmaker structures problems of evidence and on the properties he ascribes to evidence. If, however, the decisionmaker's way of configuring a problem is untrustworthy, so are the results generated by the analytical machinery he employs, no matter how elegant. This form of skepticism, however, is less cogent than it seems. Its silent premise is that the decisionmaker's way of seeing a problem is untrustworthy and must remain so as long as it remains unjustified. This premise is wrong. No form of knowledge—except tautologous knowledge—is ultimately justifiable in the sense suggested. Hence, the question of the value of theorizing about inference must be more narrowly phrased. For example, I have suggested several times that we can be confident that rational discussion of an actor's beliefs, motivations, principles, and the like is the best and most reliable way to go about making guesses and inferences about human beings, and that this kind of discussion amounts to good inferential reasoning. This way of attacking a problem, like any other way of attacking a problem, is not justifiable in any ultimate sense. Nonetheless, as I shall show next, we are entitled to believe that this method of attacking certain problems of evidence is a rational method.

C. *The Internal Perspective on the Logic of Inference and the Logic of Human Action*

Inferences about human actions have an added layer of complexity because they require guessing about human meanings, principles, and beliefs. Such guesses have unique features, and evidence about such matters has unique properties. These unique properties are not captured by either Baconian or Bayesian models of evidence and inference.

Many of us believe that human beings are actors who regulate themselves and their conduct by a kind of practical belief-logic. Human beings have various systems of beliefs, principles, and sets of mental operations and activities which they often use to regulate and control their conduct. Believing this, many of us implicitly believe that making good guesses about human beings often requires an understanding of an actor's practical belief-logic—the principles and other conceptual matters by which human actors regulate their activities. In sum, many of us believe that human beings are self-regulating actors—i.e., we use a practical belief-logic to regulate our activities—and that guesses about the activities of human actors often require an understanding of that practical belief-logic. These beliefs are not unreasonable or unwarranted.

The self-regulating character of human beings explains why generalization-based accounts of inference often founder when a problem involves guesswork about human activity. If the human activity is a real-world event, and if that behavior is caused by practical belief-logics, good predictions about such real-world activity require an understanding of the practical belief-logic generating that activity. Sets of generalizations, however, cannot describe such workings or the effects of such a belief logic. The practical logic that human beings use to determine how they act in the world involves reasoning about things such as the importance of various beliefs, the efficacy of particular kinds of activities, the classification of events requiring action, and many other things. This sort of practical reasoning cannot effectively be described by arrays of generalizations any more than sets of generalizations can describe the logical structure and operations involved in classical mechanics.

It may be objected that generalization-based inferential reasoning is usually a superior way of guessing at human conduct because human activity, unlike natural processes determinately regulated by natural laws (such as the "laws" of classical mechanics), exhibits a great deal of randomness. Hence, there is much human conduct that cannot be explained as being a necessary implication of an actor's belief-logic. However, while self-regulated human behavior does exhibit a marked degree of randomness, it does not follow that human behavior is not generated by a practical belief-logic. The randomness of human action may be a real-world feature of the practical belief-logic of self-regulating human actors. This theoretical perspective on the random qualities of human behavior is roughly similar to the way that the decay of a single radioactive atom or other randomly-behaving events may be seen as a product of certain laws of physics. It follows that guesswork about human actions may be similar to the sort of guesswork that occurs when our guesses concern stochastic processes like radioactive decay.

Like the random properties of certain natural phenomena, the random character of human action may also be attributable to a practical belief-logic that generates random outputs. Such outputs differ from the non-random outputs associated with causation-based theories such as classical mechanics.[65] Hence, the mere fact that human activity seems to occur in random fashion does not prove that generalization-based models of inference are generally informative. This is because, as the analogue of random natural processes suggests, it remains possible that human activities are caused by logical systems that somehow inhere in random processes themselves. The random properties of natural phenomena do not obviate the importance of understanding and describing the logical structure of that random behavior. Similarly, the random properties of human activity may also rest on a logical scaffolding, as in the case of stochastic processes. In many instances there may be something like a logical structure that explains or generates the random features of human behavior. If so, generalization-based theories of inference are of little value in establishing the probability of an unknown human action because the logical structure underlying the actions of human beings, just like the logical structure of stochastic processes, generally cannot be presented by sets or collections of relative frequency statements. Instead, it is often very helpful and important to know an actor's beliefs and the rules by which those belief systems are applied and interpreted. In short, it is sometimes important to take an internal perspective when making guesses and inferences about human action and behavior.[66]

Consider a simple example. Suppose we wish to guess what Ronald Reagan will do or has done in Nicaragua. To make that guess it helps to know that Reagan thinks communism is evil and it helps to know how he reasons, if at all, about the evil of communism. It is also helpful, of course, to know what Reagan believes about a great many other things—e.g., military power, NATO, socialism, war—and how he reasons about *those* things. It is likely, although not indubitable, that general beliefs of this sort play a significant part in shaping Reagan's behavior.[67] If we know what Reagan believes, what values he has, the priorities he has, how he analyzes or assesses information (and "disinformation"), and the like, we may be able to construct a fair facsimile of his practical logic in relation to communism. If we can do so, we may be able to make better guesses about what he has done or will do in Nicaragua. (If this thesis still seems doubtful, test it by assuming that Reagan is secretly a communist.)

Suppose we conclude that we have a pretty good handle on Reagan's beliefs and on the way he reasons about them. Suppose, further, that we find that Reagan has voluntarily acted in a way that seems inconsistent with his practical anti-communist and pro-capitalist logic. Must we conclude that knowledge of his anti-communist logic is worthless? Not necessarily. It is quite possible that Reagan's practical logic does not logically mandate particular actions under particular conditions; it is possible instead that Reagan's logic may generate a probability of a certain set of actions under a certain set of conditions. In short, it is possible that an element of randomness is built into Reagan's own logic; it is possible that Reagan's practical logic is a fuzzy logic.[68] The fuzziness of Reagan's logic and the consequent randomness of his actions, however, does not mean that it is pointless to try to understand his practical logic. To say that Reagan's actions have some random properties is not to say that his actions are entirely random. To say that his logic is fuzzy is not to say that it is featureless or entirely incoherent. Therefore, knowledge of the workings of his fuzzy logic gives us a better basis for making guesses about his actions because we can then say some things about the kinds of variations found in his self-regulated behavior and their probable frequency. Hence, even though Reagan's practical reasoning about communism may be fuzzy, and even if his interpretations of what regimes are truly communist may also be fuzzy and random to a degree, the quality of our guesses about his actions in Nicaragua is almost certainly improved if we know that he thinks communism is evil.[69] Correlatively, we will be worse guessers if we do not know this.

Note that Reagan's logic, no matter how fuzzy, does not consist of mere dispositions and tendencies. Reagan probably deliberates in a complex way about how to act toward communist regimes. If so, it is wrong to portray our inferences about Reagan's behavior by picturing them as involving a variety of qualified generalizations about Reagan's dispositions toward communism. This portrait translates *logic* into *generalizations*. Much is lost by this translation since it diverts attention away from the logic that informs the qualifications that Reagan will attach to such generalizations.[70] To be sure, the operations and principles by which a person combines, interprets, and applies his beliefs are sometimes otiose. Under such conditions of ignorance, we may be tempted to explain our inferences by talking about the dispositions, propensities, and inclinations of persons. Since these characteristics lend themselves to being described by generalizations and relative frequency statements,[71] it seems to follow that generalization-based theories of inference can model the inferential thinking that relies on these concepts. Yet, although limitations on our knowledge often license this way of talking and thinking about human actions, it remains important[72] not to mistake dispositions, propensities, and the like for belief systems.

While it may be true that there are some instances where people simply have certain dispositions, it is also true that in most instances what we refer to as a disposition or propensity is a product of a belief system. Hence, when the workings of a person's belief system are discernible, it is inappropriate and misleading to talk merely about that person's dispositions and inclinations. If talk about dispositions is to be useful when discussing self-generated human behavior, that discussion must be informed by an understanding of the way that the person in question thinks about and applies his principles and sentiments.

The difficulty involved in talking about human dispositions and proclivities is akin to the difficulty of explaining and predicting natural phenomena by talking about relative frequencies. Observed relative frequencies in both the human and the natural context convey significant information only to the extent that an observer has some sense of the logical scaffolding on which such frequencies are constructed. Without some sense of the logic that *informs* such relative frequencies, an observer has no way of determining the field within which those relative frequencies hold, let alone of deciding to what extent and under what circumstances it is permissible to extrapolate from those observed relative frequencies in novel situations.[73]

To the extent that a disposition refers to the frequency of human behavior under certain conditions, the same problem inheres in discussion about dispositions.[74] It is therefore important to talk explicitly about the character and workings of a person's practical belief-logic when we believe that a practical logic of this sort significantly regulates and determines conduct.

For example, suppose that the trier's task is to make an intelligent guess, given the trier's knowledge of one fragment of a phrase, about what a person said in another portion of the phrase.[75] If the trier firmly believes that this person's use of language is regulated and structured by a specific type of grammar, it is obvious that an effort should be made to grasp the character and workings of that grammar. To talk instead about a person's disposition to use particular words in particular situations is not likely to be as helpful since this way of describing that person's linguistic practice does not focus on the principles and rules the speaker uses to decide which particular linguistic dispositions are triggered in novel situations.

Consider once again a variant of the smile problem involving the murder of T. Suppose that further evidence discloses, to our surprise, that both S and R belonged to a secret society. That society's written charter proclaims, "The purpose of this society is the extirpation of idiocy and witlessness and the extirpation of those who propagate such things." The significance of this evidence might be either great or paltry. It may be great if the trier believes that the charter states the actual purpose of the society and that extirpation refers to the killing of those persons identified as being witless. But the new evidence may also be of little significance[76] if the trier believes that the society and the charter were a farce. To appreciate and assess the force of the evidence of the charter, it is necessary to enter the mind of S and to imagine how he thinks. Did S take the charter seriously? Did S believe that T exhibited the kind of witlessness referred to by the charter? Why did S join the society? What sort of person is he and how was he likely to think about a society of the sort described by the charter?

These sorts of inferential processes are not adequately addressed by a model of inference which speaks about dispositions or generalizations of human conduct. The effort to describe S's thought pattern by talking about his dispositions falls into the error of confusing the meaning of a rule or principle with a catalogue of its past applications. As Ludwig Wittgenstein once suggested, principles and rules have a curious and important quality: they guide and regulate action and thought in novel situations. A principle somehow stands outside of the situation to which it has been, or may be, applied. Principles having this sort of independent meaning do not serve a purely lexical function. Relative frequency statements and generalizations fail to capture this dimension of the principles that regulate human conduct.

III. Ordinary Inference as a Conventional Standard of Performance

I have argued that trusted forms of ordinary inference are rational and that adequate models of inference must grow out of descriptions of the theoretical structures and processes embedded in ordinary inference. Theorists of inference, however, usually distinguish between descriptive and prescriptive theories of inference. Hence, my argument appears open to the objection that I have conflated the two types of theories—that no account of the way that inference actually works can possibly tell us how rational inference should work. This objection, however, lacks merit. We have things backwards if we say that we should devise models of rational inference and then try to make our inferential processes conform to them. The object of a theory of inference is to elucidate the logical properties of inferential techniques that actually work.

Any effort to construct purely a priori theories of rational inference amounts to an effort to replace the trusted forms of information processing that we now use. Such an effort, however, is bound to fail because (1) there are in fact no available transcendental criteria for valid empirical knowledge, (2) the assumption that conventional information processing techniques are irrational because they diverge from a particular a priori theory is unwarranted since information processing techniques that actually work are fully rational, and (3) the use of an a priori theory of inference that is not rooted in ordinary human information processing techniques would require the reconstruction of the human organism. A purely transcendental theory of inference is, by definition, suitable only for angels. Hence, an adequate theory of inference must have some correspondence or connection with ordinary forms of human information processing. Moreover, it makes no sense to charge human beings with basic irrationality in dealing with uncertainty. If one supposes, as we all do, that uncertainty pervades human existence, it cannot be said that human beings lack all ability to manage uncertainty. Most people live their daily lives without the aid of a rational theory of inference and probability! Hence, it seems likely that the main service of theories of inference is to improve the ability of human beings to deploy their proven information processing techniques. If it is objected that those techniques are irrational—then rationality be damned! In short, in our theorizing about inference we must begin and end with what we know and with the way we presently know because the aim of theorizing about inference is to try to portray the logic at work in inference. If we can offer such a portrait, we can more often use our proven inferential techniques effectively.

The distinction between descriptive and prescriptive theories of inference is misconceived if it is thought to exactly parallel the distinction between fact and value. The fact-value distinction is properly construed as the distinction between what happens to be the case and what morally or ethically ought to be the case. It should not be interpreted to mean that knowledge of what is the case can never yield conditional imperatives. I have described the role that nomological structures play in the configuration of evidence and in problems of inference. Hence, in describing ordinary inference, I have described what is the case, not just the bare facts that exist in space and time. I have described some of the logical properties of conventional forms of inference. While it is true that I see models as amounting to descriptions of inferential processes, these are descriptions of a most peculiar sort—they are descriptions of conceptual and nomological processes. Hence, in describing ordinary inference I may be describing a fact but, if so, the thing described—the logic of conventional inference—is a peculiar kind of fact. It is a fact that amounts to a standard.[77] In short, my sort of description of ordinary inference generates a kind of prescription because a portrait of the logical properties of ordinary inference differs from a purely empirical description of a thing or event in space and time.

It is quite beside the point that no one may justifiably be able to tell you that you should try to comply with such standards. The discussion here assumes that a person has an interest in being a reliable inferential machine. Hence, my analysis of the proper thrust of particular configurations of inferential problems has a loose but appropriate affinity with the classic Greek notion of excellence or *arete*. My premise is that the configuration of an inferential problem is given direction by the nomological structure present in that configuration. If I were able to do all that I would like to do, if I could definitively establish the particular excellence of every conceivable kind of theory of inference, I still would have said almost nothing about the relative importance of reliable factfinding in the hierarchy of societal and individual values.

Theories of Inference and Their Verification

I have stressed the idea that ordinary inferential processes serve as a kind of benchmark for the adequacy of theorizing about inference. This emphasis on the rational character of conventional inference raises the question of how it is possible that a description of what we already do when we draw inferences can yield lessons on how to manage uncertainty in a rational fashion. If ordinary inference is already rational, does it not follow that theorizing about inference, while perhaps interesting, is redundant and uninformative as a practical matter? This question is little more than a reformulation of an old philosophical chestnut—the question of how we can come to know something if we did not know it all along. Nonetheless, this chestnut is worth considering, not so much for its own sake, but for the light it sheds on the peculiar and important role that a mapping perspective plays in making us think more clearly about our inferential thinking.

Theories of inference are peculiar kinds of theories because they describe thought, or something very much like thought. Hence, whatever may be the explanation for the ability of human beings to acquire knowledge, formal theories of inference confer knowledge of inference to the extent that they provide rigorous descriptions of the logical processes that govern inference. Theories of inference, therefore, are in one respect very much like meta-mathematical systems that have been designed to explain the properties of a separate set of mathematical systems. Meta-theoretical perspectives of this sort advance our knowledge by elucidating the properties of valid logical systems. Abstractly considered, the philosophical chestnut about the difficulty of explaining the acquistion of knowledge applies just as much to meta-mathematical perspectives as to theories of inference. In fact, however, meta-mathematical perspectives do sometimes advance mathematical knowledge. If they do so, they do so by making the logic of a mathematical system fully transparent. Theories of inference may perform a similar service by articulating in an explicit way the logic that sits within ordinary inference; they articulate the logical properties of ordinary inference by making explicit the implicit logic of inference.

The function of a general theory of inference is to regulate the use of lower order theories of inference, which I call models. This sort of general theory of inference has certain affinities with second order, third order, and nth order mathematical theories. One of the main purposes of this paper is to inquire into what it means to have a general theory of inference, and I have concluded that a general theory aims at specifying the domains of inference. Moreover, a general theory allocates different domains of inference to different theories of inference which, taken individually, describe but one type of logic in inference. Hence, a general theory of inference is theorizing about theories (as opposed to theorizing within and through the terms of particular theories). Moreover, a general mapping theory—which is a theory about the management of theory—is meta-theoretical twice removed since, on my premises, lower order theories of inference are themselves theories that describe yet lower-order theoretical and logical structures in ordinary inferential processes. Hence, my view that theoretical management of lower order theories of inference is the peculiar function of a general theory of inference amounts to saying that a general theory of inference, as well as inference itself, is cascaded;[78] a general theory of inference is meta-theory twice removed from primary inference.

The analogy between theories of inference and mathematical theories, however, is imperfect. Quite often, advanced branches of mathematics constitute logical systems that encompass other branches of mathematics as subsets. This means that the subordinate mathematical systems can be logically deduced from the more general theories that serve to explain them. This sort of deduction or derivation, however, is not what my sort of general theory of inference promises to supply or is capable of supplying. While a mapping theory subordinates models of inference from the outside, it views those subordinate models as logically disparate models, and does not aim to supply an overall logic that provides a unifying explanation for all of them. In short, a meta-theoretical perspective on inference is not the same as a meta-mathematical perspective because, if a meta-theory of inference manages lower-order logics, higher-order theories of inference do not manage lower-order logics in inference on the basis of purely formal logical considerations. That management involves an appeal to a non-formal criterion, the manner in which inference in the world actually works.

I have said all along that theories of inference cannot be validated, verified, or tested by purely empirical observations or experiments. I am now also asserting that theories of inference cannot be verified by purely formal and logical criteria. Theorizing about inference suffers from a peculiar affliction. It is not quite an empirical science, but it is also not a pure science of logic. As a result, theories of inference cannot be validated or verified by purely formal, logical criteria and they also cannot be verified or checked by purely empirical observations of spatio-temporal events. Nonetheless, theories of inference require some sort of verification. Those verification criteria must be of a distinctive kind, appropriate to their subject matter. My mapping theory points the way toward such verification criteria.

In this paper I have often spoken, somewhat loosely, of taking existing inferential practices as the benchmark of adequate theorizing about inference. I began by speaking of the importance of paying attention to the way that problems of evidence are constructed and, more simply, of the way that evidence is seen. I now wish to restore that earlier emphasis. The counterpart to the principle of empirical verification in the natural sciences can only be found in the messiness of ordinary inference. This messiness may well be seen instead as a richness that we can use both to check the power of our theories and to explore for the outline of new theories.

It is in the nature of formal theories of inference to tend toward a kind of logical simplicity. The aim of my meta-theorizing is to work in the opposite direction, to remind us that inference is in some sense a rather messy affair because it is not easily reducible to one logical pattern, however elegant. To be sure, elegant formal theories are rich and powerful in their own way. Nonetheless, formal theories of inference are not entirely like formal mathematical theories—or, if they are, their aims are not the same as the aims of a purely formal theory. Formal theories of inference are in an important sense abstractions—they are systems which, while explaining much detail, necessarily ignore many features of the thing which they try to explain. This is why it is always necessary to revisit the richness of the subject that we are trying to describe.

Ordinary ways of seeing and thinking about evidence are the benchmarks against which our theories must be checked. That ordinary inference can serve as a benchmark for its own theoretical explanation may seem to be a paradox. But this mystery is no deeper than the old philosophical chestnut about the difficulty of explaining the growth of theoretical knowledge. One equally hoary solution for the chestnut seems to work here. Ordinary ways of seeing and thinking about evidence are sources of knowledge about inference because in those ordinary ways of seeing and knowing there is much implicit theoretical knowledge. The fact is that we already know a great deal about reliable processing of information. We shall probably never reach the point when we can make all of that implicit knowledge explicit. Polanyi has argued that human beings have tacit knowledge that extends beyond explicit self-conscious knowledge.[79] If we assume that it is we who configure evidence, the fact that our configurations of evidence often seem to work well implies that there is tacit knowledge even in the supposedly bare evidence that we see.

A periodic return to the richness of the ways that we see and think about evidence does not deprive us of the lessons we have learned through our analyses of the logical properties of inference. If I have done anything, I have shown that evidence may be partitioned in different ways and that how we partition and see evidence depends on the theoretical perspectives with which we approach it. Rigorous theories of inference, therefore, allow us to see many things in ordinary inference that we did not see before.

NOTES

† © 1986 by Peter Tillers.
* Professor of Law & Director, Program for Evidence in Litigation, Benjamin N. Cardozo School of Law, Yeshiva University.

I am grateful for the comments of Dr. L. Jonathan Cohen and of Professors Craig Callen, David Kaye, Richard Lempert, Robert Rosen, David Schum, and William L. Twining. Special thanks go to Professor Ronald Allen, who made extensive comments on several drafts of this paper. Special thanks also go to Adrian Zuckerman and the other Fellows of University College, Oxford for their hospitality during my visit there, a visit that prompted me to work on an early forbearer of this paper. I also want to thank the New England School of Law for making the trip to Oxford possible. Thanks also go to Anne Fallon for drawing several of the diagrams in this paper.

I am also grateful to the wonderfully inquisitive faculty of the University of Miami School of Law and its fellow-traveler, William Twining, for giving me a chance to try out on them some of the ideas in this paper. I am sure I learned more from them than they did from me.

My thanks also go to Professor Neil MacCormick and the Faculty of Law of the University of Edinburgh for giving me a chance to try out an even earlier version of some of the ideas in this paper.

Much of the inspiration for the general thrust of this paper is the result of my reflections on the work of Richard Lempert, an imaginative thinker, a lucid writer, and an incomparably illuminating critic. Special thanks go to him for several letters in which he set me straight about the importance of formal theories as models of inference.

As always, I must thank my wife, Julie Glendon, for putting up with a spouse who secretes himself, if not in an ivory tower, then in a wooden tower at home to ponder the mysteries of uncertainty and probability.

[1] *See generally* Twining, *The Rationalist Tradition of Evidence Scholarship*, in WELL AND TRULY TRIED 211 (E. Campbell & L. Waller eds. 1982); W. TWINING, THEORIES OF EVIDENCE: BENTHAM & WIGMORE (1985).

[2] *See* 1A J. WIGMORE, EVIDENCE §§ 37.1-37.5 (P. Tillers rev. ed. 1983). *See also supra* note 1.

[3] There are many different accounts of Bayes's Theorem and Bayesian theory. *See* 1A J. WIGMORE, *supra* note 2, § 37.1 n.6 & § 37.6 *passim* (collecting citations). Two of the more accessible descriptions are Lempert, *Modeling Relevance*, 75 MICH. L. REV. 1021 (1975), and Tribe, *Trial by Mathematics: Precision and Ritual in the Legal Process*, 84 HARV. L. REV. 1329, 1352 (1971).

[4] The early efforts by legal scholars to use Bayes's Theorem to analyze the structure of inference were almost shattered by Professor Laurence Tribe early in the movement. *See* Tribe, *supra* note 3. Professor Vaughn Ball registered an earlier dissent to Bayesian theory. *See* Ball, *The Moment of Truth: Probability Theory and Standards of Proof*, 14 VAND. L. REV. 807 (1961). More recently Professor Charles Nesson has attacked various kinds of "mathematical evidence," and in certain respects his arguments resemble Professor Tribe's. *See* Nesson, *Reasonable Doubt and Permissive Inferences: The Value of Complexity*, 92 HARV. L. REV. 1187 (1979); Nesson, *The Evidence or the Event? On Judicial Proof and the Acceptability of*

Verdicts, 98 HARV. L. REV. 1357 (1985). *Cf.* Nesson, *Agent Orange Meets the Blue Bus: Factfinding at the Frontiers of Knowledge*, 66 B.U.L. REV. 521 (1986).

Tribe's attack on Bayesianism was provoked in part by Finkelstein & Fairley, *A Bayesian Approach to Identification Evidence*, 83 HARV. L. REV. 489 (1970). Finkelstein and Fairley replied to Tribe in *A Comment on "Trial by Mathematics,"* 84 HARV. L. REV. 1801 (1971), and Tribe countered with *A Further Critique of Mathematical Proof*, 84 HARV. L. REV. 1810 (1971). Fairley continued the argument in Fairley & Mosteller, *A Conversation about Collins*, 41 U. CHI. L. REV. 242 (1974). I believe Tribe had the better of this particular argument.

Tribe's attack, however, fell short of destroying the Bayesian movement. An American legal casebook containing a substantial amount of Bayesian analysis was published in 1977—a sure sign that Bayesian analysis in the law of evidence was here to stay. *See* R. LEMPERT & S. SALTZBURG, A MODERN APPROACH TO EVIDENCE (1977). *See also* Lempert, *Modeling Relevance*, 75 MICH. L. REV. 1021 (1975). A second edition of A MODERN APPROACH TO EVIDENCE was published in 1982. For further citations to legal literature in a Bayesian vein, see 1A J. WIGMORE, *supra* note 2, §§ 37.1-37.6. The papers at this symposium attest to the continuing and growing vitality of the Bayesian school.

Tribe's critique of Bayesianism was in fact less cogent than it first seemed. *See id.* at § 37.6 n.4. Tribe did not directly challenge the notion that Bayes's Theorem describes the structure of rational thinking about inference. Thus, Tribe's analysis of the use of Bayesianism in criminal trials had to focus on practical, moral, and social matters rather than on epistemological issues; Tribe was in effect forced to argue against the use of rational analysis in problems of evidence and inference in criminal trials. Moreover, the later work of scholars such as Anne Martin and Professors David Kaye, Richard Lempert, and David Schum suggests that Tribe may have overestimated the practical difficulties of using Bayesian analysis. *See, e.g.*, Schum & Martin, *Formal and Empirical Research on Cascaded Inference*, 17 LAW & SOC'Y REV. 105 (1982). Furthermore, Tribe maintains the controversial position that the values inhering in rituals mandate rejection of Bayesian analysis in criminal trials. *Cf.* Allen, *Rationality, Mythology, and the "Acceptability of Verdicts" Thesis*, 66 B.U.L. REV. 541 (1986) (attacking analogous claims made by C. Nesson).

Tribe's attack on Bayesianism may have rested at least in part on the importance of factual certainty in the criminal process. Although it is unclear, Tribe may have meant to say that the social values attached to the criminal process make it important to proceed as if certainty about facts were attainable. *Cf.* Nesson, *The Evidence or the Event? On Judicial Proof and the Acceptability of Verdicts*, 98 HARV. L. REV. 1357 (1985) (perhaps also suggesting this position). If Tribe was implicitly espousing this dubious thesis, then his argument is vulnerable. *See* 1A J. WIGMORE, *supra* note 2, § 4.

Arguably, Tribe actually meant to assert that factual certainty of some kind is possible, and rejected Bayesian analysis in criminal trials on that ground. It is perhaps just as well that this interpretation is not supported by the text of Tribe's article. While some philosophers seem to have suggested that some sorts of certainty are possible about certain kinds of matters, *see*, *e.g.*, L. WITTGENSTEIN, TRACTATUS LOGICO-PHILOSOPHICUS: LOGISCH-PHILOSOPHISCHE ABHANDLUNG § 6.51 (Suhrkamp ed. 1963) (arguing against radical skepticism by saying that disbelief requires grounds); L. COHEN, THE PROBABLE AND THE PROVABLE §§ 22, 23 & 72 (1971) (seeming to challenge the view that inferences from evidence are always subject to a measure of uncertainty), no scholar of the law of evidence has ever taken the view that inferences about facts can be certain, *see* 1A J. WIGMORE, *supra* note 2, § 37 (survey of modern theories of inference and relevancy). It seems clear that Tribe would have had to say a great deal more than he did to convince anyone that factual certainty is actually attainable in the ordinary criminal trial. In short, if Tribe's attack on Bayesianism is construed as an argument that factual certainty is actually attainable, the attack is very weak indeed.

If Tribe meant to say that one can be certain about a fact and still make a mistake, either he and I are using language in different ways or Tribe did not really mean to say that certainty about factual matters is attainable. It is possible, I suppose, to take the view that Bayesian portraits of the logic of uncertainty are immaterial because there can be no mistakes in factfinding. This view rests either on the theory that truth is always relative to the evidence, or on the theory that facts are only those facts which are legally declared to be the facts. While this view is suggestive of some important features of facts (e.g., the extent to which we constitute and shape evidence and facts), the thesis that mistakes cannot be made is implausible. It would hardly make sense to worry about the quality and quantity of the evidence in a trial—let alone about the criteria for making good inferences—if we did not think that mistakes could be made.

The sum of it all is that Bayesianism is not an easy theory to demolish. The Bayesian challenge to the traditional way of thinking about evidence and inference is a powerful one. The axioms of the standard calculus of probability seem intuitively appealing and "true." No serious mathematician questions the validity of the standard probability calculus, and the standard calculus has proved its mettle in innumerable fields of science. These are persuasive reasons for taking Bayesianism seriously.

A large part of the aversion to Bayesianism may rest on a distrust of numbers and numerical quantification. However, this sort of attack on Bayesianism misses the major thrust of Bayesianism.

[5] *See* L. COHEN, THE PROBABLE AND PROVABLE (1977). The theory is conveniently summarized in Cohen, *The Logic of Proof,* 1980 CRIM. L. REV. 91 (Sweet & Maxwell 1980). *See also infra* note 8.

[6] Cohen himself has drawn on legal theorizing about inference, *see*, *e.g.*, L. COHEN, *supra* note 5, at § 15. In the past, Cohen has justified his theory in part by an appeal to what he takes to be the usual judicial practice in the evaluation of problems of evidence. *See id.* §§ 14-39, 67-76. But Cohen now claims (or makes clear) that his theory of inference does not depend on the structure of juridical proof but rests entirely on independent considerations. *See* Cohen, *The Role of Evidential Weight in Criminal Proof, supra* at 113.

⁷ Baconian probability, as elaborated by Dr. Cohen, has two central components: (1) an emphasis on the place of generalizations in inference, and (2) an emphasis on the way in which the ranking of generalizations affects inference. Cohen's complex theory has many additional features and it is far from clear that they coincide with the assumptions made by legal scholars in the rationalist tradition. When I refer broadly to Baconian theories I am referring only to the two components listed above. The virtue of doing this is that those two features seem to have been embraced by many traditional legal scholars such as Edmund Morgan, George James, John Henry Wigmore, and, more recently, Judge Jack B. Weinstein, co-author of the leading multi-volume treatise on the Federal Rules of Evidence. *See* 1A J. WIGMORE, *supra* note 2, at §§ 37.2, 37.3.

⁸ See the range of commentary by more than fifty scholars in 6 BEHAV. & BRAIN SCI. J. 487-533 (1983); 4 BEHAV. & BRAIN SCI. J. 331-370 (1981).

⁹ While there were two early reviews of the THE PROBABLE AND THE PROVABLE in legal literature, *see* Schum, *A Review of a Case Against Blaise Pascal and His Heirs*, 77 MICH. L. REV. 446 (1979); Wagner, *Book Review*, 1979 DUKE L.J. 1071, the interest in Cohen's complex theory grew slowly among legal scholars. But as the symposium papers themselves attest, interest in Cohen's theory continues to grow.

¹⁰ Cohen's theory has been bitterly attacked. *See* Williams, *The Mathematics of Proof (Parts I & II)*, 1979 CRIM. L. REV. 297, 340 (Sweet & Maxwell). This critique, while making important points, is inordinately hostile to Cohen's theory; Williams has failed to discern some of the theory's most valuable features. Although it may be nonsense, Cohen's theory is important and fruitful nonsense.

¹¹ *See infra* note 17.

¹² *See generally* W. TWINING, *supra* note 1.

¹³ *See, e.g.,* Lempert, *The New Evidence Scholarship: Analyzing the Process of Proof, supra* at 61.

¹⁴ Fienberg & Schervish, *The Relevance of Bayesian Inference for the Presentation of Statistical Evidence and for Legal Decisionmaking*, 66 B.U.L. REV. 771, 772 (1986).

In this extract Fienberg and Schervish effectively say that one must accept the implications of a coherent axiomatized theory whose axioms are intuitively appealing. By saying this, they seem to be claiming that it follows that anyone who accepts those axioms must also conclude that normative Bayesian descriptions of the logical structure of problems of evidence are always preferable to any other kind of description. This claim is wrong. Professors Fienberg and Schervish do not acknowledge the important distinction between, on the one hand, the question of formal coherence and elegance of a theory of inference and, on the other hand, the question of the appropriate domain for the application of such a theory. This distinction which they ignore, is akin to the distinction between formal and applied mathematics. Moreover, it is not logically impossible or incoherent for a person to "accept" two or more disparate axiomatized systems. *See infra* note 15.

¹⁵ The argument of this paper explains why it is a mistake to assume that logically differentiated theories of inference are necessarily incompatible. However, two

preliminary points must be noted. First, formal theories that rest on different basic principles are not logically inconsistent. Such disparate formal theories are logically valid as long as they are coherent (that is to say, free from logical inconsistency). There is no logical contradiction just because disparate formal theories are incommensurable. Of course, it does not follow that one is bound to accept the axioms of every formally coherent theory.

Second, it is not true that only one set of axioms is intuitively plausible. The mistake of supposing that only one set of axioms can be intuitively appealing appears to reflect the assumption that different sets of axioms that seem to be talking about the same thing are in fact talking about the same thing. Mathematically and logically viewed, however, axioms are not representations of anything, but are merely formal premises for a logico-mathematical argument. Hence, the assessment of the intuitive appeal of two different axiomatic systems may well depend on our intuition about how well one or another axiomatic system meshes with the characteristics of the domain in which we propose to apply it.

[16] L. COHEN, *supra* note 5, § 1.

[17] Cohen's work, when properly (or imaginatively) construed, contains many important insights. Nonetheless, although I do not pretend to understand every facet of his version of Baconianism, some of the wrinkles in his theory are troubling, particularly the following:

(1) Cohen, his protestations notwithstanding, *see* Cohen, *The Logic of Proof*, 1980 CRIM. L. REV. 91 (Sweet & Maxwell), does seem to deny the basic complementarity of the relation between the probability of H and the probability of its negation (e.g., the probability of rain and the probability of not-rain). To be sure, this complementarity need not hold if H and some other state of affairs are not conceived of as being disjoint and exhaustive hypotheses. However, the hypothetical problems that Cohen has constructed to illustrate the workings of his theory, *see*, *e.g.*, Cohen, *Subjective Probability and the Paradox of the Gatecrasher*, 1981 ARIZ. ST. L.J. 627 (responding to a point made in Kaye, *The Laws of Probability and the Laws of the Land*, 47 U. CHI. L. REV. 34 (1979)), seem to involve disjoint and exhaustive hypotheses, and yet Cohen insists that an increase in the probability of one of those hypotheses does not entail a decrease in the probability of what seems to be its negation. *See* 1A J. WIGMORE, *supra* note 2, § 37.7 n.5; Kaye, *Paradoxes and Gedanken Experiments: A Response to Dr. Cohen's Reply*, 1981 ARIZ. L.J. 635; Kaye, *Book Review*, 89 YALE L.J. 601 (1980); Kaye, *The Paradox of the Gatecrasher and Other Stories*, 1979 ARIZ. ST. L.J. 101.

In taking this position, Cohen seems to be making two mistakes. First, he seems to conflate the notions of probability and quantity of information. *But see* Cohen, *The Role of Evidential Weight in Criminal Proof*, *supra* at 113. (relying on Keynes's theory of probability and weight). Second, he does not see that the lessons about the relative probability of scientific theories and hypotheses (which hypotheses are often not seen as mutually exclusive) cannot be directly transposed to questions about factual states of affairs when those alternative factual states of affairs are seen or described as being both disjoint and exhaustive of the possible states of affairs.

(2) Cohen seems to take the inherently implausible position that subordinate inferences supporting a higher inference may or must be treated as being a certainty. *See* L. COHEN, *supra* note 5, at 71. This claim seems flatly wrong. *See* 1A J. WIGMORE, *supra* note 2, § 41. Cohen would have been on stronger ground if he had said that human beings sometimes suppress or ignore uncertainty.

(3) Insofar as I can tell, Cohen does not distinguish carefully enough between (a) the shape of probabilistic thinking about evidence and (b) stochastic probability and statistical inference. This makes his arguments against the use of Bayesian analysis less encompassing than he thinks. His analysis, to a large extent, amounts to an argument against the use of certain types of statistical evidence.

[18] While multiple models of inference are necessary, not every theory of inference is coherent or useful. Moreover, we are not free to choose any coherent theoretical perspective we please when examining a problem of evidence. Reasons can and should be given for the choice of a particular theoretical perspective and this paper tries to show the nature of those reasons.

[19] My emphasis on the need to take our cue from ordinary ways of thinking about the world is the *leitmotif* that characterizes the way I search for a good theory. I maintain that the primary and ordinary inferential techniques and processes that we use daily are in every significant respect rational. The actual structure of those techniques must be a central ingredient of any general theory of inference that purports to teach us how to manage uncertainty in a rational fashion.

[20] The claim that a general theory of inference should regulate the use of lower order theories of inference gives the general theory a meta-theoretical cast. However, the meta-theory I present here is an informal one. Meta-theoretical talk can be meaningful and coherent even in the absence of a formalized metatheory. *Cf.* Brilmayer, *Second-Order Evidence and Bayesian Logic, supra* at 147.

[21] David Schum has already created an impressive map of the different functions that theories of probability and inference may play, both in inferential processes in general and in the adjudicative process in particular. *See* Schum, *Probability and the Processes of Discovery, Proof, and Choice, supra* at 213. Schum's creative imagination in the field of inference is unequalled. His paper is a major contribution to the study of probability and inference.

[22] It is appropriate to intermingle talk about theory as a part of evidence with talk about theory as part of a *problem* of evidence because evidence, in my usage, is anything that is taken as something that points to a possible conclusion. Hence, the existence of evidence is pretty much synonymous with the existence of a problem.

[23] My black box problem has affinities with black box problems that have been studied in connection with theories about artificial intelligence. My use of the black box problem, however, does not require a theoretical commitment on the question whether the construction of something like a machine that mimics or duplicates the operations of the brain effectively amounts to a physical replica of the brain. My black box problem only involves a commitment to the proposition that it is important to understand the internal logic of the brain (or whatever it is that performs the

internal nomological computations of the human being), temporarily setting aside the question of how it is, precisely, that we see inside the brain. It is nevertheless pertinent to note that my argument about the importance of human nomological structures does assume that there is more to human thinking than the physical properties of the brain. In the computer age, this proposition no longer seems startling or novel.

[24] Some transformation rules may predict the behavior of the box without describing its logic. However, if we assume that the physical characteristics of the box determine its logical characteristics, then we are excluding the possibility of a fortuitous coincidence of different sets of transformation rules. Hence any effective set of transformation rules may amount to a description of the innards of the box. If so, the distinction between external and internal perspectives may evaporate.

[25] To be sure, human beings may respond to the world in ways that do not implicate the intellectual or theoretical construction of reality. But such responses are, by hypothesis, not things that we can reconstruct in and by our thought. And if they are, it is enough to say that what occupies us in this investigation are those inferential strategies that have nomological structures—thoughts, beliefs, hypotheses, and the like—as important constituents.

[26] To say that inferential problems of evidence are constructed does not necessarily mean they can be constructed haphazardly or that we receive no objective information from the world. Our concepts and webs of belief do not leave us unconstrained in our thinking about problems of evidence. The hypothesis that human beings conceptually impose a structure on everything they see is not incompatible with the suppositions that (a) human beings do not alone generate the inputs which they process, and (b) there are some restraints—whether by reason of physiology, psychology, or something like Chomskyian "deep structures"—on the kinds of nomological structures that may be chosen to configure inferential problems.

[27] The type of theory under discussion in the above text is Theory II. Earlier I distinguished Theory I from Theory II by saying that the first theory refers to logical characteristics that are the properties of actual things (e.g., DNA) whereas the second theory is imported into a problem of evidence and inference by the decision-maker. The distinction, however, is not meant to deny the ontological and actual status of Theory II. There is a coherent way of expressing the distinction between the two types of theories without implicitly robbing Theory II of ontological status: whereas Theory I consists of the logical properties of the things that the behavior of evidence leads us to see, Theory II consists of the logical properties of evidence itself.

To talk about theories as consisting of properties is unconventional. This inelegant language emphasizes, however, the necessity of taking seriously the logical behavior and characteristics of the evidence that we think we see outside of ourselves. *See infra* note 28.

[28] This odd sounding talk about ontology is important; the enterprise of theorizing about the structure of rational inference falls apart without an attribution of at least practical reality to the theory that resides in evidence.

²⁹ The central problem facing any logically coherent description of inference from evidence is the enormous complexity and variety of inferential processes. The structure of inferential processes affects the value of any particular model of inference. The variety of inferential processes implies the need for various models of inference.

My argument that model builders should take their cues from the structure of actual inferential processes is open to the objection that models constructed out of such cloth are either redundant or merely descriptive. Such models, it is argued, are based on existing inferential processes, processes which have no capacity to improve the quality of our inferential ability.

³⁰ Theoretical simplicity is of course not intrinsically a bad thing. Consider, for example, the equation $E = mc^2$. Moreover, the apparent complexity of the variety of the data sought to be understood does not preclude a simple theoretical explanation. Often the gauge of the adequacy of a theory is its simplicity; a good theory aims to simplify by reducing apparent variety and disconnectedness to certain simple or fundamental principles. Hence, the simplicity of a formal theory such as Bayesianism is not necessarily bad. The question is whether such a theory is overly simple or simplistic. This question reduces to the question of whether or not there are sources or types of uncertainty in inference which formal descriptions of a particular sort do not address. My answer is that the two formal models of inference discussed here are simplistic in this narrow sense.

³¹ *See infra* Part III.

³² To refer to evidence having different kinds of properties, I speak of different kinds of nomological structures. This terminology is in part merely an unfamiliar way of phrasing the familiar claim that the logical implications of evidence are a function of the types of properties attributed to it. Moreover, the concept of a nomological structure, like the concept of a formal theory, makes no sharp distinction between the properties of evidence and its logic, since both concepts take for granted that it is as legitimate to say that the properties of evidence are a function of the logic ascribed to evidence as it is to say that the logical implications of evidence are a function of its properties. (One might say that the properties of evidence are but part of the grammar of a certain logic of inference.) The concept of a nomological structure, however, differs in several subtle but important ways from the concept of a formal theory.

The first difference is a matter of emphasis, the emphasis on the thesis that logic is built into evidence. This emphasis is nothing more than a useful reminder that there is an intimate and reciprocal relationship between the properties of evidence and its logical implications. The second difference, while not in conflict with the concept of formal theory, is more substantive. The talk about nomological structures being a part of the fabric of evidence is meant to serve as a reminder that, in the real world, people may infer the properties of evidence before they understand its logical structure. Notwithstanding these important differences, the concept of a nomological structure has an important affinity with the concept of a formal theory of inference because the word nomo*logical* alludes to the fact that properties are associated uniquely with a particular type of logic. Moreover, the talk about nomological configuring of evidence serves as a useful reminder that properties of evidence do not exist "in the air," but are ascribed to evidence. Evidence, therefore, is not simply given—it is conceptually constructed.

[33] It is almost certainly the case that other types of nomological structures exist and it is likely that some of the nomological structures listed above can and should be broken down into more discrete categories. Refinements of this sort do not impair, but strengthen, the general thesis of this paper.

[34] As Dr. Cohen noted in a seminar discussion, I do not list Bayesian logic as a separate nomological category. Nonetheless, a Bayesian nomological pattern is a subset of the second nomological structure. The second nomological category includes nomological systems that describe random natural processes. Hence, it includes theoretical systems that incorporate the logic of the standard probability calculus and the logic of Bayes's Theorem.

Dr. Cohen's observation, however, raises a broader issue. If it is true that the legitimacy of models of inference and probability depends on the nomological configuration of a problem of evidence, can I coherently take the position that Bayesian logic applies to problems of evidence that are not seen as involving random processes in nature?

I leave this last question open for the time being since the focus of this paper is not on Bayesian theory as such. Nonetheless, the question is simultaneously a difficult, important, and fruitful one. It is a *difficult* question because I am strongly inclined to think that Bayesian analysis does have value in portraying some problems of inference that are conventionally described as being purely or primarily problems of epistemic uncertainty rather than stochastic uncertainty. The question is *important* because of the importance of the proper scope of Bayesian reasoning about uncertainty. The question is *fruitful* because it offers a test of my thesis about the relationship between nomological structures in evidence and the uses of theoretical models of the logic of inference and uncertainty.

My present inclination is to think that the answer to the difficulty suggested by Dr. Cohen's observation lies in the fact that my analysis tends to dissolve any sharp distinction between epistemic and stochastic uncertainty. *But cf. supra* note 27. Hence, my argument may force me to take the position that inferential problems of epistemic uncertainty which seem amenable to Bayesian analysis are in many respects the same as problems involving stochastic uncertainty or, vice-versa, that cases of stochastic uncertainty may equally well be seen as cases of epistemic uncertainty.

Bayesianism is not the only theory raising the questions presented in this note. Thus, for example, many of the above questions could be asked about Zadeh's fuzzy set theory.

[35] In saying that three of the nomological structures are always potentially present in any problem of evidence and inference, I make certain assumptions about the field on which events that are probabilistically assessed take place. Specifically, for example, I assume that every event must take place in a space-time continuum. It is for that reason that I think something like an "imaging nomological structure" is

always potentially present in any inferential problem; the person who assesses evidence must situate his inferential problem in space and time.

It is of course true that there are imaging processes that occur involuntarily. On this, it may be urged that a nomological imaging structure is not always potentially present in an inferential problem since a nomological structure is one that is generated by a conceptual process rather than by something like a neural or other causal process. The objection does not withstand analysis. My claim is only that a nomological imaging process is always potentially present. If it is assumed that all inferential problems must be pictured as occurring in the space-time continuum, an imaging nomological structure is always potentially at hand since the decisionmaker, when trying to draw an inference, is always free to reject something like a neurally-generated image as a true image of the thing in question. To state the argument in another way, acceptance of a neurally-generated image may be seen as involving, at least implicitly, a nomological acceptance of the neurally-generated image. The point is that a decisionmaker is not bound to fashion a sector of the space-time continuum after the images generated by neural or other processes.

Although I argue that at least several different types of nomological structures are always involved in inference, I am not implying that the nomological criteria leading to acceptance of images generated by other types of processes are always transparently comprehensible. In fact, I maintain that exactly the contrary is true—that often these nomological processes are opaque even though we must assume that they occur. It is precisely this opaqueness which, as a theoretical matter, explains why a limited set of nomological structures often gives a satisfactory or informative model of the inferential problem at hand. *See infra* note 55. It is interesting to note that even neural imaging often seems to involve thought-like, albeit involuntary mental operations. *See* I. ROCK, THE LOGIC OF PERCEPTION (1983).

[36] I should add that everything I have said in the text also applies to a Bayesian account of inference. Therefore, (a) each of the nomological structures depends, in principle, on certain constraints or assumptions (e.g., the complementalness of the respective probabilities of state of affairs X and its negation, not-X) whose implications are examined by the standard probability calculus; (b) the outcomes of Bayesian calculations have the capacity to undermine the calculations based on any of the remaining nomological structures, and vice-versa; but (c) the incommensurability of Bayesian computations with the processing techniques suggested by the remaining nomological structures makes it impossible to formulate any invariant relationships between Bayesian processing techniques and those of other nomological structures. Therefore, it is impossible to give a stable description of interdependence described in clause (b).

[37] The peculiar thing about the fourth structure is that it is an interpretation or explanation of another nomological structure. All theories of inference are theoretical models of nomological structures. When inferences involve guesses about human beings, however, this double layer of theory often acquires a third layer. In these cases the modeling of inference involves a configuration of evidence which itself attributes ontological status to a part of the configured problem. *See infra* Part II § C.

[38] *See* T. KUHN, THE STRUCTURE OF SCIENTIFIC REVOLUTIONS (2d ed. 1969) (assigning great importance to theoretical paradigms and questioning the importance of crucial experiments). Although Kuhn's work is the best known, much literature on the topic of theory-dependence and related matters exists. *See also* I. HACKING, REPRESENTING AND INTERVENING: INTRODUCTORY TOPICS IN THE PHILOSOPHY OF NATURAL SCIENCE (1983) (surveying the views of C.S. Peirce, H. Putman, T.S. Kuhn, P. Feyerabend, I. Lakatos, N.R. Hanson, and others on the relationship between theory and matters such as phenomena, observations, and experiments).

[39] G. SHAFER, A MATHEMATICAL THEORY OF EVIDENCE (1976).

[40] For example, after an airplane passes through his visual field, a person may reconstruct his estimate of the actual shape of the airplane from the information acquired through highly sensitive radar. This radar data may show with great precision the distance of various portions of the airplane from the point of observation. On the basis of this information, the observer can calculate the actual shape of the plane by varying the apparent shape of different parts of the plane.

[41] No experience or evidence is presented in its pure form, free from theoretical shaping of information. Nevertheless, the thesis that all data, information, and experience are merely functions of pre-existing biases, presuppositions, theories, and the like, is unwarranted and unfounded since it remains possible that external events, as well as subjective theories, shape the evidence we see. Moreover, no extra-logical considerations speak in favor of epistemological solipsism. Belief in theoretical solipsism necessarily leads to the conclusion that carefully structured human encounters with data, information, and experience have no capacity to advance human knowledge, and experimental methods have not advanced scientific knowledge. While skeptical theorists question whether scientific knowledge has progressed, most theorists do not entertain such skepticism. The subjectivity or objectivity of human knowledge has a decisive bearing on my general thesis that theorizing about inference must be rooted in human introspection about the properties of evidence in general. Theories of inference, like other theories, require verification or validation. The verification of a particular theoretical account of an evidence and inference problem ultimately depends on how the observer sees and understands that evidence. A concession that the theoretical perspectives of the observer shape the observer's experience of the world does not render the verification procedure hopelessly subjective. In the following discussion, hypothetical problems show that inference is not entirely a matter of the mind's eye. The final section of the paper articulates how ordinary forms of inference may serve as an effective method for the validation of theories of inference.

[42] This statement requires qualification. If time allowed, I would argue that the stated interdependence among diverse nomological structures is actually a dependence on theoretically constituted information that may have, for example, Bayesian or Baconian properties. In short, all evidence has logical properties, and image-generating nomological constructs of the sort discussed in the text are dependent on evidence that falls within a different nomological classification. Although not elaborated upon here, a related argument reappears in my analysis of inferences involving guesses about human aims and meanings. *See infra* text accompanying notes 45-46.

[43] It is arguable that the thesis of coordinate determination of inference by theory and evidence is inconsistent with the quasi-idealist hypothesis that theory shapes evidence since the latter hypothesis prevents stating, as an abstract matter, exactly what constitutes basic or primitive information or data. The thought-experiments presented in the text demonstrate, however, that there is no inconsistency because we are ultimately compelled, as a practical matter, to see and experience some things as actual and unchallengeable, as genuine information and data. It may be theoretically permissible or necessary to treat some of the things we see and experience as genuine, actual, and real. Otherwise, we have no foothold for any inference, and we are left awash in rootless imaginings. We may not understand how we know information when we see it, but people have managed to improve their understanding and management of natural and social phenomena by treating some information as genuine information. There is no reason for taking all human experience as merely the expression of the conceptual psyche of the observer.

Nonetheless, decisionmakers are not precluded from revising their opinions about the quality, direction, or existence of particular pieces of evidence. Stating that some events or states must ultimately be taken as containing genuine information recognizes that an observer can revise his opinion of what putative data or information actually contain. Theoretical reconstruction and reinterpretation of problems of evidence problems can in effect alter the available information and evidence. John Jackson has suggested, quite correctly, that the interaction between theory and evidence should be seen as a dialectical process, with successive revisions of each on the basis of the other. Jackson, Two Methods of Truth-Finding and the Evolution of Criminal Procedure § 6 (July 1986) (unpublished).

I do not deny that revisions of opinion and perspective by a decisionmaker can effectively transform data, evidence, and information. In fact, one of the claims made in this paper is that a quasi-idealist perspective on theorizing is valuable because it increases sensitivity to the different possible formulations and configurations of an evidence problem. Nothing in this quasi-idealist claim, however, is inconsistent with my thesis that epistemological solipsism is an inadequate hypothesis. While an observer's ruminations about a problem may make him revise his opinion of what constitutes data and its significance, the observer will eventually arrive at information that seems solid and effectively unchallengeable. *See infra* text accompanying note 48.

The question whether any data can be seen as genuine or authentic has an important bearing on the argument of this paper. Introspection into the ways we ordinarily see problems of evidence validates the use of particular formal models of inference. Introspection of this sort, however, can serve as a verification principle only if it is assumed that human beings can reach a point where they can see a body of evidence in only one particular (if complex) way.

[44] *See* 1A J.WIGMORE, *supra* note 2, § 37.6.

[45] My argument that Bayesianism cannot offer a useful description of the smile problem described in the text above draws on an earlier work of mine about the limitations of Bayesian analysis. *See* 1A J. WIGMORE, *supra* note 2, §§ 37.1, 37.6, 37.7. There I asserted that subjectivist Bayesian theory offers no explanation of the

relatively primitive and atomic probability estimates that are the grist for the Bayesian logical mill, i.e., the mill that describes how such primitive assessments are to be coherently integrated. In the smile problem posed by Figure 3, the trier is not concerned with the conditional relationships among events and various pieces of information, but rather is trying to assess the logic of humor. The trier's focus is on the question whether the logic of this person's humor is such that he would decide to make the facial expression shown in Figure 3 if in a mirthful mood. The result of this analysis is a conclusion about the relationship between two items of information, e.g., facial shape and mirth. This is a rational process which Bayesianism does not address.

[46] To render Bayesian accounts of the structure of inference almost entirely immaterial in this problem, I stipulate that my hypothetical trier takes the configuration of the face shown in Diagram No. 4 as a whole as being indicative of humor. This trier believes that dissecting the face in order to explain his inference of humor would be a contrived explanation. This is an irrebuttable rejoinder since Bayes's Theorem itself has nothing to say about the manner in which evidence should be partitioned and dissected or, for that matter, about whether any evidence should be partitioned at all. *See* 1A J. WIGMORE, *supra* note 2, § 37.6.

[47] Note that this particular inferential problem can be forced into a Bayesian structure, but the ability to use Bayesianism to describe this (and every other) inferential problem demonstrate only that reality may always be reconstituted to fit a formal logical system.

[48] Bayes's Theorem mandates no particular partition of evidence, and I have argued that it does not even mandate that evidence be partitioned. *See* 1A J. WIGMORE, *supra* note 2, § 37.6. That argument has been extended here, and I now assert that a priori abstract epistemological considerations do not force anyone to configure evidence and inference in a specified manner. This argument implies that there are no benchmarks by which it can be objectively and authoritatively determined whether one person's approach to a problem is superior to another's. But if it is not assumed that a particular person, e.g., a juror, has a right to frame a problem as he sees it, it is not necessarily irrational to let someone else, e.g., society, determine how a problem of evidence is configured.

Nonetheless, granting society the right to structure factual problems in adjudication does not ensure that jurors will assess those factual problems more reliably (from society's standpoint) than if left to their own devices. There are intrinsic difficulties involved when a person partitions and dissects evidence for consideration by another since the other person, e.g., a juror, may find that chosen configurations of evidence present him with little meaningful information—given the way that he tends to see things and given the relationships and questions that he regards as significant. Significance lies in the eye of the beholder. *See* 1A J. WIGMORE, *supra* note 2, § 37.7.

[49] On the one hand, the observer must decide if humor implies a smile and, if so, whether the logic of a smile implies a very specific type of facial expression. On the other hand, he must examine the facial expression and ask whether that sort of facial expression is logically indicative of a smile and, hence, of humor. In short, his second

job is to discern what meanings are normally conveyed by the facial expression he takes as having existed.

[50] The following formula expresses the likelihood ratio where P = probability, E = evidence or information, and H = the event, thing, or hypothesis whose probability is to be computed by means of Bayesian calculations.

$$\frac{P(E|H)}{P(E|\text{not-}H)}$$

[51] Thus, if S_1 represents a particular shape of the mouth, D represents the information consisting of the dots of light, $O(S_1)$ represents the prior odds that the shape of the mouth is S_1 in the absence of the information D, and $O(S_1|D)$ represents the posterior odds that the shape of the mouth is S_1 given information D, then the Bayesian equation might look like this:

$$O(S_1|D) = O(S_1) \cdot \frac{P(D|S_1)}{P(D|\text{not-}S_1)}$$

[52] In this paper I usually assume that generalizations amount to relative frequency statements and I further stipulate that the phrase relative frequency statement denotes only statements asserting that events of one type occur with a certain frequency when events of another type occur. I also assume that it is appropriate to construe Cohen's generalizations as relative frequency statements. My reading of Cohen's theory is probably both problematic and controversial, particularly since Cohen himself insists that there is a marked difference between relative frequency statements and generalizations. Hence, if my interpretation of Cohen's effective meaning is wrong, many of my comments about the limitations of generalization-based theories of inference do not apply to Cohen's theory (although they do apply to the theories of scholars like Wigmore).

Nonetheless, it is appropriate, for present purposes, to think of Cohen's generalizations as a certain kind of relative frequency statement. Cohen, the person primarily responsible for the rejuvenation (in non-legal circles) of a Baconian theory of inference and probability, does not espouse a frequentist theory of probability, a theory that in general views probability statements as extrapolations from observed regularities and frequencies. *See* 1A J. WIGMORE, *supra* note 2, § 37.6. Instead, he emphasizes that observed regularities and observed frequencies do not somehow speak for themselves, and that non-frequentist theoretical considerations determine what kinds of extrapolations may be made from, for example, statistical data. *See* L. COHEN, *supra* note 5, §§ 6-7, 81-85.

But Cohen's aversion to frequentist theories of inference and probability does not mean that the sorts of generalizations that he discusses are not relative frequency statements in the sense in which that phrase is used here. Cohen believes it is necessary to use generalizations to extrapolate properly from information and evidence. Insofar as I can tell, these generalizations (offering a warrant for our inferences from evidence) amount to statements taking the form of propositions about relative frequencies. That these backing generalizations are not mere reports of observed relative frequencies is immaterial to my argument here.

Granting this distinction, Cohen nonetheless continues to deny that the generalizations he addresses are relative frequency statements even in this diminished sense. Thus, for example, he construes statistical explanations as statements describing "propensities." L. COHEN, *supra* note 5, § 85. The motive for doing this appears to lie in Cohen's desire to emphasize the notion that statistical explanations can be hedged and qualified. *Id.* at §§ 77-80. Hence, Cohen essentially denies that statements setting forth either observed or assumed relative frequencies convey any significant information by themselves, at least when an effort is made to extrapolate from such relative frequency statements in novel situations.

Cohen has done a great service in pointing out the inherent limitations of all efforts to explain statistical inference on the basis of relative frequencies alone. Still, for present purposes, his generalizations remain relative frequency statements in a significant sense. Regardless of how hedged and limited the generalizations may be, and however subject those generalizations may be to further hedging, qualification, and limitation, they convey meaningful information, and have discernible sense to the extent that they amount to the proposition that there is at least a propensity for events X to occur with a stipulated frequency in relation to events Y.

[53] The premise of this argument is that real world states such as humor can have a logical fabric of their own. To be in a mirthful mood may, perhaps, sometimes be little more than a mere inclination, but in many instances a mirthful, witty, and jocular mood is an inclination to engage in certain sets or types of complex mental operations, calculations, and deliberations. Hence, humor is often more than a propensity or a pattern of engaging in mirthful events with a certain frequency in a set of trials over a specified period of time. The mental and conceptual patterns involved in being witty are quite complex. Witty activity has a kind of logic to it. In fact, many witticisms depend on a twisting of ordinary logic or patterns of thought or speech, and the twisting itself involves a kind of surprising logic.

[54] *See supra* note 44.

[55] While generalization-based models of inference are often defective portraits of rational guessing about human beings, it is still true that there is no one correct way of structuring inference. In fact, even good guesses about the behavior of human beings sometimes rely primarily on generalizations, even when it seems that the true explanation for a human action lies in the internal belief system of the actor. Resort to generalizations seems appropriate when the internal logic of human actors is impenetrable to an outsider. Baconian reasoning will also often be appropriate when that internal logic is simply absent. From the standpoint of the outsider the explanation for his lack of knowledge of that internal logic is immaterial. Thus, for example, sometimes it may be appropriate to invoke the generalization (if we believe it) that people who smoke cigarettes lie more often than non-smokers. If we have no idea of the dynamics of the human personality that generate this behavior, this may be the only way we can talk about the habit of smoking.

It can be argued that generalization-based accounts of inference are neither descriptive nor enlightening. In a criticism reminiscent of the criticism of frequentist probability theories, *see* 1A J. WIGMORE, *supra* note 2, § 37.6, it can be argued that the mere observation of regularities in nature does not explain or specify the conditions under which the generalization becomes inapplicable or the circumstances under which the generalization must be modified. For example, neither the data in the smoking problem or the generalization based on the data answer the question whether or not, and to what extent, the generalization about lying applies to women, children, Native Americans, etc. Moreover, there is no refuge in narrow and specific generalizations that seem to be well-supported by data. Problems of inference are infinitely varied, and inevitably we find we lack data that bears on the significance of the touted generalization.

This criticism of generalization-based models of inference is legitimate since much more is involved in the applications of a relative frequency statement than its content. *See* 1A J. WIGMORE, *supra* note 2, § 37.6 n.6. In a practical sense, however, the criticism may be immaterial for the analysis of a particular evidence problem which incorporates a type of reasoning whose logical processes we are unable to determine. In these situations generalization-based accounts of inference are sufficiently enlightening because the decisionmaker's ignorance does not allow any further explanation of his inferential processes.

[56] It is evident, however, that the theoretical constructs found in humor are not fully encompassed by the second nomological structure. First, the logic of humor is less systematic and rigorous than the logic of classical mechanics or relativity. Second, humor is an example of the fourth nomological category, a category which involves the dimension of things such as meanings, intentions, and aims. *See infra* Part II §C.

[57] *See supra* note 23.

[58] For example, whether the presence of a stranger in a neighbor's garden at night supports the conclusion that the stranger is up to no good depends in part on whether that generalization is more plausible than competing generalizations (such as that strangers under such circumstances are lost).

[59] It seems arguable that the analogy is defective because the judge, when addressing questions of weight, focuses more on general hypotheses supported by the evidence than on the extent to which general hypotheses support a particular inference. This analysis, however, is wrong. The judge, when inquiring into the relative strength of guilt hypotheses, as opposed to innocence hypotheses, is not a scientist trying to determine which general laws are most strongly supported by the available data. The judge typically asks which factual hypotheses are most strongly supported by the evidence in the case.

[60] Although certain facts have been presented, the judge now must draw appropriate conclusions from those facts, raising the difficult task of choosing a theoretical or conceptual lens to determine what those facts suggest. The judge is looking for a story that gives him a sensible way of interpreting the facts. This search, however, is not necessarily a search for hypotheses about the sequences of events in time. *See infra* note 62.

[61] This section is not limited to stories that arrange events in a chronological sequence. The discussion does not, however, canvass every kind of story. Several of

the papers presented at the symposium examine other types of stories in inference. *See, e.g.,* Allen, *A Reconceptualization of Civil Trials, supra* at 21; Kaye, *Do We Need a Calculus of Weight to Understand Proof Beyond a Reasonable Doubt?, supra* at 129; Schum, *Probability and the Processes of Discovery, Proof, and Choice, supra* at 213.

[62] The two discordant hypotheses about the events on May 1 illustrate my thesis of the interdependence of evidence and theory. *See supra* text accompanying note 36. The characterization of the shape of the mouth on the basis of one of the two hypotheses is not solely an interpretation flowing from the subjective theoretical perspective of the trier. First, the hypothesis of mirth is suggested by certain evidence, the testimony of W about the oral exchange he had with S on May 1. Second, the trier may ask whether the shape of the mouth offers a good fit with the hypothesis rather than perform what he sees as radical surgery on his interpretation of the meaning of the mouth's shape. Third, the trier's ability to characterize the shape of the mouth as a smile adds evidence pertinent to the ultimate issue since the smile characterization should have the effect of strengthening the mirth hypothesis.

In response, it might be repeated that all information acquired by the trier is filtered through the trier's predispositions and inclinations. This argument may be true (as I believe it to be), but it is largely irrelevant since it fails to demonstrate that the trier does not acquire information in part by encounters in the world, with evidence and information. If there is a need for subjective information-processing techniques, the only necessary conclusion is that the information is processed at least partly as a function of predispositions. It does not demonstrate that all information is solely a function of such predispositions, nor does it demonstrate that predispositions cannot be altered or affected by experience, information, and like matters.

[63] *See supra* note 58 and accompanying text.

[64] Society will not necessarily allow the trier to do so. Most societies implicitly structure the inferential activities of triers in formal adjudication in quite specific ways (even if that structure allows, within its borders, a degree of variation). If this is correct, the question arises as to the extent to which the inferential processes favored by specific adjudicative processes reflect societal views about the right way to discover the facts and, alternatively, the extent to which societally favored inferential processes reflect collateral considerations and values, e.g., the conservation of resources, the value of ritual in trials, and like matters.

Professor Charles Nesson has suggested that social and cultural values may be implicated in mundane rules of evidence which seem to have nothing to do with values. *See generally* Nesson, *supra* note 4. But we should not leap to the conclusion, as Professor Nesson has, that procedural and evidentiary rules geared toward reliable factfinding reflect either social values or intellectual and epistemological dogmatism. It is not objectively demonstrable that one person's characterization of an evidence problem is better than another's. Hence, it does not inexorably follow that the way a society structures particular problems of evidence and inference is epistemologically worse than how a particular juror does.

The difficulty in the American or English trial system is more specific. In these systems triers of fact—jurors and judges—are all expected to use their own common sense to evaluate the evidence presented by the parties. Thus, the issue is whether the societal structuring of inferential activity by its legal system is incompatible, epistemologically speaking, with the hypothesis that judges and jurors generally use their common sense to evaluate the evidence.

The answers to these questions may depend on the extent to which the triers of fact have epistemic beliefs corresponding to beliefs implicitly contained in the applicable evidentiary and procedural rules. A tentative guess is that English and American trial systems do not structure evidence and information in a way that assuredly satisfies the epistemic interests of the triers of fact. It should be noted that in an adversary system rules of procedure and evidence may generate a wide variety of evidentiary presentations at trial. *See generally* 1A J. WIGMORE, *supra* note 2, § 37.7. John Jackson has reached a similar conclusion in an intriguing article adding a comparative perspective to his epistemological approach. Jackson, Two Methods of Truth-Finding and the Evolution of Criminal Procedure (July 1986) (unpublished).

[65] It is permissible to think of equations explaining random processes (such as radioactive decay) as laws that generate, govern, or predict random processes even though such equations, by their own terms, produce probabilistic or frequentistic outputs. Thus, for example, equations describing radioactive decay do not imply that any particular atom must decay at a particular time, but they express the probability that a mass of atoms with certain properties will decay at a given rate. These equations express the character of the real-world forces that cause radioactive decay. Similarly, it is not inconsistent to say that human actions are governed by belief-logics and that human actions often have random qualities.

[66] There are instances in which an understanding of the subject's internal world seems unnecessary; the subject is sometimes regulated by exterior principles of which he is not aware. For example, the actions of automobile drivers (even taken individually) may be better predicted by various formulae which no one supposes the driver self-consciously has in mind. However, it is irrelevant that self-regulated activity is sometimes better predicted by propositions that make no reference to an internal point of view because in some other situations an internal point of view affords a superior basis for inference.

[67] It is possible that Reagan does not in fact believe that communism is a bad thing. This is immaterial, however. I am only assuming that Reagan holds and acts upon some set of beliefs, whatever they may be.

[68] Reagan's actions may appear random even though we believe he is regulating his activities entirely or substantially by his belief-system. Beliefs, of course, change. More significantly (in Reagan's case, in any event), the rules for drawing conclusions from belief systems, and the rules for identifying and characterizing various conditions in pertinent ways, are not precisely defined.

[69] While this author does not like Reagan's contra policy in Nicaragua, he fully shares the sentiment that many forms of communism are horrible and detestable.

[70] *See supra* text accompanying notes 52-57.

[71] *See supra* note 55.

[72] *See supra* note 44.

[73] The novelty of every situation makes this objection fatal. *See* 1A J. WIGMORE, *supra* note 2, § 37.6.

[74] It is immaterial that talk about dispositions may refer only to hypothesized relative frequencies. It remains necessary to determine how dispositions and proclivities work in novel situations.

[75] *See* 1A J. WIGMORE, *supra* note 2, § 37.7.

[76] Evidence of the charter and the secret society, however, has probative force on another matter. The membership of S and W in the same small voluntary association may support the inference that W has warm feelings for S that color his testimony in a way favorable to S.

[77] This description of ordinary inference is similar to a description of a mathematical theory rather than a description of a spatio-temporal event. A description of a mathematical system yields rational prescriptions because it consists of rules whose precise function is to constrain, regulate, and structure thought. The structure of conventional inference is equivalent to the conventional rules used to draw inferences from evidence. I take such rules to be standards.

[78] This terminology comes from Professor David Schum. *See* Schum & Martin, *supra* note 4. I have given his terminology an entirely different meaning.

[79] *See* M. POLANYI, PERSONAL KNOWLEDGE: TOWARDS A POST-CRITICAL PHILOSOPHY 95-100 (1964).

Peter Tillers,
Professor of Law and Director,
Program for Evidence in Litigation,
Benjamin N. Cardozo School of Law, Yeshiva University.

WARD EDWARDS*

SUMMING UP: THE SOCIETY OF BAYESIAN TRIAL LAWYERS†

Professor David Schum, whose work is reported elsewhere in this volume, has spent many years applying Bayesian ideas to inference from evidence in legal contexts. I have come to realize recently that, thanks to Schum and others, the Bayesian position has become a kind of orthodoxy among evidence law scholars. Reading and attendance at meetings has persuaded me (a) that Bayesian theory is indeed an orthodoxy in such circles, (b) that the dominance of this orthodoxy within evidence scholarship has had little impact on anything that happens in legal practice, and (c) that a central purpose of this book is to provide an opportunity for assorted opponents of this orthodoxy to take their best shots at it.

I am scarcely qualified to be a scorekeeper, since I myself am an advocate of the orthodoxy. Indeed, I see no usable alternatives to Bayesian methodology among the assorted heterodoxies presented in this book, or among the Maginot-Line resistances to it. But I can express a simple bias—no one laid a glove on us![1] While many participants challenged the orthodox Bayesian view, none of them backed us in to a corner ducking; for each punch there seems to be a counterpunch.

Parenthetically, it fascinates me that those who resist the Bayesian position in evidence law do not express similar resistance to it in medical diagnosis and treatment, where it is becoming widely accepted. At the Massachusetts General Hospital, Dr. Stephen Pauker makes his hospital rounds pulling a personal computer with him on a wheeled cart. When he encounters serious diagnostic or treatment decision problems, he fires up the computer, enters his own assessments of probability, obtains utility judgments from the patient, and does whatever form of analysis seems appropriate to the problem at hand. I doubt that those who oppose the Bayesian position in evidence law would resist this kind of application if they were patients in the hospital with their own lives at stake. (Perhaps I should exempt Professor Shafer from this comment; I am less clear about what he would want in such a context than I am about the others.)

Omitted Ideas.

Two important ideas are virtually omitted from the papers presented here. For the sake of completeness, I would like to mention them briefly. One is a fairly straightforward theoretical point from the Bayesian perspective. Any decision, including the decision not to decide, is, if made under conditions of uncertainty, equivalent to an assessment of a vector of probabilities. Given a payoff or utility structure, the boundary between any two acts can be regarded as a plane or hyperplane in a probability space. Where more than two acts are being considered, and in particular, where some acts are "hedging" acts (which would not be taken if the true state of the world were known with certainty), the region within a probability space best suited for a given act is a closed convex region.

That esoteric bit of information bears on our discussion because it effectively says that a set of probability estimates is itself a decision and, more important, that such a decision is equivalent to a set of probability estimates. From a Bayesian point of view, the idea that decisionmaking can occur in the absence of probability assessments is not merely unwise, it is impossible. If one takes that position it seems reasonable that these probability assessments should be explicit and open to inspection and evaluation.

The second missing idea is sensitivity analysis. A decision analysis that does not include a set of sensitivity analyses would be regarded by virtually any practitioner as woefully incomplete if the issue at hand were of any importance or difficulty. The basic idea of sensitivity analysis is to vary the numbers entered into the "base case analysis" to see if changing them makes much difference. Typically one finds that for a reasonable variation of the numbers it does not make much difference. The rare case in which numbers are important invites a second careful elicitation of the key numbers in the analysis.

Those who apply decision analysis know that exploring a change in the numbers within a given structure is less valuable than exploring and trying to simplify the structure itself. In decision situations generally, new and dominant options occasionally emerge from sensitivity analysis; when this happens, it is cause for jubilation. Often, sensitivity analysis permits considerable simplification of a given structure without changing its conclusions. This simplification may not make much difference to the analysis, but it makes a lot of difference to the analyst: it better reveals the argument's analysis and thus facilitates the analyst's justification of the conclusions reached. The spirit of sensitivity analysis is like the spirit of summation in legal argumentation. No discussion of the application of decision theory to legal decisionmaking should omit this important step.

The Society of Bayesian Trial Lawyers

In generating some thoughts about the first meeting of the highly imaginary Society of Bayesian Trial Lawyers, my fevered imagination combined a degree of concern for the rather academic flavor of our discussions with some regret that no one laid a glove on us Bayesians (perhaps because they didn't know where to aim their punches). These Bayesian Trial Lawyers, I suppose, would be experts at decision analysis. In addition, there would be a more receptive environment for them to bring such analysis to bear on legal decisionmaking than currently exists. Like all good trial lawyers, the Bayesians would be strictly practitioners, making points and money from cases won rather than from papers published. What problems might surface at that first meeting?

1. The first issue here, as anywhere in decision analysis, is structuring. The Trial Lawyers would need models of their cases. If a case largely depended on inference from evidence (which would be unusual), the lawyers would have to develop an inferential structure for the issue at hand. They would then have to defend that structure against competing structures. I deeply regret to say it, but structuring in general is art, not algorithm (and this applies more forcefully and pejoratively to the structuring of hierarchical inferences than to any other form of structuring).

Then, as now, an appropriate approach to argumentation would be to develop a competing structure leading to different elicitations and different conclusions. So far as I know, Bayesians have no experience conducting arguments about the appropriateness of inferential structures in adversarial settings. Procedural rules of debate would have to be developed for the task. Fortunately, these rules would be essentially the same procedural rules already in use. A more difficult problem would be to develop rules for determining who wins such a debate.

It would be unfair not to add that the problem of structuring a case is hardly unique to the Bayesian Trial Lawyers; it applies to any lawyer's thinking about any legal process. Perhaps the demands imposed by Bayesian formalism are more severe, but the nature of the problem is no different.

2. The Bayesian Trial Lawyers, despite pre-Bayesian experience with depositions and similar tedious discovery procedures, would experience new heights of frustration as they learned about the problems of elicitation of numbers. Theoreticians worry hard about whether the numbers elicited are in some sense "right." Practitioners worry even more about whether the task of eliciting them is tolerable. The problem is one of combinatorial

explosion. Consider a single probability expression, $P(A|B,C,D,E)$. To make life as simple as possible, let us suppose that A, B, C, D, and E can each have two and only two states. Then the expression stands for 32 probabilities, of which at least 16 must be assessed. This kind of combinatorial explosion problem dominates any attempt actually to apply hierarchical Bayesian inference. Yet I am inclined to believe it soluble, essentially because the numbers to be assessed are no more than reformulations of questions that should be examined in any case. But the drudgery likely to be involved solving it is beyond my present threshold of tolerance.

3. If numerical values were substituted for existing metaphors, the notions of "presumption of innocence," proof "beyond a reasonable doubt," and proof by a "preponderance of the evidence," would lose much of their metaphoric character. The notion of presumption of innocence is truly a guide to inference: it simply says that the evidence to be assessed is all the evidence that distinguishes this person from others who might be accused. That statement, though sensible, is amazingly hard to put into operation. Such notions as proof beyond a reasonable doubt, and even perhaps proof by a preponderance of the evidence, smuggle utility considerations into a probability oriented procedure. While this is sensible, it simply makes the combinatorial-explosion problem even worse, by raising a host of other elicitation issues. Thus, making legal decisionmaking *explicitly* dependent on values requires a major reorientation in the nature of the legal enterprise. That reorientation should not be too difficult to conceptualize, since its traces are ubiquitous in the legal literature. But turning those traces into an explicit view, and making that view numerical, are radical steps. Thus, factfinding would be contaminated by the values behind the actions to which the facts may lead. The problem exists now, but it is likely to be exacerbated by dealing with the values more explicitly.

4. Vagueness is one of the most valuable intellectual tools we have. We rarely require ourselves to be more explicit than is necessary. Bayesian Trial Lawyers, who would have more commerce with numbers than do present trial lawyers, would constantly find themselves dealing with complaints from those who felt pressed to be more explicit than they could be. The Trial Lawyers, good Bayesians all, would know that this was not the case. An empirically derived number is never completely precise—it need be no more precise than the words it replaces. But the popular idea that numbers are precise is deeply ingrained. Sensitivity analysis, by exhibiting the near ubiquity of insensitivity, might be helpful. Still, it would fall on the Trial Lawyers, as it currently falls on statisticians, to remind the world that the

change in language from words to numbers does not imply a change in thought from vague to precise. This problem is especially ironic because probabilities, of all numbers, are most deliberately designed to express vagueness, and do so very well. But this would be incessantly galling to the Bayesian Trial Lawyers, as it is now to decision analysts.

5. The Bayesian Trial Lawyers would be jubilant—if, as I am assuming, Bayesian competence was as available on the bench as among the attorneys—because formal procedures are easier to explain and justify than informal ones. Since arguments would be more explicit, bases for decision should be likewise more explicit. Consequently, both appeal, and especially reversal on appeal, should be rarer than they are now.

This review of the practical problems inherent in actually using, instead of merely advocating, a Bayesian approach to legal inference and decision is clearly intended to portray such an enterprise as difficult and demanding. I fully believe this, and doubt any Bayesian would disagree. Is the enterprise, then, worthwhile?

We won't know until we try, but a lot of past experience bears on the question. The proposal amounts to applying formal procedures to an important human endeavor. Human beings have done this extensively in the past; the sciences, engineering, and medicine are the result. Most of the authors would agree that such formalization has been useful in those contexts. If one is willing to take those examples as sources for a prior, the prior odds that the difficult enterprise ahead will be worthwhile seem to me to be very high.

Indeed, I might be willing to follow the lead of that noted legal scholar W.S. Gilbert. In *Iolanthe*, the Lord High Chancellor of England, after attributing his rise to eminence to strict adherence to the rules of evidence sings:

In other professions in which men engage
  (Said I to myself—said I),
The Army, the Navy, the Church, and the Stage
  (Said I to myself—said I),
Professional license, if carried too far,
Your chance of promotion will certainly mar—
And I fancy the rule might apply to the Bar
  (Said I to myself—said I!).[1]

Probably just that kind of professional license is needed to introduce a full-fledged Bayesian argument into a real case as a matter of course; in fact, the history of *People v. Collins*[2] rather suggests it. But we now have another profession in which men engage—Science. When it comes to rules of thought, its successes have redefined what is and is not licentious. Scholars of evidence, following the Chancellor's line of thought, might well want to conclude that the rules that make science work so well apply to the Bar. Human beings in general find it useful, not licentious, to make vague ideas less vague. I consider it likely that practitioners of the law, as well as legal scholars, will have a similar experience.

NOTES

† © 1986 by Ward Edwards.
* Director, Social Science Research Institute, University of Southern California.
[1] W. GILBERT & A. SULLIVAN, *When I Went to the Bar*, IOLANTHE act. 1, line 489 (1882), *reprinted in* 1 THE ANNOTATED GILBERT AND SULLIVAN 195 (I. Bradley ed. 1982).
[2] People v. Collins, 68 Cal. 2d 319, 438 P.2d 33, 66 Cal. Rptr. 497 (1968) (prejudicial error to admit testimony of probability theorist on issue of identity).

Ward Edwards,
Professor of Psychology and Industrial and
Systems Engineering and Director,
Social Science Research Institute,
University of Southern California.

INDEX OF NAMES

Allen, Ronald J.
 21–60, 61, 62, 63, 66, 68, 71, 79–87, 103–111, 129, 147, 178, 481, 673, 704
Anderson, Terry
 224, 230, 244

Bacon, Francis
 214, 215, 228, 241, 243, 277
Barclay, S.
 263
Barnes, David W.
 33
Bergman, Paul
 219–21, 230, 248, 257
Berenson, F. C.
 137
Bernoulli, Jacque
 186
Binder, David A.
 219–21, 230, 248, 257
Brilmayer, Lea R.
 68, 147–67, 169–74, 177–81, 265
Brook, James
 151, 178
Burbank, Stephen
 61
Burks, Arthur W.
 163

Callen, Richard W.
 22, 39
Carnap, Rudolf
 115, 118
Cassirer, Ernst
 285
Cohen, L. Jonathan
 21, 22–24, 29, 33, 35–43, 48, 50, 62, 63, 66, 68, 79, 80, 113–28, 129–30, 132, 134, 136, 137, 157, 187, 194–95, 215–16, 228, 235, 240–50, 257, 262–66, 271, 273–74, 277–79, 284, 298–99, 331–32

Cohen, Neil B.
 21, 135
Cohen, Robert
 x
Cohn, Donald
 135

Doyle, Arthur Conan
 228–230

Eco, Umberto
 228, 229
Ellman, Ira
 129, 133
Edwards, Ward
 224, 231–32, 234, 244, 246, 271–76, 337–42
Elekof, Per Olof
 187, 259–60
Epstein, Richard
 50

Fairley, William B.
 62–63, 213
Fienberg, Stephen
 61, 71, 177, 278
Finkelstein, Michael O.
 62–63, 213
Freeling, A.
 260
Freund,
 191, 208

Garber, Daniel
 186
Gilbert, W. S.
 341
Grady, Mark F.
 24
Graham, Michael
 x

343

NAME INDEX

Green, Eric
  ix–xi
Green, Michael
  21

Haack, Susan
  163–64
Hacking, Ian M.
  135, 227–28
Hale, Matthew
  205
Hanson, N. R.
  227
Hastie, Reid
  21, 233
Hempel, Carl
  115, 118
Hunter, Geoffrey
  150

Jeffrey, Richard C.
  130, 163

Kant, Immanuel
  285
Kaplan, John
  21, 28
Kay, Wendell
  133
Kaye, David H.
  1–19, 21, 27, 29–31, 43–45, 50, 61, 62, 71–72, 74–75, 124–26, 129–145, 150–54, 159, 164, 177–83, 202
Kelly,
  263
Keynes, John Maynard
  117–19, 129–130, 132, 134
Kent, Sherman
  275–76
Kernstein, David
  62
King, Joseph H., Jr.
  39

Leibniz, Gottfried W. F.
  186

Lempert, Richard
  x, 21, 61–102, 103–11, 132, 177, 178–79, 202
Levi, Isaac
  227

Martin, Anne
  65, 68, 70–71, 169–175, 178, 250
Mellor, D. H.
  124–25, 137
Melsa, James
  135–36
Mill, John Stuart
  214, 215, 228, 241, 243, 246
Morgenstern,
  241

Nesson, Charles R.
  32–33, 46–47, 64, 130, 132, 147, 178, 307
Neyman, James
  125

Pascal, Blaise
  214–17, 234–36, 241–48, 251–55, 257–58, 260, 263–64
Pauker, Stephen
  337
Peirce, Charles S.
  227, 228
Pennington, Nancy
  233
Penrod, Steven D.
  233
Phillips, L. D.
  232
Polanyi, Michael
  317

Regan, Donald
  61
Rubinfeild, Daniel L.
  46

Sahlin,
  260

NAME INDEX

Salmon, Wesley C.
227, 235
Saltzburg, Stephen
109, 134
Savage, Leonard J.
68, 187, 202, 271
Schervish, Mark
278–79
Schmidt, Marlene
ix
Schum, David A.
x, 21, 34–35, 40, 61, 62, 65, 186, 205–12, 213–70, 271–76, 279, 337
Schwartz, William
x
Sebeok, T.
228, 229
Shafer, Glenn
22–23, 49, 66, 68, 71, 77, 80, 185–204, 205–10, 211, 214, 215–16, 235–41, 246–47, 249, 253, 257, 260, 262–66, 271, 273, 278, 285–86, 337
Skyrms, Brian
134–35
Steir, Serena
21
Suppes, Patrick
130

Thompson, Judith
61
Tillers, Peter
ix–xi, 21, 61–62, 67, 261, 277–336

Toulmin, Stephen
231
Tribe, Laurence
23, 32, 62, 65, 83, 214, 277
Tversky, Amos
177, 186, 188, 201, 205
Twining, William L.
x, 224, 230, 244, 277–78

Van Quine, Willard
286
Von Neumann, John
241
Von Winterfeldt, Detlof
232

White, Michael
129, 134
Wigmore, John Henry
208, 213, 220, 223–24, 228, 231, 244–46, 248–50, 258–60, 272, 277
Winter, Lawrence
129
Wittgenstein, Ludwig
312

Zabell, Sandy
186
Zadeh, Lofti
214, 216–217, 254–58, 263–64, 271–72, 274–75, 278
Zuckerman, Adrian
x